D1545846

Amphibians of East Africa

Comstock Books in Herpetology
Aaron M. Bauer, Consulting Editor

Amphibians of East Africa

Alan Channing and Kim M. Howell

Comstock Publishing Associates
a division of

CORNELL UNIVERSITY PRESS
ITHACA, NEW YORK

Cornell University Press gratefully acknowledges the Royal Norwegian Embassy, Tanzania, the Norwegian Agency for Development Cooperation, and the Wildlife Conservation Society for their generous support of this publication.

The Wildlife Conservation Society saves wildlife and wild lands through careful science, international conservation, education, and the management of the world's largest system of urban wildlife parks. These activities change attitudes toward nature and help people imagine wildlife and humans living in sustainable interaction on both a local and a global scale. WCS is committed to this work because we believe it essential to the integrity of life on Earth.

First published 2006 by Cornell University Press

Color plates printed in Hong Kong

Library of Congress Cataloging-in-Publication Data

Channing, A.
Amphibians of East Africa / Alan Channing and Kim Howell.
p. cm. — (Comstock books in herpetology)
One chapter in Swahili.
Includes bibliographical references and indexes.
ISBN-13: 978-0-8014-4374-9 (cloth : alk. paper)
ISBN-10: 0-8014-4374-1 (cloth : alk. paper)
1. Amphibians—Africa, East. I. Howell, Kim. II. Title. III. Series.

QL662.A353C48 2005
597.8'09676—dc22

2005015832

Cornell University Press strives to use environmentally responsible suppliers and materials to the fullest extent possible in the publishing of its books. Such materials include vegetable-based, low-VOC inks and acid-free papers that are recycled, totally chlorine-free, or partly composed of nonwood fibers. For further information, visit our website at www.cornellpress.cornell.edu.

Cloth printing 10 9 8 7 6 5 4 3 2 1

Contents

Color plates follow page 226

Preface

When two people set out on a journey that will last a number of years in almost unknown territory, one would expect a respectable reason for such a venture and hope for a fruitful outcome. The writing of this book was such a journey.

This book has two separate origins. One of us, Alan Channing, based in Cape Town and interested in the life histories and evolution of amphibians had studied these animals during a number of short visits and a sabbatical year in Tanzania, as well as during short trips to Uganda and Kenya. He was struck by the difficulty of identifying the frogs and by the lack of up-to-date reference material. Often this required him to delve into the old literature and seek assistance from others interested in East African amphibians.

The second of us, Kim Howell, has been based at the University of Dar es Salaam for 32 years. He had been collecting information on East African amphibians and was keen to see a book on them produced, as identification of specimens collected in the region was difficult, with many needing to be sent to large museums in the United States and Europe for assistance. He received a large number of queries concerning the names and biology of frogs from many different areas and was encouraged by Arne Schiøtz to "write a book."

At a herpetological meeting in 1994 the two of us met and bemoaned the lack of a suitable reference for East African amphibians. We each encouraged the other, and some years later began the next step of the journey. It seemed natural to combine our efforts, as we both had made progress compiling data.

Our task was complicated by the changes to the names of localities where specimens had been collected over many years, and even to a series of name changes of the species themselves. Many of the smaller species, such as the leaf-litter frogs, are remarkably similar, making the task of explaining how to distinguish them more difficult. The last checklist of all the species compiled by a scientist who had worked in East Africa was written nearly half a century ago, in 1957, by A. Loveridge.

The writing process is never easy; preparing text that is accurate while being readable and useful to a broad audience, when many details of life history remain unknown, is taxing. We were excited that during this process a number of frogs new to science were discovered.

The journey has ended. The once unknown territory is beginning to be recognizable, and the outcomes are presented here in words and pictures. Both of us hope that the book generates as much satisfaction in the reading and using as it did in the completing, and starts others on their own journeys enjoying, observing, and recording information on amphibians in East Africa.

Alan Channing and Kim M. Howell

Stellenbosch 2003

Acknowledgments

We thank our wives, Jenny and Imani, whose contributions and understanding allowed us to do the fieldwork and spend the time writing this book.

A volume such as this results from the efforts of many herpetologists and other field workers who have made important discoveries and published them in the scientific literature. Both of us have benefited from opportunities for fieldwork, museum visits, and advice from Bob Drewes, who has been studying amphibians in East Africa for many years. John Poynton made available drafts of work in progress on the taxonomy of Tanzanian amphibians. D. E. (Eddie) van Dijk kindly made his literature collection available and found many obscure details we needed.

We would like to recognize and thank the following who assisted us in our East African endeavors: Beryl Akoth, Mathias Behangana, Joe Beraducci, David Bygott, Alison Channing, Jenny Channing, Barry Clarke, Tim Davenport, Hartwig Dell'Mour, Siobain Finlow-Bates, Bunty Grandison, Svein Erik Haarklau, Jeanette Hanby, Peter Hawkes, Michael Klemens, Stefan Lötters, Simon Mduma, Felix Mkonyi, Charles Msuya, Wilirk Ngalason, Damaris Rotich, Arne Schiøtz, Tony Sinclair, Bill Stanley, Chris and Tilde Stuart, and Mills Tandy. Alison Channing, Jenny Channing, Rafael De Sá, Michele Menegon, Felix Mkonyi, David Moyer, Charles Msuya, Wilirk Ngalason, Bill Stanley, and Chris and Tilde Stuart provided recordings and/or specimens. A special word of thanks to three generations of the Phillips family of Iringa, who made us welcome in so many ways. Martin Walsh kindly corrected our use of local language names. John Measey, Simon Loader, Mark Wilkinson, and Dave Gower generously shared their field experiences of little-known caecilians.

David Moyer of the Wildlife Conservation Society was always willing to offer advice, ease administrative procedures, and use his skills in the field of animal sounds generously. He recorded many of the calls reported on here and played an important role in the discovery of many new species by discovering unknown calls. Elia Mulungu brought in many useful frogs and recordings and undertook difficult fieldwork, often in miserable weather, always with a smile.

The organizations that made it possible to undertake this work include our respective employers, the University of the Western Cape in South Africa and the University of Dar es Salaam in Tanzania. COSTECH (The Tanzania Commission for Science and Technology) issued research permits. Funding for Alan Channing was provided by the Senate Research Committee of the University of the Western Cape, the Royal Society, the National Research Foundation of South Africa, National Geographic Society, the Royal Norwegian Embassy, and Wildlife Conservation Society. Kim Howell thanks the Department of Zoology and Marine Biology at the University of Dar es Salaam, the Biodiversity Database housed there, and the director of Wildlife, Ministry of Natural Resources, and Tourism as the CITES management authority.

The Danish ENRECA program (to enhance research capacity) provided funding for Kim Howell for a museum visit and fieldwork in Tanzania as well as continuing support to the biodiversity database at the University of Dar es Salaam.

Many museums kindly permitted us access to their specimens and data on East African amphibians and we are grateful to those in charge of the collections: R. C. Drewes of the California Academy of Sciences, San Francisco; S. K. Rogers of the Carnegie Museum, Pittsburgh; A. Resetar of the Field Museum of Natural History, Chicago; K. Beamen of the Los Angeles County Museum, Los Angeles; D. Rotich of the National Museums of Kenya, Nairobi; and J. B. Rasmussen of the Zoological Museum, Copenhagen. J. C. Poynton of the Natural History Museum in London identified all specimens collected by Frontier Tanzania. Tadpoles for illustration were loaned by M. Bates of the National Museum, Bloemfontein; L. du Preez of the University of Potchefstroom; and M.-O. Rödel of the University of Würzburg.

Participants and staff of Frontier Tanzania, a collaborative project between the Society for Environmental Exploration in London and the Faculty of Science at the University of Dar es Salaam, collected specimens at many localities in Tanzania and their efforts are greatly appreciated.

This book would not have been possible without the generous photographic contributions of a number of friends and colleagues: Bill Branch, Marius Burger, Tim Davenport, Elizabeth Harper, Malcolm

Largen, Stefan Lötters, John Measey, Michele Menegon, Martin Pickersgill, Edoardo Razzetti, Mark-Oliver Rödel, Bill Stanley, Chris and Tilde Stuart, Mills Tandy, Lorenzo Vinciguerra, John Visser, and James Vonesh. The photographers are credited in the relevant captions. All other photographs are by Alan Channing.

We wish to acknowledge the assistance of Mr. S. Mwansasu, Geographic Information System (GIS) specialist at the Institute of Resource Assessment, in the final map preparation, and Mr. N. Gwacha of the University College of Land and Architectural Studies at the University of Dar es Salaam in earlier phases of map design.

The tadpole drawings were expertly undertaken by Jenny Channing, and the line drawings that illustrate the keys were mostly prepared by Greg Crutsinger. We thank them both for their efforts. Imani Swilla, David Moyer, Charles Msuya, and Henry Ndangalasi kindly assisted with translating the information presented in Swahili.

The staff at Cornell University Press have efficiently and expertly guided the production of this book. Their assistance made the task easy for us. Mark-Oliver Rödel made many useful comments on the text, including providing unpublished observations.

Amphibians of East Africa

Introduction

East Africa consists of Tanzania, Kenya, and Uganda. The region is well known for its wealth of plants and animals, represented in game reserves such as the Serengeti, the Maasai-Mara, and the Bwindi Impenetrable Forest Nature Reserve. These reserves and others like them have been the localities of wildlife television documentaries that cover the well-known large mammal predators, the crocodiles, the herds of migrating buffalo, and, of course, the gorillas (Fig. 1). Less well known, but playing an important role in the East African ecosystems, are the amphibians. The ancestors of the present-day frogs and caecilians were in attendance before the rift valley started to form, and long before the

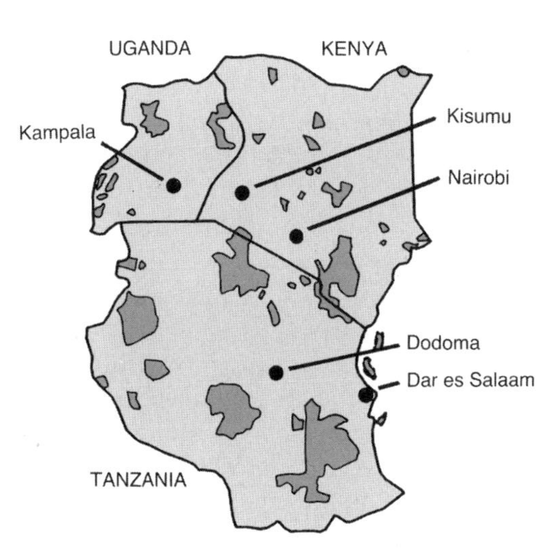

Fig. 1 Map showing the nature reserves and other protected areas in East Africa, indicated in dark shading.

first true mammals shook off the last vestiges of their reptilian ancestry.

There are more than 5500 species of amphibians known worldwide. In East Africa there are just over 200 species, with more being discovered every year. Amphibians have been studied in this region for more than a century, but for the student there exists no comprehensive overview, other than an online world list of species, since the publication of Arthur Loveridge's East African checklist in 1957. In this book we summarize and compare all that is known about the biology of amphibians found in East Africa. To the west, the Congo Basin is largely unexplored for amphibians, and the countries south of Tanzania have been covered in the book *Amphibians of Central and Southern Africa*, published in 2001.

One of the most common reactions we get in the field when explaining that we are looking for frogs, often at night and in the rain, is complete disbelief, and possibly some concern for our mental well-being. However, amphibians, and frogs in particular, are fascinating as they have a range of different body shapes, color patterns, and behaviors. Some can survive in harsh arid conditions, while others may spend much of their lives in permanent ponds. Even the montane forests have tree-living species that rarely come down to the ground. Frogs are very vocal; the males produce calls that advertise their presence to receptive females. These calls are useful indicators of the presence of particular species.

While many frogs lay eggs in water that develop into tadpoles, some lay eggs in nests out of water, where they develop directly into small frogs. In East Africa there is a unique reproductive mode for African frogs; the forest toads in Tanzania (and relatives on Mt. Nimba in West Africa) have internal fertilization, and the eggs are retained in the oviducts, where they develop through the tadpole stage before being "born" as small frogs.

There is a huge variation in the habitats where frogs are found; these range from the arid interior to the moist forests of the mountains. The tadpole stage, where it is present, permits the larvae to grow in an aquatic environment, feeding mostly on algae, thus being ecologically separated from the terrestrial adults that largely eat insects.

Frogs are useful to humans and have the potential to become very important in the future. They eat large numbers of insects, reducing the level of insect populations. The frogs are, in turn, eaten by birds, snakes, mammals, and other frogs. Frogs also are known to be very sensitive to many chemicals that have the potential to harm humans, and the loss of frog populations is evidence that the environment is declining in quality—they thus act as an early-warning system.

Apart from their environmental value, frogs can be considered as virtual pharmaceutical factories. Their skins secrete a range of pharmacologically active substances, including antibiotics and painkillers. Frogs do not have to be killed to harvest the active compounds in the skin, and these substances can be made in the laboratory once they have been discovered.

The concern, worldwide, that amphibian populations are declining has not yet been supported or documented for East Africa. Amphibian-monitoring projects are presently ongoing in Kenya, and data are available for some localities in Tanzania.

A common question concerns the meaning of the words *frog* and *toad*. Here we use the word *frog* to refer to all anurans. Toads mostly belong to the family Bufonidae and are therefore a subset of frogs. Members of the family Pipidae are sometimes called *clawed toads* or *clawed frogs*, which may add to the confusion. The word *amphibian* refers to the anurans and caecilians in sub-Saharan Africa but includes salamanders in other areas of the world where they occur.

Arrangement

The book is arranged to be useful to anyone interested in the natural history of Africa, but it also includes details for the professional biologist. Just over two hundred (203) species are covered, consisting of 194 frog species and 9 caecilians. The meanings of the scientific names are explained, and local names and some names used in the literature are listed. Each species is briefly described, along with a statement of its known habitat preferences and distribution, its advertisement call, breeding biology, and tadpole. A section of notes and a guide to key references for further reading complete each species account.

We did not attempt to review taxonomic changes. The scientific names here were complete and accurate at the time of publication. Many new names are expected as more detailed work is carried out in East Africa. Studies using molecular data, such as DNA sequences, and new software capable of resolving phylogenetic relationships with large data sets will undoubtedly lead to changes in the classification and naming system, as a better understanding of the relationships between species develops.

Identification

Keys are provided to aid the identification of adults and tadpoles. These keys are a series of pairs of options concerning details of the animal to be identified. A choice of one of the options will lead either to the name of the animal or to a further set of options. Practically all the adults can be identified using the keys. Not all of the tadpoles are known, however, so the tadpole key only leads to identification at the level of the family. Since only a small number of species are usually found at each locality, the distribution maps will suggest the presence of species, which can then be checked using the descriptions of adults or tadpoles. See the preamble to the keys for more explanation.

Layout of Species Accounts

Each species is listed by its common and scientific name, followed by alternative common names where they are known.

Common Names

Most languages spoken in Africa make no distinction between various species of frogs, except where they have some importance to the speakers, for example, as food items, like the bullfrog. Few features would make frog species readily identifiable by a layperson, so it is not surprising that there are only a few common names, such as *butwa* for a large frog, *kibutwa* for a small toad, and *mabutwa* for many large toads. The Maasai call all frogs *endwaen'gare*. *Shembumi* is the Sambaa equivalent. *Jula* is Shambala for all frogs. In Nandi a toad is called *kuchwa*, while the name *mororoch* is used for all frogs. In Luo *ogwal* refers to a frog. In Kiikoma frogs are *ebyora*, while toads are *chakuru*. The Bantu languages in Uganda use the general name *ekikele* or variations like *ekikere* and *ekichere* for all frogs, while in Gisu it is *isodo*, in Madi *lichaliro*, and in Atesot *adodoko*. Many of the local names recorded by Arthur Loveridge are *chura* or variations on this word—Swahili for "frog." Rain frogs are known as *chura bila pua* in Swahili, "frog without a nose." Frogs are not always well known: the occupants of one village were amused at the thought that frogs might be found in trees!

Description

We have selected features that should assist in the identification of the species, and give them in an abbreviated form, rather than a complete technical description. The references in the bibliography should enable interested readers to find the original literature and other detailed descriptions.

Habitat and Distribution

This is a brief statement of the habitat where the species is found and its known distribution. A map showing the distribution in East Africa is available for each species. Please read the section below on maps to appreciate the information presented and to understand the limitations of distribution maps.

Advertisement Call

Male frogs use the advertisement call to attract females. This is a convenient tool for identifying a species, as each species has a unique call. The calls that are known have been described. Recordings of frog calls for some parts of Africa are commercially available and enable comparisons of calls heard in the field. Many frog species are so similar in shape and color that only the different calls show they are separate species. Many computer programs are capable of producing a sound spectrogram from a call, which is a graphical representation of the call, with time represented by a horizontal axis, frequency by a vertical axis, and amplitude (loudness) by darker areas. The descriptions of the calls are based on sound spectrograms generated using software such as RAVEN, available from the Cornell Laboratory of Ornithology (Macaulay Library of Natural Sounds, Cornell Laboratory of Ornithology, 159 Sapsucker Woods Road, Ithaca, NY 14850 USA. Web site: www.birds.cornell.edu).

Breeding

This section describes the eggs, where they are laid, and other associated information. This information should not be regarded as definitive, as the variation in breeding systems is not well known. Rather, regard these details as a starting point. If you can, check what is presented. Only in this way will we begin to understand the reproductive variation in natural populations.

Tadpoles

Relatively few tadpoles are known, despite the fact that tadpoles are easy to catch and to identify to family and often to genus. Tadpole behavior is presented where it is known. Only 40% of the tadpoles from East African frogs have been described. We are hopeful the gaps here will encourage more work on the tadpoles of this part of Africa. The tadpole information is presented in a separate chapter.

Notes

We use this section to present extra information on the biology of the species. This has been subdivided in a few cases where there is a lot of information.

Key References

This section cites important sources of information that will serve as an introduction to the literature. Each source is listed by author and date of publication. For example, "Drewes & Perret 2000" can be looked up in the bibliography to give the complete citation: "Drewes, R. C., and J.-L. Perret. 2000. A new species of giant, montane *Phrynobatrachus* (Anura: Ranidae) from the central mountains of Kenya. *Proc. Cal. Acad. Sci.* 52: 55–64." Any library should be able to locate the original sources listed.

Maps

The maps have been compiled from three sources: recent literature, the collections of many major museums, and our field notes and those of many of our colleagues. Each symbol on a map represents one specimen or more.

Map interpretation is open to abuse. It is important to realize that the maps show only *localities where the species is known to occur*, and not *all the localities where the species does occur.*

A number of possible errors can occur when mapping distributions. These include misidentifying the specimen, confusing the locality, and introducing errors when transferring the data from a catalogue to the map. These errors can sometimes be detected as points far removed from the others on the map. We have been particularly cautious and have ignored doubtful records or presumed misidentifications. Unrecognized new species, for example, are always mapped incorrectly as some other species. The newly described southern torrent frog *Arthroleptides yakusini* was initially confused with Martienssen's torrent frog *Arthroleptides martiensseni.* Earlier maps of the distribution of Martienssen's torrent frog consisted of records of both species. In cases where two species are similar in color and shape, it may be possible to tell them apart only when they are alive and calling. Of course, female frogs do not call, so they have to be identified by positive association with the males.

Gaps in distributions are real only if the species does not occur there. As mentioned earlier, however, the gaps may be due to a lack of collecting. Negative collections are rarely indicated, as collectors are always hopeful they will find range extensions or new species in the future.

Undercollected species can be recognized by their patterns of distribution. The map symbols will be plotted in lines that coincide with the roads along which casual collecting has occurred. Casual collection on roadsides or while driving on roads at night is a useful technique, but further extensive collecting should follow. Unusual species are often collected repeatedly at the same locality, as the collector wishes to be certain that the animals will be available, or the accommodations or other facilities are convenient. Many remote areas in East Africa remain poorly surveyed.

Bibliography

The bibliography is a list of useful reports and publications concerning the amphibians represented in East Africa. We follow the standard citation format for the sources of information used.

Future Work

As this book is being prepared, many additional species are in the process of being described, and many little-known areas are sure to yield other undescribed frogs. Local universities are starting to utilize approaches like DNA sequencing, and cladistic analyses are commonplace. The next few years are expected to produce a surge in amphibian systematic publications from this area.

Adding to the Knowledge of Amphibians

There are many gaps in our knowledge. Most of these can be filled in by careful observation and recording. People living in remote areas are in a unique position to assist, as they can be present at the start of the rains, when most frogs are active. Many scientific societies cater to people interested in frogs and provide newsletters for the presentation of casual observations, as well as scientific journals for the publication of detailed studies. The Herpetological Association of Africa offers membership to anyone interested in African frogs (and reptiles). The association produces the *African Journal of Herpetology* and a newsletter; any library or museum should be able to trace the current address of the association. Recent books or chapters in books (listed in full in the bibliography) that will be useful for anyone interested in African frogs include Passmore & Carruthers 1995, Schiøtz 1999, Poynton 1999, Tinsley & Kobel 1996, and Channing 2001. Libraries may have copies of some of the older books listed in the bibliography.

Natural History Museums

Many natural history museums house collections of frogs and serve as a source of information about amphibian biodiversity. The collections include the specimens, along with important information about each. The specimens are preserved so as to be useful in the future, and may be the source of comparative material for purposes of identification, or as material for examining the diet of amphibians and their reproductive biology, among many other uses. The information associated with each specimen is critical—for example, the date and place where it was collected—as without this the material is all but worthless. Animals collected for a particular research purpose—for example, a study of parasite transmission—are placed in a collection to serve as voucher specimens. Voucher specimens serve as a reference so that other workers can confirm the findings of the original investigation. New species and frogs from unusual places are particularly valuable as voucher specimens.

East Africa has many unique amphibians, and the amateur herpetologist, as well as the general public, can play an important role in contributing to our knowledge. Once the amphibians in an area are identified, their distributions and biology can be compared to what is reported in this book. New findings can be conveyed to the institutions listed below, along with tape recordings of the advertisement call, a voucher specimen (properly labeled), and detailed observations.

At the time of writing, the following institutions had amphibian collections and staff to look after them. Other museums may have collections, but unless they are being actively curated, it is better to ensure that the specimens get to one of the institutions listed. The telephone numbers are not given as they can change, but it should be simple to find the current contact details. Zoology or biology departments at other universities and conservation authorities will also be able to advise.

Addresses of East African Institutions with Amphibian Collections

Zoology Department, Makerere University, P.O. Box 7062, Kampala, Uganda

Herpetology Department, National Museums of Kenya, P.O. Box 44486, Nairobi, Kenya

Department of Zoology and Marine Biology, P.O. Box 35064, University of Dar es Salaam, Dar es Salaam, Tanzania

Collecting Amphibians and Information for Museums

Frogs are usually caught at night by using a strong light. The males can be located by their calls in the breeding season. It is important not to collect too many, or to collect species for no reason. Collections should be made of what appear to be new distribution records, or specimens that cannot easily be identified, as these may be new to science. Often sound recordings are adequate to identify a species.

The specimens can be kept alive in a plastic bag, one to each bag. Keep the bags cool and add a little water to prevent the frog from desiccating. If possible, take the frog alive to a museum or university. If it is not possible to deliver the frog alive within a day or so, then the specimen can be killed, using a little teething gel rubbed onto the top of the head. Alternatives include MS 222 (3-aminobenzoic acid ethyl ester), a little powder dissolved in enough water to be 5 mm deep in the container, or 5% ethanol, or 1% formalin. In extreme cases the specimen can be placed in deep freeze for a few hours until frozen. After the specimen is killed, it must be fixed by laying it out, with the fingers and toes spread, between layers of cheesecloth or paper towel dampened with 5% formalin, inside a closed plastic container. Large specimens, over 25 mm long, will require a small incision in the side to permit the fixing solution to get to the internal organs. After 24 hours the specimen can be rinsed and placed in 70% ethanol for storage.

Each specimen must be labeled. This is done by tying a good-quality paper or plastic label with thread around the waist of the specimen, with the following details written in pencil or alcohol-proof ink: exact locality, given so others can find the place; date of collection; name of collector; identification; other useful information. Museum expeditions assign a collector's number to each specimen, and this is linked to detailed notes in a permanent catalogue. Global Positioning System devices are readily available, and these will give detailed latitude and longitude coordinates.

Sound recordings can be made using anything from a simple cassette recorder, to digital tape or mini-disk recorders. In all cases it is essential to use a good microphone positioned as close to the calling animal as possible, to minimize background noise. The recording details, such as the temperature, exact locality, date, time, and species identification, should be added to the end of each recording, to prevent these details from being lost.

Museum staff will be willing to advise on procedures and technical details like labeling paper. Be aware that you may need a permit to collect amphibians.

KEY REFERENCES: Loveridge 1957, Heyer et al. 1994, Frost 2002.

History of Amphibian Studies in East Africa

The unique animals and plants of Africa have long attracted the attention of explorers and scientists. Amphibians, and frogs in particular, were collected during the earliest expeditions to the continent. The first frog formally named from Africa was a member of the genus *Breviceps* from near Cape Town, which was described by Linnaeus in 1758.

During the period leading up to the early 1900s, the collecting of amphibians in East Africa was undertaken largely by explorers, colonial officers, and missionaries from Britain and Germany. Because of the "scramble for Africa" among colonial powers, the area now known as East Africa was administered partly by Germany and partly by Britain. Kenya was a British colony and Uganda, a protectorate under Britain. In contrast, Tanganyika (the mainland portion of the United Republic of Tanzania) was part of German East Africa. The islands of Zanzibar and Pemba were administered as a British protectorate but ruled by a sultan.

German explorers and scientists provided some of the earliest accounts of the area, and often the first descriptions of many of the amphibian species. Specimens collected in the field were shipped to museums in Europe, where they were studied by specialists. For example, F. Steindachner from Vienna, G. Pfeffer of Hamburg, and G. Tornier from Berlin described material collected between 1867 and 1897. Other collectors who made important contributions in this period include H. H. Johnston and W. H. Nutt. Both traveled in the Lake Rukwa area of southern Tanzania. Nutt provided the type material for two new species of amphibians described by G. A. Boulenger at the British Museum in London. W. Peters, based in Berlin, was one of the first scientists to spend time in the field, mostly in Mozambique and on the islands off Tanzania, during a six-year stay in the middle of the nineteenth century. At this time the island of Unguja, also known as Zanzibar, served as a base and staging point for many travelers and explorers who wished to make expeditions into the remote areas of the hinterland. Zanzibar was an important first stop for many collectors, partly because the British consul there from 1866 to 1880, Sir J. Kirk, not only was a naturalist but also encouraged others to collect scientific specimens on their journeys.

Museum workers based in Europe described many species of amphibians sent to them from East Africa. It is easy to understand how exciting it must have been for them to receive shipments of specimens from places they knew of only vaguely and could not hope of visiting. Their early studies are critically important to our understanding of modern taxonomy. Foremost among these workers was G. A. Boulenger, of the British Museum of Natural History in London. From 1882 to 1889 he published a "catalogue," which was effectively the list of world amphibians and reptiles at that time and which set the baseline for future systematic work. One of the earliest collections was made by a Swedish expedition that worked on Mt. Kilimanjaro, Mt. Meru, and the surrounding Maasai Steppe between 1905 and 1906.

The East Usambara Mountains in Tanzania provided a cool break for those employed at the tropical coastal town of Tanga. The mountain settlement of Amani became an important center of botanical research, and later became an important center for malaria research. F. Nieden described some amphibians in 1911 from the Amani area of the East Usambara Mountains.

In 1914 an event changed the shape of East African herpetology. A Welshman, Arthur Loveridge, took up a position at a newly formed museum in Nairobi, Kenya. Unfortunately, the hostilities between the Germans and their adversaries in World War I began in East Africa in the same year. Loveridge joined the British East African Mounted Rifles in 1915 and for three years, in addition to serving in the military, collected amphibians and reptiles. After the war, Loveridge returned to the museum in Nairobi before becoming an assistant game warden in Tanganyika. In 1924 he moved to the Museum of Comparative Zoology at Harvard University in the United States, where he was responsible for the herpetological collections.

Fortunately for East African herpetology, Loveridge not only was an excellent museum worker but also led collecting expeditions to some of the most important sites in East Africa for amphibians. These included the Uluguru and Usambara mountains of Tanzania in 1926–1927, the highlands of southwestern Tanzania in 1929–1930, Kenya and eastern Uganda in 1933–1934, and Kenya, Tanzania, and Uganda in 1938–1939. Of all the scientists who have contributed to amphibian studies in East

Africa, Loveridge must be regarded as the most important. He published large numbers of scientific papers on the distribution of species as well as descriptions of new species and documentation of their biology. Loveridge popularized his East African fieldwork in *Many Happy Days I've Squandered* published in 1944, *Tomorrow's a Holiday* in 1947, and *Forest Safari* in 1956.

An important development for amphibian studies during this period in Kenya was the establishment of the Coryndon Museum in Nairobi in 1930. This institution continues today as the only natural history museum in the East African region capable of curating large numbers of specimens, and serves as a regional center of expertise. C. A. Du Toit from Stellenbosch, South Africa, collected frogs on the Kenyan slopes of Mt. Elgon in 1934 that Loveridge later described as a new species, *Arthroleptides dutoiti*. In 1936, H. W. Parker listed the amphibians collected by an expedition to Lake Rudolf collected in 1934. D. R. Buxton included amphibians in his account of the fauna of the dry Turkana area of northern Kenya in 1936. R. Moreau and R. Pakenham published a zoogeographic analysis of the fauna of Mafia, Pemba, and Zanzibar islands in 1942.

The years of World War II, 1939–1945, saw only a few publications on East African amphibians.

Except for the continuing work of Loveridge, the period after World War II was characterized by a decline in publications from scientists based in Europe, particularly in Germany and Britain. Reasons for this included the destruction of collections in museums in Europe during the war, and the reduced funds available for research.

However, in Belgium, G.-F. de Witte and R. Laurent were actively reporting on Central African amphibians, many of which are also found in East Africa. In 1941, De Witte prepared a detailed report on the amphibians of the Albert National Park in what is now the Democratic Republic of Congo, and many of his findings are relevant to East Africa. Laurent reviewed the systematics of frogs from this area in a series of papers. In the United States, Loveridge continued to publish and in 1957 summarized many of his taxonomic findings in a checklist of amphibians and reptiles for East Africa.

In the 1950s and 1960s, other workers began to conduct field studies and thus continued Loveridge's tradition of fieldwork, which for many years had been neglected. In Tanzania in 1958, B. M. and R. F. Chapman undertook one of the first field studies in East Africa on *Bufo* toads in the Lake Rukwa area. Important collections were made in southeastern Tanzania by A. Rees of the Game Department. D. Vesey-FitzGerald collected extensively in the Lake Rukwa area while serving with a red locust control program and later on Mt. Meru when he was a staff member of Tanzania National Parks. Few field studies seem to have been conducted during this time in either Kenya or Uganda.

The 1970s saw the beginning of renewed interest in East African amphibians. Working in Uganda, R. Tinsley studied the ecology of African clawed frogs in the genus *Xenopus* and described a new species of this genus in 1973. He and his colleagues continue to study this genus. A. Schiøtz, Ronalda Keith, and A. Duff-Mackay undertook fieldwork in East Africa; a major innovation in their reports was the inclusion of sound spectrograms made from tape recordings of the advertisement calls of male frogs. The results of some of these studies were summarized in a guide to the tree frogs by Schiøtz in 1975. P. Van der Elzen and D. Kreulen published a study of vocalizations of some of the amphibians of Serengeti National Park in 1979. R. G. and M. H. Bowker conducted one of the first studies on the use of a seasonal breeding site in Kenya, the results of which were published in 1979.

From 1972 to the end of the twentieth century, African amphibian studies have benefited from both an increased interest in the group and major improvements in technology. Frog advertisement calls are known to be species-specific, and in this period sound recording and analytical tools became widely available.

Although sound spectrograms of African frogs had been published in the 1960s, the study of the systematics of eastern African amphibians using this technique started when M. Tandy and R. Keith used these analyses of calls to group species within the genus *Bufo* in 1972. M. Tandy, J. Tandy, R. Keith, and A. Duff-Mackay described a new species of toad from the dry country of East Africa, *Bufo xeros*, in 1976, which was recognized on the basis of its unique advertisement call. C. Vigny reported on a study of the vocalizations of various species of *Xenopus*, including many from East Africa, in 1979.

Not only did technology provide characters useful in distinguishing species, but improvements in computer speed and memory permitted the application of rigorous methodology to analyze the data and determine relationships. The rapid rise in phylogenetic systematics, or cladistics, is reflected in the use of this approach in contemporary frog studies.

Laboratory work in the form of detailed anatomical studies can provide many data suitable for cladistic analyses. B. T. Clarke used osteological characters for a generic-level phylogeny of African ranids in 1981. The systematic treatment of the family Hyperoliidae by R. C.

Drewes in 1984 involved a cladistic analysis of data derived largely from a detailed study of osteology and muscle characters. By the mid-1980s it had become apparent that changing species concepts and new analytical tools would lead to large-scale revisions of taxa, both at the species level and in terms of systematic arrangements. The important systematic review of the amphibians of the world, edited by Frost in 1985, served as the starting point for this process. An updated version of this list is currently available online from the American Museum of Natural History at http://research.amnh.org/herpetology/amphibia/index.html. One of the problems facing workers in East Africa and elsewhere has been the confusion caused by repeated descriptions of the same species in different areas, sometimes with the type specimens no longer available. An important series of papers by J. C. Poynton and D. G. Broadley from 1985 to 1991 on the amphibians in the neighboring Zambezi catchment has gone a long way toward sorting out the taxonomy and systematics of a number of these problem species.

Poynton discussed the distributions of sub-Saharan amphibians in 1999, in a book on the patterns of amphibians of the world edited by W. E. Duellman.

Molecular analyses have become increasingly important in taxonomic studies. The use of simplified biochemical techniques has recently placed DNA sequencing within the reach of many laboratories. The application of this technique to phylogenetic studies of East African amphibians is exemplified by C. Richards and W. Moore, who reported on the relationships of the tree frog family Hyperoliidae in 1996. More recently, A. Wieczorek and others have used DNA sequence data to investigate the *Hyperolius viridiflavus* species complex, and demonstrated that many of the earlier reported species and subspecies can be placed together as members of one large, extremely polymorphic species, while others can be recognized as distinct species.

The late 1980s and 1990s saw an increase in fieldwork on East African amphibians. In Kenya, R. C. Drewes continued studies on coastal forest amphibians at Arabuko-Sokoke. A. Grandison and S. Ashe described the reproductive biology of *Mertensophryne micranotis* in 1983, which was shown to have internal fertilization, to lay eggs in small volumes of water found in snail shells, and to produce an unusual tadpole. M. Tandy and D. Feener described a new species of toad from the Lake Turkana area in 1985.

In Tanzania, R. Pakenham updated his work published more than 40 years previously with a report on the amphibians of Zanzibar and Pemba islands in 1983. Frontier Tanzania, a cooperative research project between the London-based Society for Environmental Exploration and the University of Dar es Salaam, began surveying coastal forests and other habitats in 1989. Over 50 sites, mostly in forest, have been surveyed to date. K. Howell surveyed various forests in Tanzania for amphibians and in 1993 reviewed the forest amphibian fauna.

In Uganda, R. C. Drewes and J. Vindum published a study in 1994 that documented the amphibians of the Impenetrable Forest. C. Msuya completed a study on habitat use by Tanzanian coastal forest amphibians in 2001; this resulted in the first doctorate awarded to an East African for a field study of amphibians. The following year J. Vonesh published the effects of predation on the eggs, larvae, and adults of amphibians found in Kibale National Park in Uganda.

Although in comparison to Europe or the Americas the studies on East African amphibians are relatively few, they have already generated further interest in the fields of amphibian biology, ecology, and taxonomy. The number of local herpetologists in East Africa is small but growing, and a core of nationals trained at the postgraduate level are now actively studying amphibians and their conservation. Furthermore, long-term cooperation with institutions and individuals outside the East African region continues to expand. With the increased global interest and concern for amphibians, there is unlimited scope for studies on amphibians by the interested amateur as well as the professional biologist.

Geography and Environment

Amphibians are indirectly and directly affected by three important factors: topography, climate, and vegetation. The effects of these factors on amphibian distribution involve historical and recent events, as well as the ongoing interactions between them. We briefly outline these three factors for East Africa.

Topography

The landforms, such as rift valleys and high mountains, influence the climate and produce habitats that are often isolated. This has led in many cases to the evolution of endemic species of animals and plants associated with these habitats. The rugged scenery of East Africa is the central reason for the very high numbers of endemic amphibians found there.

The topography of East Africa is dominated by a series of volcanoes. Major features include a coastal strip, central plains, the rift valley, volcanoes, and ancient block faulted mountains (Fig. 2). The altitudinal range is great, stretching from sea level to the summit of Mt. Kilimanjaro at 5896 m.

Among the most notable East African rivers are the Victoria Nile, Albert Nile, Katonga, and Aswa in Uganda, which eventually drain northward into the Nile; the Tana and Galana in Kenya, which drain into the Indian Ocean from the highlands; and the Pangani, Ruaha, Kilombero, Malagarasi, Rufiji, and Rovuma in Tanzania, which drain eastward from the inland plateau. The catchment areas of these rivers are vast, and their banks serve as preferred habitat for many amphibians.

Some of the most striking features of the East African landscape are landforms resulting from the tectonic processes of faulting and volcanic

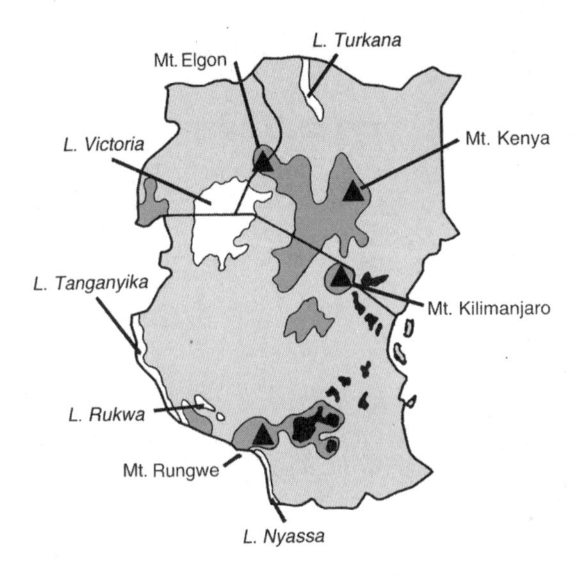

Fig. 2 Map showing the major lakes and mountains of East Africa. The Eastern Arc Mountains are indicated in black. Areas above 1500 m are dark gray.

activity. These include escarpments as well as volcanic peaks of the East African rift system.

The rift valley was formed about 17 million years ago. Its eastern branch is known as the Great or Gregory Rift, while the western branch is referred to as the Albertine Rift. The western rift includes Lakes Malawi, Rukwa, and Tanganyika in Tanzania, and Lakes Edward and Albert in Uganda. The Great Rift includes Lakes Eyasi and Natron in Tanzania (the latter borders Kenya) and Lakes Naivasha, Elmenteita, Nakuru, Bogoria, Baringo, and Turkana in Kenya. These lakes have varied in size down to relatively small puddles over geological time. Lake Victoria, the largest natural lake in Africa, is found between the two rift valleys. It is presently up to 80 m deep, but in the past has often dried up into a few small isolated water bodies. Rift valleys serve both as barriers to species on either side and as a dispersal route for species associated with the riverside vegetation.

Volcanic mountains associated with the rift valley are of two ages. The older includes Mt. Elgon, formed about 20 million years ago. Other mountains of volcanic origin such as Mt. Kilimanjaro, the Ngorongoro crater highlands, and Mt. Kenya are more recent and originated about 1 to 2 million years ago. Volcanic soils are typical of the Kenyan highlands;

the Kilimanjaro, Meru, and Mt. Rungwe areas of Tanzania; and the Kigezi area of Uganda. These soils are fertile, and many interesting amphibians are found in the forests that grow there.

A major feature in the eastern portions of Kenya and Tanzania is a chain of crystalline, block faulted mountains known as the Eastern Arc Mountains. These include the Teita Hills in Kenya and, moving southward, the North and South Pare, West and East Usambara, Nguu, Nguru, Ukaguru, Uluguru, Rubeho, and Udzungwa mountain ranges in Tanzania. The southern highlands of Tanzania are separated from the Eastern Arc by the Makambako gap. The various mountains comprising the Eastern Arc have been isolated for a long time, resulting in the evolution of endemic amphibians, such as the species of forest toad *Nectophrynoides*, on many of them.

The coastal strip of Kenya and Tanzania is generally narrow and rises to higher ground farther inland. This strip has served as a route for amphibian movement up and down the coast. Some species, such as the tinker reed frog *Hyperolius tuberilinguis* are found from Kenya southward along the coastal strip to the tropical parts of South Africa.

Major offshore islands include Zanzibar, Pemba, and Mafia. Numerous small islands dot the coast, but these remain largely uninvestigated by biologists and many lack permanent sources of freshwater. The coastal islands become part of the mainland when the sea level is low, and are distinct when the sea rises. The islands have been separated by high sea levels for the last 8000 years, which has led to the development of some interesting endemic amphibians. Islands also occur in the larger freshwater lakes.

All of these features, past and present, affect the distributions of amphibians on both a larger and a smaller, more local scale. Topography influences climate, and the two together determine the patterns of vegetation and animal distribution.

Climate

Most amphibians depend on relatively moist conditions and need water in which to deposit eggs and for the successful development of larvae. Therefore, the amount of rainfall and its distribution are critical aspects of the environment. The single most noticeable climatic feature in the tropics is the alternation between dry and wet seasons. The majority of amphibians are seasonal breeders during the wet periods.

Although often unpredictable and extremely local, the rainfall in East Africa nevertheless tends to follow an annual cyclic pattern (Fig. 3). This pattern often has two peaks but may have only one, depending on a number of factors.

Two periods of maximum rainfall, one beginning in October (short rains, or *vuli* in Swahili) and another in April (long rains, or *masika*), occur. But East Africa extends from 5° north latitude to 12° south latitude, and this bimodal rainy season pattern is most distinct in the areas nearest the equator; to the south, the two rainy seasons tend to merge into one.

Despite these typical patterns, one location may receive rain and another only 50 or 100 m away may remain entirely dry, as frequently occurs along the coast.

The general rainfall pattern may be further modified by the effects of physical features, especially mountain ranges. One side of the mountains may be in an extremely dry "rain shadow," with resulting differences in vegetation and microclimate conditions. In contrast, the areas immediately surrounding Lake Victoria tend to experience more regular and pre-

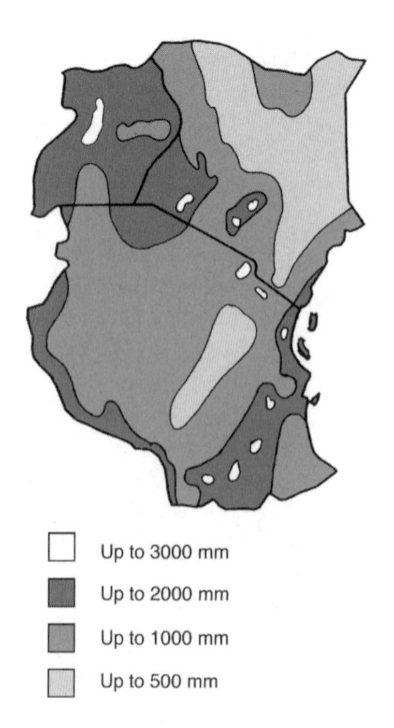

Fig. 3 Map showing the average annual rainfall patterns of East Africa.

dictable rainfall. Climate is modified by elevation; highland and mountain areas are much cooler than lower-lying areas. While the coastal plains and other lower areas may be extremely hot, highlands and mountains regularly receive frost. These same montane areas experience large daily variations in temperatures, with high insolation during the day but freezing conditions at night. In western Uganda, glaciers top the Ruwenzori range at over 5000 m. On the highest peaks of Mt. Kilimanjaro and Mt. Kenya, snow is present year-round.

Vegetation

We use a simplified vegetation scheme to describe the habitats of East Africa (Fig. 4). Vegetation reflects the complex interaction of climate,

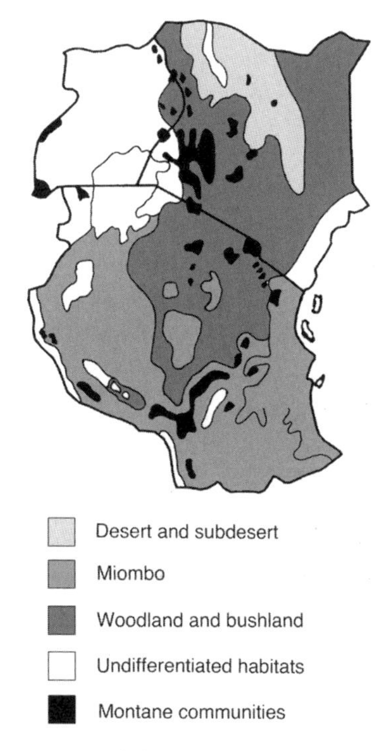

Fig. 4 A simplified vegetation map of East Africa. Miombo is dominated by *Brachystegia* and *Julbernardia* species. Wood and bushland are dominated by *Acacia* and *Commiphora* species. The undifferentiated habitats are mainly wooded and bushed, including tracts of forest, grassland, and swamp. The vegetation changes caused by subsistence agriculture are not indicated.

soils, and altitude. In many cases it is possible to associate amphibian distribution with broad vegetation types and zones.

Forest

In East Africa as a whole, only a relatively small portion of the total land area is covered with closed natural forest. In Tanzania and Kenya, about 2% or slightly less of the land is covered by closed forest, while in Uganda the figure is about 3.6%.

Traditionally, forest is often categorized as lowland (up to 800 m), submontane (800 to 1200 m), and montane (1200 to 2000 m), but there is some disagreement among specialists as to the exact elevational boundary limits of the categories and the terminology applied to each type. The particular amphibians associated with a type of forest will also depend on the geological history of the area.

For example, amphibian species typical of forest in western Kenya, Uganda, and (although no studies have been done there) presumably northwestern Tanzania west of Lake Victoria show affinities with the adjacent larger block of Guinea-Congolean forest to the west. An example is the Congo spiny reed frog *Afrixalus osorioi* in Uganda and Kenya.

Species that are found in forest on mountains in western Uganda and that extend westward to the eastern part of the Democratic Republic of Congo include Christy's tree frog *Leptopelis christyi*, the Congo wot-wot *Phylictimantis verrucosus*, and Lang's reed frog *Hyperolius langi*.

Species on the very high mountains in western Uganda that extend farther west include the Kivu tree frog *Leptopelis kivuensis* and the brown reed frog *Hyperolius castaneus*.

Forest species found on mountains and highlands of volcanic origin include the montane reed frog *Hyperolius montanus* and bladder reed frog *Hyperolius cystocandicans* in Kenya and the Ruwenzori river frog *Afrana ruwenzorica* in Kenya and Uganda.

Coastal forest is a specialized category of dry lowland forest. Typical species include the forest spiny reed frog *Afrixalus sylvaticus*; the woodland toad *Mertensophryne micranotis*, which lays its eggs in the water of empty land snail shells; and the Mrora toad *Stephopaedes howelli*, endemic to Mafia and Zanzibar islands, which probably lays its eggs in water in crevices of trees and buttress roots.

Because they have been isolated for millions of years from the Guinea-Congolean forests, the Eastern Arc Mountains differ greatly from the west African forests in plant species composition and in amphibian fauna. The forests found on mountains of relatively recent volcanic origin have yet another set of amphibians.

Typical submontane and montane forest on the Eastern Arc Mountains is habitat for a number of species that are endemic to that chain of mountains, including the Uluguru tree frog *Leptopelis uluguruensis* and the long-fingered forest frog *Probreviceps macrodactylus*. Genera that are endemic to particular mountain blocks within the Eastern Arc include *Parhoplophryne* (black banded frog), *Hoplophryne* (three-fingered frogs), and *Callulina* (warty frogs).

The coastal forests of Tanzania and Kenya have been recognized as having strong affinities with those of the Eastern Arc Mountains, including high species diversity and endemism. Their importance has been emphasized by their recognition as the Eastern Arc/Coastal Forest Global Biodiversity Hotspot.

Specialized Montane Habitats

In montane habitats, bamboo is often found on the moister slopes at elevations of 2500 to 3000 m. In Tanzania, the robust forest toad *Nectophrynoides viviparus* is found in bamboo at these elevations, although it is not restricted solely to this vegetation type. Grassland may also be present at high elevations, and the Karissimbi tree frog *Leptopelis karissimbensis* is known from high-altitude grassland and heather in Uganda. At 3000 to 3500 m, a tree heath or ericaceous zone often occurs as part of a distinct Afro-Alpine vegetation association.

Species found at these high elevations often have special adaptations of physiology and behavior. For example, the Uluguru forest frog *Probreviceps uluguruensis* is a high-elevation species in Tanzania's Uluguru Mountains that lives in the litter and lays its eggs in burrows. *Nectophrynoides* species found at high elevations are apparently all live-bearing. The Mt. Meru stream frog *Strongylopus merumontanus* is a high-altitude species associated with wetlands.

Woodland

In contrast to forest, woodland refers to stands of trees at least 8 m (up to about 18 m) in height, with an open or continuous but not thickly interlaced canopy with a cover of more than 20%. The Swahili term *miombo* is used to indicate woodlands dominated by *Brachystegia* and *Julbernardia* species, which are widespread on much of the plateau country in Tanzania.

Bushland

Bushland is a term used for groups of woody plants with a shrub canopy less than 6m in height with a few emergents, and a canopy cover of more than 20%. It may be present in extremely dry areas of low rainfall, but other types of bushland are also found in seasonally flooded areas where the black cotton soil is waterlogged for long periods of time, conditions that do not permit the growth of large trees.

In East Africa, the term *savanna* has often been loosely applied to vegetation which includes both grassy plains and open woodland and bushland. Many of the more widespread species occurring in East Africa are found in woodland and bushland. The silvery tree frog *Leptopelis argenteus*, the marbled snout-burrower *Hemisus marmoratus*, the Senegal kassina *Kassina senegalensis*, and the southern foam-nest frog *Chiromantis xerampelina* are examples.

Grassland

Grasslands are those areas mainly covered by grasses, either tall or short, and some other herbs. In East Africa extensive grassland areas are often periodically burnt. Seasonally inundated low-lying areas, including floodplains, are often grasslands.

The Nairobi toad *Bufo nairobiensis*, African bullfrog *Pyxicephalus adspersus*, and the Guinea snout-burrower *Hemisus guineensis* are typical species of grassland that is seasonally flooded. Montane grasslands are a specialized habitat of relatively small extent in East Africa.

Seasonal Wetlands

Seasonal wetlands could be termed *grasslands* or *bushland* in the dry season, but during the rainy seasons they are important breeding sites for amphibians when they are inundated as pools. Even small pools, ponds, and pans may be extremely important for local populations of amphibians. In contrast, vast expanses of floodplain are inundated annually in East Africa; such areas are often dominated by black cotton soils and are known as *mbuga* in Swahili. Rees's toad *Bufo reesi* is known only from the Kilombero valley floodplain in Tanzania, but many widespread species of frogs make use of seasonal breeding pools in and at the edge of floodplains.

Permanent Wetlands

Permanent wetlands include the larger freshwater lakes and smaller water bodies like swamps, pools, and ponds. Permanent swamps, including papyrus swamps, which are often extensive, are a major feature of the larger East African lakes and water bodies. African clawed frogs (genus *Xenopus*), the eastern groove-crowned bullfrog *Hoplobatrachus occipitalis*, and several species of puddle frogs (genus *Phrynobatrachus*) are typical of such habitats. Permanently wet areas may be found at the edges of streams and rivers and in the deltas of large rivers, and these are the only sites at which the mud caecilian *Schistometopum gregorii* has been found in Kenya and Tanzania. The Kihansi Gorge waterfall wetland is a specialized habitat that is dependent on the spray from a waterfall in the Kihansi River of Tanzania. This is the only locality where the endemic Kihansi spray toad *Nectophrynoides asperginis* is found.

Human-Modified Habitats

Habitats such as drainage and irrigation ditches, dams, fish and garden ponds, and gardens may be locally important habitats for amphibians. A widely distributed species that is able to tolerate urban environments as well as those of modern agriculture is the guttural toad *Bufo gutturalis*.

In East Africa, human-modified habitats also include extensive monocultures such as maize, ornamental flower, pyrethrum, sisal, tea, and coffee plantations and plantation forests. In most cases, the species under cultivation are exotic, and generally speaking, these large-scale agricultural areas have been created at the expense of natural habitats more suited to amphibians. Even a crop such as cardamom, grown by small-scale farmers, is planted at the cost of removing the understory forest vegetation and associated amphibians.

Semidesert and Desert

Kenya is the driest of the three East African countries, with over 75% of its area classified as arid or semiarid and receiving less than 200 mm of rain annually. The north and northeast of Uganda is also relatively dry. Evidence suggests that there was once a dry corridor from the northeast of Africa to southern Africa as well as an extension of the Somali arid zone from Tsavo in Kenya down through central Tanzania. These dry portions of East Africa are generally sparsely vegetated. Species found under these conditions include the Lugh toad *Bufo lughensis* and the desert toad *Bufo xeros*.

Conservation

Amphibians are usually only protected indirectly, when they happen to occur within an area set aside for the conservation of other animals, more often than not those that are large and hairy. To appreciate the need for amphibian conservation, we must ask what role frogs and caecilians play in the scheme of things, and why it is desirable to conserve them.

The Importance of Amphibians in the Ecosystem

Amphibians are an important but often overlooked component of most terrestrial and freshwater aquatic ecosystems. During the breeding season, extremely high densities of adults and tadpoles may occur. They consume large amounts of food. Adult frogs are important predators on invertebrates; the positive role of frogs in feeding on insects such as mosquitoes as well as insects that feed on crops is often overlooked. Tadpoles may be filter feeders, grazers on algae, or carnivorous. The ecological role of the limbless gymnophionans that live mostly underground is not well known, but they appear to be important in aerating the soil and circulating nutrients that maintain soil quality and fertility.

Amphibians themselves are important food items for other animals, including other amphibians, at all stages of their life cycles. Frog eggs are fed on by invertebrates such as crabs and fly larvae as well as by frogs and other vertebrates. Tadpoles are eaten by insects, fish, adult amphibians, and reptiles such as monitor lizards. In addition, they may form an abundant food source that attracts birds such as storks and herons. Adult frogs are eaten by a wide variety of predators, including birds and reptiles, especially snakes and monitor lizards. Some bats are also known to feed on frogs. Details of known predators are included under the species accounts.

It has been recognized that on a global basis, some amphibian populations are in decline, and rapid extinctions have been observed for certain species. Only a few populations of East African amphibians have been monitored, and thus it is difficult to say whether or not populations in the region are generally stable or declining.

Threats to Amphibians

A number of threats may affect populations of East African amphibians. These include habitat destruction, pollution, the pet trade, and others.

Habitat Destruction

Habitat alteration and destruction is the single most important threat to East African amphibians. Some species can tolerate a wide variety of environmental conditions and vegetation types but are threatened by agriculture, particularly when large areas are planted with a single crop. Many of the endemic species are limited to a particular habitat such as forest or high grassland, and are unable to exist where these have been modified, such as when natural forest is removed to plant pine trees.

Species such as Parker's tree frog *Leptopelis parkeri* and the long-fingered forest frog *Probreviceps macrodactylus* are dependent on forest. They are restricted to closed forest habitat and are most at risk from any change in forest quality or decrease in the relatively small area of remaining forests.

Because relatively little is known about the detailed life history and ecological requirements of these species, it is difficult to say exactly why and how they are dependent on forest. Members of the genus *Probreviceps* (the forest frogs) do not need free water to breed but require moist litter and soil in which to burrow and lay eggs. Two members of the genus *Stephopaedes* (the forest toads) are found only in forest, one in a few patches of the East Usambara Mountains and the other on remnant forest on Mafia and Zanzibar islands. Christy's tree frog *L. chrysti* is known from forest in Uganda and the plain tree frog *L. modestus* from forest in western Kenya. These forest species are not found in other habitats and need to be protected by maintaining sufficient natural forest.

It is not clear if altering the physical structure of a forest—for example, removing most of the straight-stemmed saplings of a few particular tree species used for building poles—might affect amphibians directly. Studies suggest that such removals will alter the physical features and species characteristics of the forest in the long term. These changes in tree species composition and physical structure may affect amphibians directly by eliminating their hiding places in tree holes and under bark, or the calling sites for males. Changes to a forest that affect its size, alter the leaf-litter composition and therefore the invertebrates available for food, and the physical and chemical composition of the soil will likely have direct and indirect negative effects on amphibians.

Many forested areas are associated with high rainfall and are therefore prime agricultural areas. In East Africa, large tracts of land have been cleared for coffee and tea plantations. In addition to the physical loss and fragmentation of forest, these types of agriculture often involve the use of agrochemicals such as insecticides, which may themselves have negative effects on amphibians.

The remaining smaller forests are often under pressure from agriculturists as well as for the felling of timber and the extraction of other wood products, especially building poles. The coastal forests of Kenya and Tanzania are for the most part small and highly fragmented and are especially vulnerable. Examples of amphibians at risk from coastal forest destruction include the Mrora forest toad *Stephopaedes howelli* and the Usambara forest toad *S. usambarae.*

In some East African forested areas, more than 40% of forest cover has been lost in the last 50 years. In the Eastern Arc forests, 76% of the cover has been lost in the last 200 years. However, this reduction in forest cover has been happening for a long time. In some places, considerable areas of "virgin" forest on the Eastern Arc Mountains had been cleared by humans in connection with iron smelting some 2000 years ago.

Much of East Africa is woodland of varying richness and quality. Many areas are a mosaic of woodland and grassland, which have been traditionally used for seasonal agriculture and grazing. In some places, such habitats have been taken up by large-scale farming and ranching. Sisal plantations form one type of large-scale production that renders much of the habitat unsuitable for any natural fauna.

In the highland areas of Kenya, for example, natural vegetation, mostly woodland and grassland, was cleared to plant eucalyptus and conifer plantations of pine and cypress. Wattle, valued for the tannins in its bark, was planted in extensive areas around Njombe and Mufindi in Tanzania. Conditions in these plantation forests are largely unsuitable for natural fauna and flora, and while no quantitative data are available specifically for amphibians, plantations must be seen as having a negative effect.

In the lowland regions of East Africa, large areas of forest and other vegetation have been cleared to plant sugar cane. Burning and the use of agrochemicals associated with the production of sugar cane is generally perceived as unfavorable to local fauna and flora. However, a study in southern Africa indicates that frog species that tolerate the original clearing of the natural habitat may be able to survive in sugar cane plantations.

Recently, large blocks of rich miombo woodland have been cleared for teak, a nonindigenous tree of high value for its timber. The effect of this is still being assessed, but it is likely that the removal of a number of species of plants and animals will result in a deterioration of the habitat for amphibians.

Because many amphibians require either seasonal or permanent water bodies for reproduction, wetlands are especially important. Any human activities that negatively affect wetlands and their surrounding vegetation will also affect amphibians. Such activities include clearing of surrounding vegetation, changes in water flow, and pollution.

Development Projects

The construction of hydropower dams, hotels, lodges, roads, and mines may create conditions that are harmful to amphibians, many of which must undergo seasonal movements to and from breeding sites. Species such as the Senegal kassina *Kassina senegalensis* and the edible bullfrog *Pyxicephalus edulis* may move hundreds of meters from breeding ponds to dry-season refuges such as forest and woodland patches. If their route is blocked or altered, they may be unable to complete the normal movement patterns necessary for successful breeding and dispersion.

An East African species that is known to have decreased greatly in number due to the destruction of its highly specialized habitat is the Kihansi spray toad. The Kihansi River in Tanzania flows off the escarpment of the Udzungwa Mountains, known for their globally high biodiversity values. The river flows through a narrow gorge, and spray from large falls has resulted in the formation of a specialized spray-dependent wetland. A dam that reduced the flow of water over the falls and in the river gorge was built for the Kihansi hydropower project. Although an environmental impact assessment had been conducted, it was not sufficiently detailed in its fieldwork. In 1996, after construction was started, a species of live-bearing toad new to science, the Kihansi spray toad *Nectophrynoides asperginis*, was discovered in the spray wetland. Studies and further fieldwork indicated that the species was endemic only to the spray-drenched portions of the Kihansi river gorge. The funders and proponents of the project were unwilling to consider other options to address the conservation of the toad or, indeed, the entire fauna and flora of the gorge. Once the dam was built and in operation, the spray-drenched wetland on which the toad is dependent decreased in size by 98% due to the removal of water for power generation. This resulted in a well-publicized environmental disaster. In December 2000, five hundred of the toads were moved to specialized captive breeding facilities in the United States, and later an artificial sprinkler system was installed in the gorge in an attempt to maintain what little remained of the habitat. By mid-2003 the population of spray toads in one of the largest wetlands in the gorge was showing signs of recovery, having increased more than tenfold over its lowest point. Late in 2003 it was reported that the population had crashed, with a loss of 95% of the animals in the wild. Agrochemicals in use upstream of the dam were suspected as the cause, although the ubiquitous chytrid fungus has also been implicated.

The tribulation of this tiny toad is probably an indication of things to come in the East African region as the pressures for "development" and the need for hydropower increases due to a growing and increasingly urbanized population.

The Kihansi spray toad has come to serve as a flagship species and has raised national, regional, and international awareness of the need to include amphibians in environmental studies for development projects.

Pollution

Amphibians may be vulnerable to pollution at all stages of their life cycle. Agricultural chemicals such as insecticides, herbicides, and fertilizers can be harmful to them. These chemicals are used by both large farms and smallholders. Some agrochemicals act as hormones that interfere with reproduction in amphibians by disrupting the normal development of the reproductive organs. If the use of agrochemicals is not tightly controlled, their entrance into the natural ecosystem may have negative effects on amphibians, especially in the aquatic environment.

The insecticide DDT has been banned in Europe, North America, and elsewhere largely because of its negative effects on the environment, especially its concentration in the food chain. Experience in other parts of the world suggests that the use of DDT, especially if aerial spraying is deployed, may cause serious environmental problems relating to the food chains of freshwater ecosystems. Until recently it had been gradually phased out in East Africa, but its use is now being reconsidered to help in the battle against the malaria mosquito.

There has been a general trend for an increase in the use of pesticides in East Africa. However, there is often not the concomitant effort at education on how to use them effectively and safely. In countries where a large portion of the population lives as peasant farmers, often with little education and scientific training, there is considerable risk to the environment. Literally tons of old pesticides are stored improperly. Simply handling the associated cleanup is an expensive operation fraught with dangers to humans and to the environment.

Improper usage of pesticides that can affect the environment, and directly or indirectly amphibians, includes the application of excessive pesticide to crops, which leads to runoff into water bodies; the employment of insecticides to poison grain-eating waterfowl, such as ducks and geese; and the use of insecticides as a fish poison in the Lake Victoria basin.

Pet and Live-Animal Trade

In recent years a worldwide trade in amphibians has developed, and in East Africa thousands of some species have been shipped abroad. Most species in the pet and live-animal trade are large or colorful, such as the banded rubber frog *Phrynomantis bifasciatus*, some 12,000 of which were exported from Tanzania in the period 1989–1998. Other species exported in large numbers over the same time period were reed frogs (genus *Hyperolius*), 10,100, and bullfrogs (genus *Pyxicephalus*), 4600.

The Convention on the International Trade in Endangered Species of Wild Fauna and Flora (CITES) controls the international trade in endangered species. It establishes lists of species for which there is a need for monitoring and controlling trade. For species in Appendix I, those threatened with extinction, trade is permitted only in exceptional circumstances. Appendix II lists species not necessarily threatened with extinction, but for which trade is controlled and permitted under restriction, including that of recording and reporting all animals imported and exported. Appendix III contains species that are protected in at least one country which has asked other CITES parties for cooperation and assistance in controlling the trade.

The only East African frogs on a CITES appendix are members of the genus *Nectophrynoides* (forest toads), all of which are listed in Appendix I. The most up-to-date listings of the CITES appendices can be found at http://www.cites.org.

Collection of Amphibians for Food and for Traditional Medicine

In some parts of East Africa, the larger species of frogs such as bullfrogs (genus *Pyxicephalus*) are eaten. In most places this is not likely to have a serious affect on the species as a whole, but could result in the depletion of local populations.

Fungal and Parasitic Infections

A new potential threat to amphibians has recently been identified in Africa. Some frogs have been found to be infected with a chytrid fungus, which attacks the skin. This fungus is believed to be responsible for the demise of frog populations elsewhere in the world, so its presence here is of concern. It has been identified in clawed frogs and ridged frogs in East Africa. A variety of parasitic worms also infect amphibians, but usually these alone do not cause a threat to populations. But if the amphibians are already experiencing other forms of stress—for example, from pollution—the effects of the parasites might be more serious.

Conservation Efforts

Although East Africa is renowned for its large mammal biodiversity and its extensive protected-area network of national parks and game reserves, such as the Serengeti, a number of important habitats and the animals associated with them are not protected by these parks, which were established to preserve large mammals.

In general, most of the East African amphibian species of limited distribution are those found in mountain forests, especially those of the Eastern Arc Mountains in Kenya and Tanzania. However, the coastal forests also contain endemic species of amphibians. Since the highest number of species, and the highest number of endemic species, occur in forests, those of the Eastern Arc and coastal forests have been designated among the world's top-ten biodiversity hotspots. The Albertine Rift forests are also recognized for their high biodiversity values.

A major challenge for the future is to see that habitats which contain high amphibian biodiversity are included within the East African protected-area network of national parks and other protected-area categories such as nature reserves.

Conservation Status of East African Amphibians

The World Conservation Union (IUCN) has established a set of criteria that objectively assess the threat of extinction for each species of animal. The IUCN categories for organisms that are threatened with extinction are Critically Endangered, CR; Endangered, EN; or Vulnerable, VU. Those species not considered threatened are regarded as of Least Concern, LC; or Near Threatened, NT. Species that have not been assessed are listed as DD, Data Deficient.

In the East African region, according to the Global Amphibian Assessment (GAA), 52 species are regarded as threatened and are placed in the following categories: Critically Endangered (6), Endangered (17), and Vulnerable (29). These are listed in Table 1. We have omitted species of uncertain status found on the GAA website. Some of the newer species in this book are not yet evaluated by the GAA. At least 30% of the East African amphibian species are under threat of extinction.

Table 1

Scientific Name	Common Name	IUCN Status
Order Anura		
Family Arthroleptidae		
Arthroleptis nikeae	Nike's squeaker	EN
Arthroleptis tanneri	Tanner's squeaker	VU
Schoutedenella xenodactyla	Eastern squeaker	VU
Family Bufonidae		
Bufo brauni	Braun's toad	EN
Bufo uzunguensis	Udzungwa toad	VU
Churamiti maridadi	Beautiful tree toad	CR
Nectophrynoides asperginis	Kihansi spray toad	CR
Nectophrynoides cryptus	Uluguru forest toad	EN
Nectophrynoides minutus	Dwarf forest toad	EN
Nectophrynoides poyntoni	Poynton's forest toad	CR
Nectophrynoides pseudotornieri	Pseudo forest toad	EN
Nectophrynoides vestergaardi	Vestergaard's forest toad	EN
Nectophrynoides viviparus	Robust forest toad	VU
Nectophrynoides wendyae	Wendy's forest toad	CR
Stephopaedes howelli	Mrora forest toad	EN
Stephopaedes usambarae	Usambara forest toad	EN
Family Hyperoliidae		
Afrixalus morerei	Morere's spiny reed frog	VU
Afrixalus orophilus	Montane spiny reed frog	VU

Table 1 (*Continued*)

Scientific Name	Common Name	IUCN Status
Afrixalus sylvaticus	Forest spiny reed frog	EN
Afrixalus uluguruensis	Uluguru spiny reed frog	VU
Hyperolius castaneus	Brown reed frog	VU
Hyperolius cystocandicans	Bladder reed frog	VU
Hyperolius discodactylus	Highland reed frog	VU
Hyperolius frontalis	Pale-snouted reed frog	VU
Hyperolius kihangensis	Kihanga reed frog	EN
Hyperolius minutissimus	Dwarf reed frog	VU
Hyperolius tannerorum	Tanner's reed frog	EN
Leptopelis barbouri	Barbour's tree frog	VU
Leptopelis karissimbensis	Karissimbi tree frog	EN
Leptopelis parkeri	Parker's tree frog	VU
Leptopelis uluguruensis	Uluguru tree frog	VU
Leptopelis vermiculatus	Vermiculated tree frog	VU
Phlyctimantis keithae	Keith's wot-wot	VU
Family Microhylidae		
Hoplophryne rogersi	Roger's three-fingered frog	EN
Hoplophryne uluguruensis	Uluguru three-fingered frog	VU
Parhoplophryne usambarica	Usambara black-banded frog	CR
Probreviceps macrodactylus	Long-fingered forest frog	VU
Probreviceps rungwensis	Snouted forest frog	VU
Probreviceps uluguruensis	Uluguru forest frog	VU
Spelaeophryne methneri	Scarlet-snouted frog	VU
Family Pipidae		
Xenopus vestitus	Jacketted clawed frog	VU
Xenopus wittei	De Witte's clawed frog	VU
Family Ranidae		
Arthroleptides dutoiti	Mt. Elgon torrent frog	CR
Arthroleptides martiensseni	Martienssen's torrent frog	EN
Arthroleptides yakusini	Southern torrent frog	EN
Phrynobatarchus irangi	Irangi puddle frog	EN
Phrynobatrachus krefftii	Krefft's puddle frog	EN
Phrynobatrachus ukingensis	Ukinga puddle frog	VU
Phrynobatrachus uzungwensis	Udzungwa puddle frog	VU
Phrynobatrachus versicolor	Green puddle frog	VU
Strongylopus kitumbeine	Kitumbeine stream frog	VU
Strongylopus merumontanus	Mt. Meru stream frog	VU

Strategies for Conserving Amphibians

The mounting concern over what appear to be global declines in amphibian populations has led to calls for a reassessment of the threats to East African amphibians. The IUCN Red Data book and its online Web version (http://www.redlist.org) are a positive step. Each of the three East African countries has different categories of protected areas, and each has its own priorities; nevertheless, actions can be taken to improve the conservation status of amphibians in the region.

1. Ensure that amphibians continue to receive attention and are kept in the public eye. In conjunction with these efforts, use indigenous knowledge and names to help popularize amphibians and make them more familiar to the general public.
2. Analyze existing distributional data to determine which species are included within existing protected areas such as national parks, game reserves, and nature reserves. This can be done using existing data and should serve as a focal point for establishing an amphibian atlas mapping project. Once an analysis of distributional data has been completed, steps can be taken to ensure that populations currently outside of protected areas receive the appropriate conservation measures.
3. Involve villagers, community-based organizations, and private-land owners in conserving habitats for amphibians.
4. Ensure that amphibians receive the necessary attention and consideration when planning and implementing development projects, especially those involving infrastructure.
5. Initiate-long term studies of breeding sites and identify breeding areas that are critical.
6. Monitor populations to detect short- and long-term changes.
7. Monitor the trade in amphibian species as well as the trade process; where appropriate, use existing national legislation or implement CITES regulations to ensure that trade in amphibians is sustainable.

The long-term future of East African amphibians depends on the will of those employed in the public and private agencies involved with conservation. The readers of this book have an important role to play in initiating and maintaining discussion to keep amphibians on the conservation agenda.

Classification

Frogs, salamanders, and caecilians are placed in the class Amphibia, which is one of the divisions of the vertebrate animals, or Vertebrata, which includes fishes, amphibians, reptiles, birds, and mammals. All the amphibians are soft-skinned, most have lungs, and most have free-swimming larvae, which develop into adults.

The class Amphibia is divided into three orders.

Order Caudata

Caudata includes the salamanders and their relatives, all of which have a long tail and usually two pairs of limbs. They are found in North and South America, Europe, Asia, and Africa north of the Sahara.

Order Gymnophiona

This order includes all the caecilians, which are elongated, limbless animals that superficially resemble large earthworms. They also possess annuli that resemble the body segments of earthworms. They occur worldwide in the tropics. Two families occur in East Africa.

Family Caeciliidae

The two genera in East Africa are *Schistometopum* with one species and *Boulengerula* with five. This is a group of widespread but little-known animals.

Family Scolecomorphidae

These caecilians have small tentacles that fit into sockets on each side of the head. The eyes are attached to the base of the tentacles. There is only one genus, *Scolecomorphus*, with three species.

Order Anura

This order consists of the tailless amphibians and includes all the common frogs. They occur worldwide except in the Arctic and Antarctic. There are over 20 families of frogs, of which 8 occur in East Africa.

Family Arthroleptidae

Commonly called squeakers, these forest-floor frogs are usually small and do not have free-swimming tadpoles. Two genera are known, *Arthroleptis* and *Schoutedenella*, with six and five species in East Africa.

Family Bufonidae

This family consists of six genera: the toads *Bufo* (24 species), red toad *Schismaderma* (1 species), forest toads *Nectophrynoides* (12 described and 2 undescribed species), and *Stephopaedes* (3 species), and the tree toad *Churamiti* and the woodland toad *Mertensophryne* (1 species each). Small to large species of terrestrial or arboreal frogs, they usually have thick glandular skin and small dark tadpoles. One group is ovoviviparus, retaining the eggs in the oviduct, where they develop into small frogs.

Family Hemisotidae

Commonly known as snout-burrowers, these smooth-skinned frogs have strong arms and a hard pointed snout. They burrow snout-first. They lay eggs terrestrially in burrows. The tadpole has a large fin, with the base of the tail muscle covered by a sheath. Only one genus, *Hemisus*, is recognized in the family, with three species in East Africa.

Family Hyperoliidae

There are five genera: the reed frogs *Hyperolius* (32 species), the spiny reed frogs *Afrixalus* (12 species), the tree frogs *Leptopelis* (14 species), the running frogs *Kassina* (3 species), and the wot-wots *Phlyctimantis* (2 species). These mostly arboreal frogs have large discs on the fingers and toes. Many are brightly colored. The tadpoles vary from large-finned pond types (*Kassina*, *Phlyctimantis*), to slender forms with a little fin (*Leptopelis*).

Family Microhylidae

Seven genera are found in East Africa. Five are only known from the Eastern Arc Mountains of Tanzania and Kenya: the warty frogs *Callulina* (two species), the three-fingered frogs *Hoplophryne* (two species), the black-banded frog *Parhoplophryne*, the forest frogs *Probreviceps* (four species), and the scarlet-snouted frog *Spaeleophryne*. The two widespread genera are the rain frogs *Breviceps*, with two species, and the rubber frog *Phrynomantis*. These frogs have a narrow head and a small mouth. The rain frogs burrow backward and have eggs that develop directly into small frogs, without a free-swimming tadpole stage. The rubber frogs climb into crevices and have gregarious tadpoles that possess tentacles.

Family Pipidae

These streamlined frogs spend their lives in water. They are commonly known as clawed frogs. There is only one genus, *Xenopus*, with six species in East Africa. The tadpoles are transparent and gregarious.

Family Ranidae

These common frogs have a wide range of different body shapes and biology. Some have direct development, but most have typical tadpoles. There are 11 genera in East Africa: the river frogs *Afrana* (3 species), the white-lipped frogs *Amnirana* (2 species), the torrent frogs *Arthroleptides* (3 species), the dainty frog in the genus *Cacosternum*, the ornate frogs *Hildebrandtia* (2 species), the groove-crowned bullfrog *Hoplobatrachus*, the puddle frogs *Phrynobatrachus* (21 species), the ridged frogs *Ptychadena* (13 species), the bullfrogs *Pyxicephalus* (2 species), the stream frogs *Strongylopus* (2 species), and the sand frogs in the genus *Tomopterna* (4 species).

Family Rhacophoridae

These large, camouflaged frogs have big adhesive discs on the fingers and toes for climbing on trees. The eggs are laid in a foam nest, and the young tadpoles drop into the water to continue their development. There are four species in a single genus, *Chiromantis*.

Identification

Introduction to the Keys

The following information is concerned with ways of identifying adult amphibians. The same principles apply to tadpole identification, and more detail is presented in the chapter on tadpoles.

One of the aims of this book is to enable readers to identify amphibians reliably. This can be difficult, however, as frogs and caecilians have very few characters that are easily seen and that are useful for this purpose. Caecilians are rather similar to each other, and a stereomicroscope aids in viewing the details needed for identification.

Generally, all frogs within a family or genus have similarly shaped bodies, webbing, tubercles, and habits. Frogs do differ considerably in color pattern, but nearly all species are highly variable in this respect; each species possesses a number of possible patterns, and there is often overlap between related species.

The surest way to identify frogs is by the advertisement call, a vocalization made by the male to attract a female of the same species. Female frogs have been shown experimentally to be very discriminating, so that only females of the same species are attracted to the male call for breeding.

Readers are advised to follow five steps to identify an adult frog:

1. Follow the identification keys as far as possible.
2. Compare the photograph and brief description.
3. Compare the description of the advertisement call.
4. Check the known distribution of the candidate species.
5. Compare the description of the habits and behavior of the species.

Each species of frog will have a unique combination of these features. Any one by itself can be equivocal, but using a combination should lead to a correct identification. Problems include the fact that, like all other organisms, individual frogs vary within populations.

It has not been possible to provide keys to all the species, nor to include photographs of all the color patterns, but we have tried to illustrate the common ones rather than the most colorful. Recordings of many frog calls are commercially available, and the serious reader will no doubt wish to obtain them. Calls can be found on the Web site for the California Academy of Sciences (http://www.calacademy.org/research/herpetology/frogs/list.html), and on the CD *South African Frog Calls* produced by J. Zähringer, which includes many species that reach East Africa. The distribution maps reflect only the known distribution, as opposed to the real distribution. It would not be unexpected to discover a species just outside its area of distribution, but if a frog believed to be confined to the high peaks of the Eastern Arc Mountains is suddenly identified from the tropical east coast, then the identification must be in doubt. Each species occupies a particular range of habitats and normally can be characterized by certain behavior. Many squeakers (genus *Arthroleptis*) live in or near forest and conceal themselves under leaf litter. If one tentatively identifies as *Arthroleptis* an individual that has been found in temporary puddles in the dry savanna, then a misidentification may be indicated. Perhaps there has been confusion with a superficially similar genus, like the puddle frogs (genus *Phrynobatrachus*).

Characters Used in the Keys

The figures with the relevant parts of the key will serve to illustrate the features discussed. Below are some brief definitions or explanations of terminology used to describe adult amphibians.

Claws on foot. Dark claws on three of the toes on each foot, and sometimes an extra clawlike structure on the prehallux, as in the genus *Xenopus*.

Cusps on lower jaw. Two sharp, toothlike projections at the front of the lower jaw, as in bullfrogs.

Digital discs. Flattened adhesive structures on the tips of the fingers and toes. These usually have distinct grooves around the edge.

Distal subarticular tubercle. See Tubercle.

Dorsal skin ridge. A very fine ridge along the midline of the back, characteristic of frogs in the family Arthroleptidae.

Dorsolateral stripe. A stripe along the back, between the side and the midline.

Dorsum. Back.

Fingers in opposing pairs. Four fingers arranged in two pairs, enabling them to grasp branches with one pair on each side of the grip, as in the foam-nest frogs *Chiromantis.*

Foot length. The measurement from the tip of the longest (fourth) toe to include the metatarsal tubercle.

Fourth metatarsus. The foot bone extending back from the fourth toe.

Gular disc. Elastic glandular tissue under the throat of some male frogs.

Inner metatarsal tubercle. See Tubercle.

Inner toe. The first toe. The longest toe is the fourth, for orientation.

Internarial distance. The distance between the nostrils.

Interocular bar. A marking, often dark, running from eye to eye over the head.

Interorbital distance. The distance between the eye bulges, taken across the top of the head.

Joints free of web. The phalanges counted from the tip that are not webbed. See also Web reaching tubercle.

Last phalanx out of alignment. The last joint of the finger that has an extra piece of cartilage causing the tip to be below the rest of the finger, as in tree frogs.

Lumbar pattern. The pattern on the lower back.

Metatarsal. The base of the toe.

Metatarsal tubercle. See Tubercle.

Outer metatarsal. The metatarsal on the outside, or the side with the longest toe.

Outer metatarsal tubercle. See Tubercle.

Palate with transverse folds. The inside of the roof of the mouth with ridges running from side to side.

Palmar tubercle. See Tubercle.

Paravertebral band. A longitudinal stripe next to the midline of the back.

Parotid. See Parotid gland.

Parotid gland. A large skin gland behind the eye in many toads in the family Bufonidae.

Pectoral markings. A pattern, usually dark, on the chest near the arms.

Phalanges. Individual bony elements of the fingers and toes. See also Joints free of web.

Posterior face of thigh. The back of the thigh.

Prehallux. An extra digit on the inside of the foot. This is essentially a sixth toe.

Proximal subarticular tubercle. See Tubercle.

Pupil shape. The configuration of pupil in live frogs, when closed in bright light. It may be a vertical slit, a horizontal slit, or rectangular.

Snout hardened for digging. Snout with a protruding tip, as in some *Probreviceps* species.

Snout-urostyle length. The body length measured from the tip of the snout to the end of the urostyle, the bony rod at the end of the vertebral column.

Snout-vent length. The body length measured as the distance between the tip of the snout and the end of the body.

Subarticular tubercle. See Tubercle.

Subocular tentacle. A small tentacle below the eye, found in the genus *Xenopus.*

Tarsal fold. A longitudinal ridge running from the heel to the ankle, especially in some toads of the genus *Bufo.*

Tarsus. The lower leg between the ankle and the heel.

Teeth on upper jaw. Fine teeth along the edge of the upper jaw. These are best felt by running a fingernail softly along the jaw, as they are difficult to see.

Throat flap. A glandular flap of skin covering the throat in males of some frogs.

Throat gland. See Throat flap.

Tibia. The lower leg between the knee and the ankle.

Tongue free behind. The tongue attached only in front, loose at the back.

Transverse skin groove. A groove or fold in the skin running transversely behind the eyes.

Tubercle. A raised hardened skin projection, often in the form of a flattened cone or ridge. Tubercles are frequently diagnostic and serve as reference points to determine the amount of webbing. *Subarticular* tubercles are found below the joints of the fingers and toes. *Metatarsal* tubercles are found at the back of the foot and can be *inner* or *outer.* Subarticular tubercles at the tip of the toe or finger are *distal,* while those at the base are *proximal.*

Tympanum. The eardrum. This is not always visible.

Urostyle. Part of the "hip" in frogs. This part is usually used as the most posterior bony rod for measuring body length, as snout-urostyle length.

Ventral surface. The lower surface, including the chest and belly.

Vertebral band. A wide stripe running along the midline of the back.

Vertebral line. A thin line running along the midline of the back.

Vocal sac. The folded skin pouch of a male frog that expands during calling. These are single under the throat, or double with one at each side of the jaw.

Vomerine teeth. Small toothlike projections on the roof of the mouth near the internal nostrils.

Web reaching tubercle. A way of measuring the amount of webbing. The web notch midway between two toes is measured against the subarticular tubercles, or the margin of web along the toe is used.

How to Use the Identification Keys

The keys consist of a series of choices. Start at 1a/1b and select the statement that best fits the frog you are trying to identify. Each choice will lead you either to a new pair of statements, to a group name with a new subset of keys, or to a species identification. For example, in the key to the species of *Xenopus*, choice 1b will identify the specimen as *X. ruwenzoriensis*, while choice 1a will lead to step 2, where you will be asked to choose between 2a and 2b, and so on.

The keys have been arranged so that the specimen is first identified to family. Within the family there is a key identifying the specimen to genus, and then another key to arrive at the species name.

KEY TO EAST AFRICAN AMPHIBIANS

1a. Adults without limbs, wormlike — Gymnophiona
1b. Adults with limbs—frogs — Anura

Species Accounts

Order Anura

KEY TO THE FAMILIES

1a. Tongue absent, three or four claws on each foot (Fig. 5) Pipidae
1b. Tongue present 2

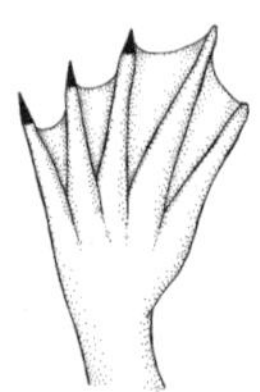

Fig. 5 Three claws on foot.

2a. Upper jaw toothless 3
2b. Upper jaw with fine teeth (Fig. 6) 5

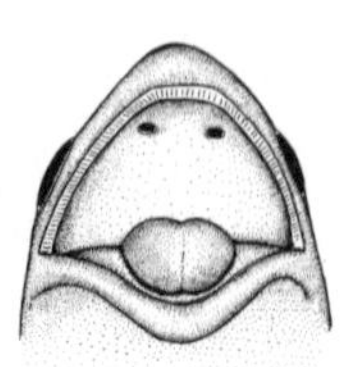

Fig. 6 Upper jaw with teeth.

3a. Transverse skin fold behind eyes, snout pointed, hardened for digging (Fig. 7) Hemisotidae
3b. No transverse skin fold behind eyes 4

Fig. 7 Transverse fold behind eyes.

4a. Angle of jaw does not reach behind eye — Microhylidae
4b. Angle of jaw behind eye — Bufonidae

5a. A fine dorsal skin ridge running along midline (Fig. 8) — Arthroleptidae
5b. No fine dorsal skin ridge — 6

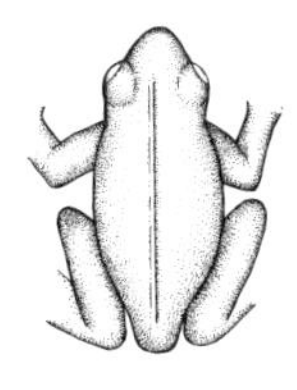

Fig. 8 Fine dorsal skin ridge.

6a. Last phalanx of fingers out of alignment (kinked) (Fig. 9) — 7
6b. Last phalanx of fingers not out of alignment — Ranidae

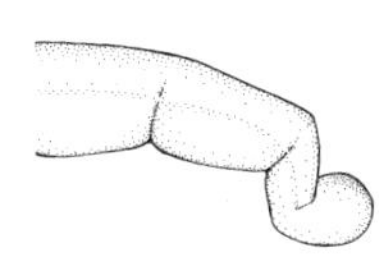

Fig. 9 Last phalanx of finger out of alignment.

7a. Fingers arranged in two opposing pairs (Fig. 10) — Rhacophoridae
7b. Fingers not arranged in two opposing pairs (Fig. 11) — Hyperoliidae

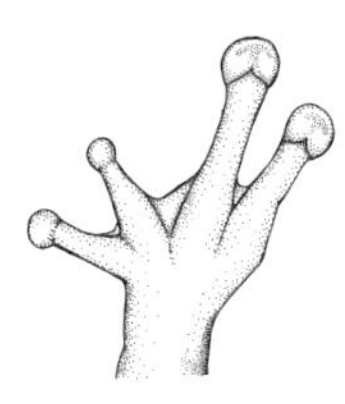

Fig. 10 Fingers arranged in opposing pairs.

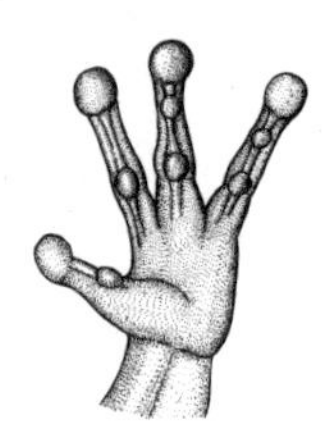

Fig. 11 Fingers not in opposing pairs.

Squeakers—Family Arthroleptidae

This family consists mostly of small forest-dwelling frogs, usually found associated with leaf litter. Members of this family often have a more or less distinct hourglass-shaped mark on the back and a bewildering array of color patterns, usually including browns and reddish browns. The frogs possess a fine skin ridge along the dorsal midline. Breeding males of many species have an elongated third finger that may reach nearly half the body length. This feature may play a role in breeding. The male protects his calling site against other males by wrestling the intruding male to drive him away. He uses the long finger during these aggressive encounters. Observations of spawning pairs suggest that the long finger plays no part in egg fertilization or deposition. In some species the finger length is variable between males. The eggs are laid in damp places under rocks or leaf litter, and development continues directly to a juvenile frog. Frogs in this group are found in the moist tropics throughout sub-Saharan Africa and breed throughout the year if conditions are suitable. Much of Africa was once covered in forest, and the remnants of this vegetation can still be found on mountain slopes and some lowlands. These forest patches are the habitat of the small leaf-litter frogs. The species vary from minute slender adults not exceeding 15 mm in length, to large robust animals over 50 mm.

The species of squeakers in East Africa are assigned to one of two genera, *Arthroleptis* and *Schoutedenella*, which differ in details of skeletal anatomy. Generally, species of *Schoutedenella* tend to be smaller and more gracile than those of *Arthroleptis*. It has been suggested that the grouping of species into these two genera should be reexamined. A phylogenetic analysis is required to determine the placement of the many small species in this group.

KEY TO THE GENERA

1a. Proportion of first finger length/distance between anterior borders of eyes = 80%–140% (Fig. 12) *Arthroleptis*

1b. Proportion of first finger length/distance between anterior borders of eyes = 38%–78% *Schoutedenella*

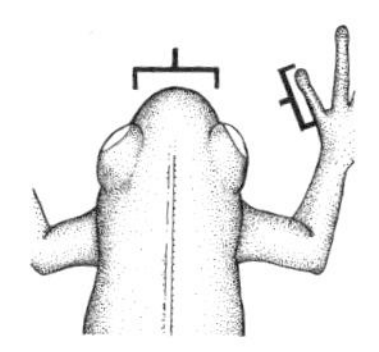

Fig. 12 Length of first finger in relation to distance between anterior borders of eyes.

Squeakers—Genus *Arthroleptis*

This group of small to medium-sized frogs is associated with forest. They are robust, distinguishable on the basis of the shape of the tips of the toes, size, and color pattern. The genus is found throughout tropical Africa, but only five species are known in East Africa. All the species are direct developers; the eggs develop into small frogs without a free-swimming tadpole stage. The taxonomy of these frogs is confusing, with some species names appearing to apply to groups of cryptic species. While this book is in preparation, other species are being described, such as a large species from the Ukagurus, in Tanzania. The following arrangement of species might be modified by a comprehensive revision of the group.

KEY TO THE SPECIES

1a. A large flangelike inner metatarsal tubercle is present (Fig. 13) *Arthroleptis stenodactylus*
1b. Inner metatarsal tubercle not flattened and flangelike 2

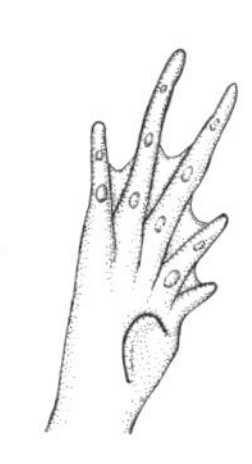

Fig. 13 Flangelike inner metatarsal tubercle.

2a. Tips of fingers clearly dilated (Fig. 14) 5
2b. Tips of fingers at most slightly swollen 3

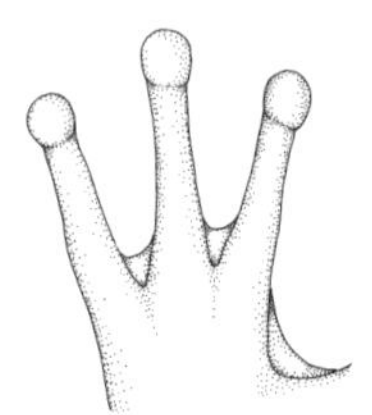

Fig. 14 Tips of fingers dilated.

3a. Many supernumerary tubercles under the metatarsals, females up to 45 mm, tibia more than half the body length (Fig. 15) *Arthroleptis affinis*
3b. A single supernumerary tubercle, or none, under the metatarsals, females up to 51 mm 4

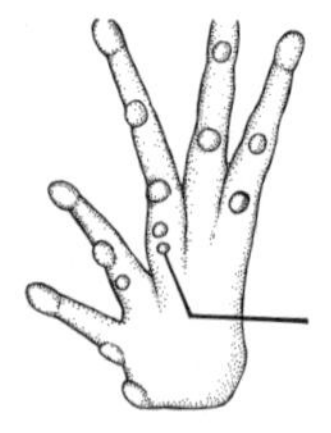

Fig. 15 Supernumerary tubercles.

4a. Tibia more than half the body length *Arthroleptis nikeae*
4b. Tibia half the body length *Arthroleptis tanneri*

5a. First finger much shorter than second (Fig. 16) *Arthroleptis reichei*
5b. First finger equal to second *Arthroleptis adolfifriederici*

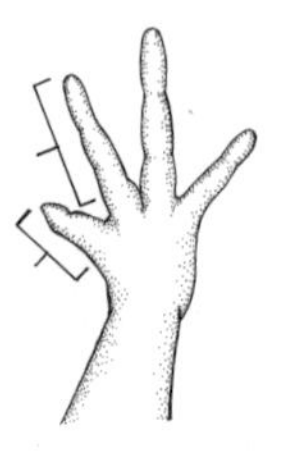

Fig. 16 First finger shorter than second.

Adolf's Squeaker

Arthroleptis adolfifriederici Nieden, 1911
(Plates 1.1, 1.2)

This species was named for the leader of the German East African scientific expedition to Central Africa in 1907–1908, Duke Adolf Friederich zu Mecklenburg.
Montane squeaker.

Fig. 17 *Arthroleptis adolfifriederici.*

DESCRIPTION: This is a stout frog. Males are known up to 32 mm long, while females may grow to 46 mm. During breeding a series of toothlike granules develop on the inner side of the second and third fingers in males. The first finger is about as long as, or slightly shorter than, the second. The inner metatarsal tubercle is about the same length as the first toe. Fingertips are slightly expanded, while the toes have large expanded tips. A median papilla is present on the tongue. The snout is one and a half times as long as the horizontal eye distance. The tympanum is distinct, half the size of the eye. The toes have a trace of webbing. Only a rounded inner metatarsal tubercle is present. Coloration is variable, with the frogs tending to be well camouflaged. A broad pale vertebral stripe is sometimes present. The underside can be white or marbled.

HABITAT AND DISTRIBUTION (FIG. 17): This species is found in leaf litter on the forest floor and is known from montane forests in eastern Democratic Republic of Congo, Rwanda, Burundi, Uganda, Kenya, and Tanzania.

ADVERTISEMENT CALL: The male calls from below or on leaf litter. The call is a series of loud short chirps, with dominant harmonics at 3.2 kHz.

BREEDING: Males call during and after wet weather. No details of breeding biology are known.

NOTES: Food items include ants, caterpillars, cockroaches, spiders, and freshwater shrimps. They are eaten by Tornier's cat snake *Crotaphopeltis tornieri*.

KEY REFERENCE: Barbour & Loveridge 1928a.

Ahl's Squeaker

Arthroleptis affinis Ahl, 1939
(Plate 1.3)

The meaning of the specific name is obscure—*affinis* refers to something related.

Fig. 18 *Arthroleptis affinis*.

DESCRIPTION: This species is large, with the male up to 35 mm in length and the female up to 45 mm. A row of supernumerary (extra) tubercles is present under the metatarsals. Adult breeding males possess small denticulations along the side of the third and second fingers. Head width is 42% of body length and 77% of tibia length in females but 75% in males. The tibia is 55%–56% of body length, and the internarial distance is 13% of body length. Some individuals have a single supernumerary tubercle at the base of the first toe below the inner metatarsal tubercle. Small discs are present, pointed, and wider than the proximal joints and phalanges. Males have elongated third fingers.

The back may have a background of various shades of brown, with a darker bar between the eyes and with irregular blotches on the back and upper legs. The phalanges and soles of the feet are dark brown with pale tubercles. A dark band passes above the tympanum.

HABITAT AND DISTRIBUTION (FIG. 18): This species is associated with leaf litter on forest floors. It is a Tanzanian endemic, known from the North and South Pare, Nguru, Rubeho, East Usambara, and Udzungwa mountains, at lower elevations, and coastal forest.

ADVERTISEMENT CALL: Unknown.

BREEDING: No details are known.

NOTES: There is some confusion concerning the identity and taxonomic status of this species, Tanner's squeaker, and Adolf's squeaker.

KEY REFERENCE: Grandison 1983.

Nike's Squeaker

Arthroleptis nikeae Poynton, 2003

This species was named for Nike Doggart who collected the first specimens.

DESCRIPTION: This species is presently only known from two specimens, both females, with a maximum body length of 56 mm. This is the largest species in the genus in East Africa. The finger tips are slightly swollen, while the toes show small indistinct discs. The tympanum is visible, with a horizontal diameter about 13% of the head width. The markings on the back consist of two dark brown distinct forward-facing V marks, in a paler broad vertebral band. A dark band with a light front edge runs between the eyes. There are darker cross bands on the legs and arms.

HABITAT AND DISTRIBUTION: The specimens were found on the forest floor in the Rubeho Mountains in central Tanzania.

ADVERTISEMENT CALL: Unknown.

BREEDING: No details available.

KEY REFERENCE: Poynton 2003b.

Reiche's Squeaker

Arthroleptis reichei Nieden, 1911
(Plate 1.4)

This species was named for Gustav Reiche, of the Königliches Zoologisches Museum in Berlin.
Large-toed squeaker, Poroto mountains screeching frog, *Buluwidi* in Kinga, *koti* in Nyakusa.

Fig. 19 *Arthroleptis reichei.*

DESCRIPTION: The male may reach 25 mm long, with the larger female reaching up to 34 mm. This slender frog has long hind limbs and expanded toe tips. The first finger is very much shorter than the second. The length of the third finger in the male is slightly more than half the head width. The tibia is about half the body length. Webbing is very slight to absent, and the second to fourth fingers, and toes, end in distinct discs. The metatarsal tubercle is very small, shorter than the first toe, while the outer metatarsal tubercle is absent. The snout is sharp. Nostrils are closer to the snout tip than to the eyes. The interorbital distance is longer than the width of upper eyelid. The tympanum is distinct, nearly two-thirds the diameter of the eye. The skin is smooth. A typical hourglass pattern in darker brown is present on the back. There is a dark, thin ridge above the tympanum. The frogs are greenish gray above, with a curved dark bar with white edge between the eyes. A brown line runs from the nostril through the eyelid terminating behind the tympanum. A pale band between the eyes is present in most specimens.

HABITAT AND DISTRIBUTION (FIG. 19): These frogs have been found in moist evergreen montane forests, where they are known from leaf litter in forest, and in areas of low shrubs and grass. They have also been found in the axils of wild banana plants. This species is seen in the mountains of northern Malawi, through the Poroto and Rungwe mountains in southern Tanzania, to the Udzungwa and Uluguru mountains. Specimens have been caught at an altitude of 2000 m.

ADVERTISEMENT CALL: Unknown.

BREEDING: No details available.

NOTES: Although Reiche's squeaker is not threatened directly, the forest habitat is in danger from clearing for agriculture.

Recorded food items include bugs, caterpillars, and forest cockroaches. Reiche's squeaker is eaten by green snakes *Philothamnus* spp. and the white-lipped snake *Crotaphopeltis hotamboeia*.

KEY REFERENCE: Loveridge 1933.

Common Squeaker

Arthroleptis stenodactylus Pfeffer, 1893
(Plate 1.5)

The specific name *stenodactylus* refers to the long third finger of the male.
Dune squeaker, shovel-footed squeaker, shovel-footed bush squeaker, Kihengo screeching frog.
At least two, if not more, species have been confused as the common squeaker. See Notes, below.

DESCRIPTION: This frog is large and stout, with the male up to 35 mm long and females up to 45 mm. The snout is rounded, with the nostril midway between the eye and the tip of the snout. The interorbital space is about equal to the width of the upper eyelid. The tympanum is distinct, smaller than the eye diameter. Finger and toe tips are slightly swollen but not with distinct discs. Webbing is absent or with only the slightest trace present. The inner metatarsal tubercle is massive, flange-

Fig. 20 *Arthroleptis stenodactylus.*

like, and as long as or longer than the first toe. This species has a relatively short leg. The tibia length is a third to less than half of the body length, and equal to the head width. However, the tibia length is less than the width of the head of a large specimen.

The male has a dark throat with loose vocal sac skin, while the female has a white throat with some speckling in the pectoral region. The male also differs from the female by having minute spines on the back and limbs, and by its narrower head and elongated third finger. The elongated third finger (up to twice as long as the first finger) grows rapidly after the male attains sexual maturity. The skin is warty, with a fold present from the arm insertion to near the vent. Coloration is very variable, with a vertebral line or band sometimes present. Ventral markings vary from immaculate to very speckled. Dorsal markings consist of a pair of dark sacral spots, with various combinations of a three-lobed dorsal band, with or without a light vertebral line. A dark line is present from the tip of the snout to the shoulder. The back is typically colored in browns, but reds and greens are common. The soles of the feet are dark.

HABITAT AND DISTRIBUTION (FIG. 20): This frog is widespread and able to live both in gardens and in natural vegetation. It is found beneath logs or leaf litter. This frog is known from southern and eastern Democratic Republic of Congo to Kenya and south to Zimbabwe, Mozambique, and northeastern South Africa, from sea level to 2000-m altitude.

ADVERTISEMENT CALL: The male calls from a concealed site in leaf litter or under vegetation, during both day and night after rain. The call is a short whistle (0.05 s), high-pitched (3.5 kHz), and repeated at half-second intervals. Males may call very rapidly in chorus. The call sounds somewhat similar to that of the common reed frog *Hyperolius viridiflavus.*

BREEDING: Eggs are deposited in hollows or burrows in damp earth, often under bushes or around the roots of trees, or under loose-leaf mold. Eggs have been found in shallow burrows guarded by a male. The 2-mm eggs are creamy white. Clutch size varies from 33 to 80. Metamorphosis occurs in the nest, and there is no free-swimming tadpole stage.

NOTES: The common squeaker appears to be a complex of species, each with particular habitat preferences and small call differences. As this book is being written, investigations are under way to resolve this problem. This frog eats a wide range of insect and other arthropod prey, as well as earthworms, snails, and even other frogs. Recorded food items include various cockroaches, beetles, grasshoppers, crickets, spiders, snails, and centipedes.

The common squeaker is in turn preyed on by many snakes, such as the olive marsh snake *Natriciteres olivacea*, the southeastern green snake *Philothamnus hoplogaster*, the eastern stripe-bellied sand snake *Psammophis orientalis*, the rufus beaked snake *Rhamphiophis rostratus*, Tornier's cat snake *Crotaphopeltis tornieri*, and the savanna vine snake *Thelotornis capensis*. This small frog is also a popular food of other larger frogs. A specimen was found in torpor in a hollow tree nearly a meter above ground level.

KEY REFERENCES: Barbour & Loveridge 1928a, Loveridge 1933.

Tanner's Squeaker

Arthroleptis tanneri Grandison, 1983
(Plate 1.6)

This species is named for John Tanner, who donated the type locality, Mazumbai Natural Forest, to the University of Dar es Salaam.

DESCRIPTION: This frog is the largest in the genus, with the males reaching 39 mm in length and the females up to 55 mm. The tympanum is distinct, with a diameter about half the internarial distance. The tips of the fingers and toes are slightly swollen, with a groove present around the toe tip. The inner metatarsal tubercle is as long as the distance from the tubercle to the tip of the first toe. No supernumerary tubercles are

Fig. 21 *Arthroleptis tanneri.*

present under the metatarsals. The tibia is half the body length, and the foot is longer than the tibia. The toes are unwebbed. The back is brown with darker markings, including a pale chevron behind the eyes. The limbs are cross barred above, with the back of the thighs mottled. The hands and feet are pinkish brown. The upper half of the iris is golden. Breeding males have small spines on the chin.

HABITAT AND DISTRIBUTION (FIG. 21): These large frogs are found on the forest floor, where they are well camouflaged against leaf litter. They conceal themselves under vegetation along streams, and they may aggregate alongside water. This is a Tanzanian endemic, known from the West and East Usambara, Nguru, and the Ukaguru mountains.

ADVERTISEMENT CALL: Unknown.

BREEDING: The eggs are unpigmented and 3.5 mm in diameter. No details of breeding are recorded.

NOTES: Tanner's squeaker is often associated with other stream-dwelling frogs such as the Angolan river frog *Afrana angolensis* and Krefft's puddle frog *Phrynobatrachus kreffti.*

KEY REFERENCE: Grandison 1983.

Squeakers—Genus *Schoutedenella*

This genus is found widely in sub-Saharan Africa. The frogs are gracile, always associated with forest or forest remnants. They are ground dwelling, often concealed in leaf litter. Five species are known in East Africa. The placement of species in the genera *Schoutedenella* and *Arthroleptis* needs to be checked using appropriate analyses of relationships.

KEY TO THE SPECIES

1a. Adult males and females 25 mm or more in length *Schoutedenella poecilonotus*?
1b. Adults less than 25 mm in length 2

2a. Discs with terminal papillate projection, head wedge-shaped (Fig. 22) *Schoutedenella xenodactyla*
2b. Discs rounded, head not wedge-shaped, hourglass pattern on back 3

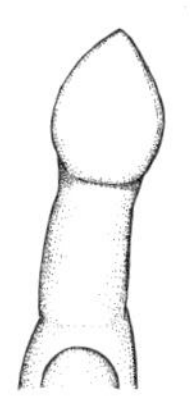

Fig. 22 Disc with papillate projection.

3a. Metatarsal tubercles less than half the length of the fifth toe, not larger than subarticular tubercles of toes (Fig. 23) 4
3b. Metatarsal tubercles longer than half the length of the fifth toe, about twice the size of the subarticular tubercles of the toes *Schoutedenella xenochirus*

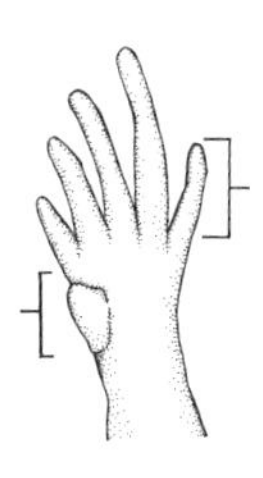

Fig. 23 Length of metatarsal tubercle in relation to fifth toe.

4a. Third finger of males less than two-thirds of head width
Schoutedenella xenodactyloides
4b. Third finger of males as long as head width
Schoutedenella schubotzi

Mottled Squeaker

Schoutedenella poecilonotus? (Peters, 1863)
(Plate 1.7)

The specific name means "mottled tympanum."

Fig. 24 *Schoutedenella poecilonotus?*

DESCRIPTION: This species is stout, with the male reaching 27 mm in length and the female 34 mm. The tympanum is about the size of the eye. The back often has scattered small warts. The fingers and toes do not have discs. The toes are not webbed. Color is variable, with the background pale to reddish brown, often with a darker hourglass pattern on a pale background, down the middle of the back. Some breeding males are very dark with no or little pattern. A pale vertebral line or band may be present. Sometimes the back is a uniform color with no pattern. Adult males have gray to violet throats.

HABITAT AND DISTRIBUTION (FIG. 24): The mottled squeaker is found in savanna, at altitudes up to 1000 m, where it is associated with leaf litter. Like other leaf-litter species, it is also known from farmlands in the forest zone. It has been recorded from the West African forests to Uganda. It is expected to occur in western Tanzania.

ADVERTISEMENT CALL: The male calls from ground level. Each call is cricket-like, consisting of 5–8 pulsed notes at a frequency of 5 kHz. The call has a duration of 0.8–0.9 s.

BREEDING: The eggs are laid in a nest below leaf litter. Each egg is 3 mm in diameter, in clutches of 10–30 eggs. The eggs develop directly into small froglets, which hatch after 15–20 days. Each female may lay 2 or 3 clutches in a season.

NOTES: This species of squeaker, like many others, may be a complex of similar-looking taxa. The East African species is different from the nominate form in West Africa. This frog may represent a form known as *Schoutedenella tuberosus*.

KEY REFERENCES: Amiet & Perret 1969, Amiet 1975, Rödel 2000.

Schubotz's Squeaker

Schoutedenella schubotzi (Nieden, 1911)

This species was named in honor of H. Schubotz, who was the zoological leader of a German scientific expedition to Central Africa in 1907–1908.
Kivu dwarf litter frog.

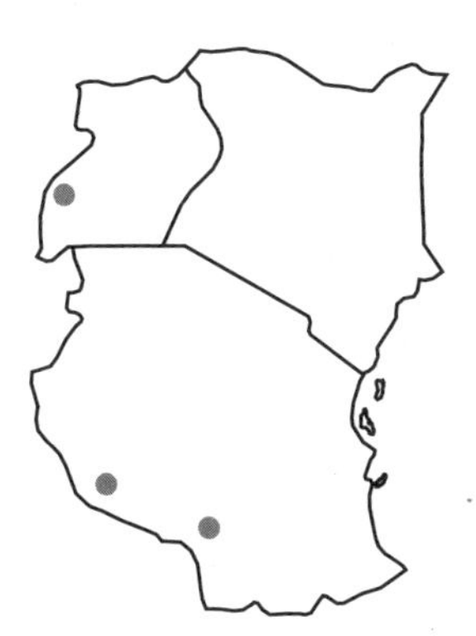

Fig. 25 *Schoutedenella schubotzi.*

DESCRIPTION: The males may obtain a length of 21 mm and the females, 23 mm. The legs are short. The tympanum is visible and half the size of the eye. The nostrils are closer to the snout tip than the eye.

The fingers and toes have distinct, although small discs. The first finger is shorter than the second. The elongated third finger of males is about as long as the width of the head, and has many spines on the inner surface. The inner metatarsal tubercle is very small and round. The back is granular, especially on the sides. The belly is pigmented. Coloration is variable in browns and grays, and many individuals have red to orange femurs. A common pattern includes a dark spot on the head. Males have black throats.

HABITAT AND DISTRIBUTION (FIG. 25): This species is found on vegetation in lowland savanna, and forest and forest remnants up to an altitude of 2800 m. It is known from southern Tanzania to Rwanda, Burundi, Western Uganda, and eastern Democratic Republic of Congo.

ADVERTISEMENT CALL: The call is described as a harsh series of double chirps. It has not been analyzed.

BREEDING: Unknown.

KEY REFERENCE: De Witte 1941 (as *Schoutedenella kivuensis*).

Plain Squeaker

Schoutedenella xenochirus (Boulenger, 1905)
(Plates 1.8, 2.1)

The specific name *xenochirus* refers to the "strange hand," with the very elongate third finger of the males.
Marimba screeching frog.

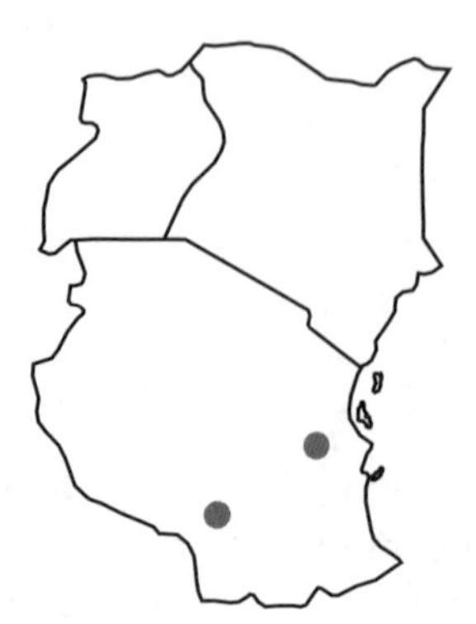

Fig. 26 *Schoutedenella xenochirus.*

DESCRIPTION: This frog is small, with the female less than 24 mm long. The tibia is between a third and a half of the body length and equal to, or slightly more than, half of the head width. The metatarsal tubercle is very large and spurlike. It is usually more than half the length of the outer toe. There are no discs, although the tips of the fingers and toes may be slightly swollen. The third finger in the male is very long, four to five times the length of the second finger. The length of the third finger relative to the snout-vent length seems to increase with age. The inner edge of the third finger possesses a row of small spines. Adults tend to have a uniform back coloration, rose to brick red, or tan to gray. The groin and forward areas of the thighs of adults are bright red, visible only when the frogs hop. Females tend to be red in the inguinal region.

HABITAT AND DISTRIBUTION (FIG. 26): This frog is associated with forest patches or streams in open grassland, between altitudes of 1800 and 2500 m. It is known from northeastern Angola, across to northern Malawi, and northward to Tanzania.

ADVERTISEMENT CALL: The call is a high-pitched trill, consisting of about 7 notes, at a rate of 15/s. The emphasized frequency is 4.6 kHz.

BREEDING: Clutches of about 9 eggs, each one white and 3 mm in diameter, are laid in shallow nests in leaf litter. Nests are constructed on mossy banks. The adults may stay near the eggs.

NOTES: These frogs burrow beneath cover or into sedge tussocks when disturbed.

KEY REFERENCE: Stewart 1967 (as *Arthroleptis xenodactyloides nyikae*).

Eastern Squeaker

Schoutedenella xenodactyla (Boulenger, 1909)
(Plate 2.2)

The specific name means "strange finger."
Koti in Nyakusa.

DESCRIPTION: The males and females reach 22 mm in length. The discs are pointed. The nostril-snout distance is less than the eye width, and the nostril is nearer the snout tip than the eye. The head is wedge-

Fig. 27 *Schoutedenella xenodactyla.*

shaped. The first finger is much shorter than the second. The sexes are dimorphic for color. The back is yellowish brown, with or without a dark hourglass pattern, with a large orange blotch above the vent. The sides are dark brown with a pale pink line above the dark brown. The thighs and inner tibia are orange or bright red in breeding males. The ventral surface is white with gray marbling, and the limbs are dark with pale speckling.

HABITAT AND DISTRIBUTION (FIG. 27): These frogs are found in leaf litter or under logs, and also in banana axils. They are widespread in Tanzania, known from the Usambara, Uluguru, Udzungwa, and Rungwe mountains, and the coastal plain and offshore islands.

ADVERTISEMENT CALL: Males call during the day and night from the forest floor, with a frantic chorus after rain. The call consists of a brief whistle, about 0.05 s long, with a dominant harmonic between 6.4 and 7.0 kHz. The call is repeated two or three times per second.

BREEDING: The yellow eggs are laid in leaf litter during the January rains.

NOTES: Recorded food items include a range of very small arthropods, such as ants, ticks, termites, amphipods, and insect larvae. Eastern squeakers are eaten by Tornier's cat snake *Crotaphopeltis tornieri*. Saw pits in the forests serve as traps for these small frogs.

KEY REFERENCES: Barbour & Loveridge 1928a, Loveridge 1933 (as *A. schubotzi*)

Dwarf Squeaker

Schoutedenella xenodactyloides (Hewitt, 1933)
(Plate 2.3)

The specific name *xenodactyloides* refers to the long third finger of the males, meaning "the shape of a strange finger."
Hewitt's bush squeaker.

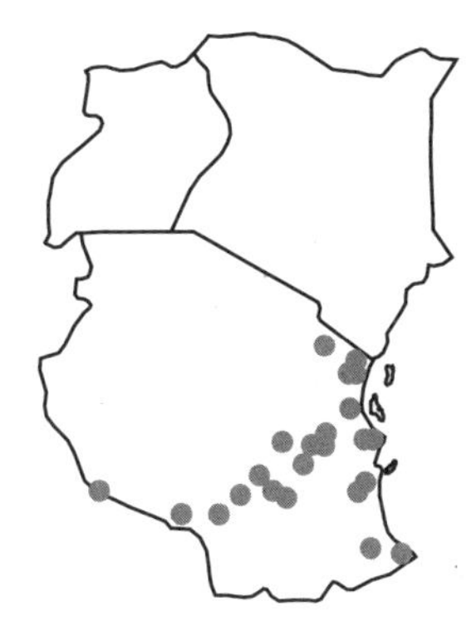

Fig. 28 *Schoutedenella xenodactyloides.*

DESCRIPTION: The adult is small, less than 22 mm long. It is slender, with the tibia between a third and a half of the body length. The metatarsal tubercle is very small, as are the weakly developed tubercles beneath the toes. The third finger in the male is less than two-thirds of the width of the head. The digits possess distinct small discs. An hourglass pattern on the brown back is typical, but a pattern of two light dorsolateral stripes is also common. The insides of the thighs and the sides of the body can be reddish brown.

HABITAT AND DISTRIBUTION (FIG. 28): This species has been found in grassland swamps at a high altitude and in forests at both high and low altitudes. It is recorded from Mozambique, Malawi, through eastern Zimbabwe, to northeastern Tanzania.

ADVERTISEMENT CALL: The male calls from under dead leaves on the forest floor and is very difficult to observe. The call is a brief cricket-like chirp, with each call consisting of about three brief clicks at 5.5 kHz. The call can be heard day and night from leaf litter.

BREEDING: The eggs are laid under leaf litter. Clutch size is 20. Each white egg is 1.5 mm within a 4-mm capsule. The males remain near the eggs.

NOTES: This extremely common frog is not easily seen, but it can be traced from its cricket-like call. It may climb in low vegetation. It has been found in the holes made by beetles in fallen rotten tree trunks. The only recorded predator is the vine snake *Thelotornis capensis*. Food records include crickets.

KEY REFERENCE: Channing 2001.

Toads—Family Bufonidae

There are two groups of toads in East Africa: terrestrial rough-skinned forms, in the genera *Bufo*, *Mertensophryne*, *Schismaderma*, and *Stephopaedes*, and mostly climbing forms in the genera *Churamiti* and *Nectophrynoides*. The large toads are common in towns, villages, and farmlands and are probably the best-known frogs in East Africa. *Ekikele* is Luganda for toad, while in Runyakole and related languages the words *ekikere* and *echichere* are used. Toads are known as *likyele* in Gisu, *ogwal-pok* in Luo, *ongwal-pok* in the related Alur, *adrukundru* in Lugbara, *lichaliro* in Madi, and *adodoko* in Atesot.

Some toads occur in very dry areas, far from any water, whereas others seem to require extremely moist conditions, such as those of moist grassland or high mountain forest. The smaller species are often found in mountainous areas. Toads are frequently observed in disturbed habitats, and the larger toads are known to hybridize in such places. These large toads are characterized by a horizontal pupil, which frequently has a lower notch. The skin is thick and warty. The skin glands, especially the enlarged parotid glands behind the eyes, secrete an irritant that serves to deter predators. Toads are caught in pit traps and eaten by the Mossi people from West Africa.

Toads eat a wide variety of insects, including cockroaches, grasshoppers, and crickets. Ants form an important part of their diet. Toads are an important link in the food chain and, despite their poison glands, are eaten by a variety of predators. When injured, *Bufo* tadpoles produce a skin secretion that stimulates the "fright reaction" in other *Bufo* tadpoles, causing them to scatter.

KEY TO THE GENERA

1a. Subarticular tubercles of fingers and toes single (Fig. 29) 2
1b. Most subarticular tubercles double 4

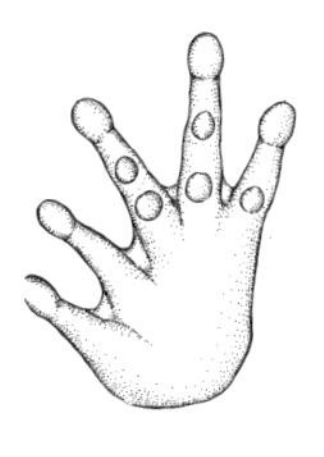

Fig. 29 Single subarticular tubercles.

2a. Jaws the widest part of the head 3
2b. Eyes the widest part of the head 7

3a. Parotid glands present, although sometimes obscured *Bufo* (part)
3b. No parotid glands *Schismaderma* (*S. carens*)

4a. Tympanum visible *Bufo* (part)
4b. No tympanum 5

5a. Parotid glands narrower than the distance between them 6
5b. Parotid glands wider than the distance between them *Stephopaedes*

6a. At least two phalanges of fifth toe free of web, fifth toe not shorter than third *Bufo* (part)
6b. Fifth toe shorter than third *Mertensophryne* (*M. micranotis*)

7a. Finger discs very large, metallic colors on back *Churamiti* (*C. maridadi*)
7b. Fingertips not expanded or only with slightly swollen tips, not metallic *Nectophrynoides*

Genus *Bufo*

This genus is large with species in most parts of the world. The species range from less than 20 mm to over 120 mm in length. All share a rough, thick glandular skin. The skin serves as efficient waterproofing, and the skin glands produce defensive substances, discussed below. Toads are found in forests, in alpine grasslands, and in very arid areas. The eggs are laid in strings in water and develop into small black tadpoles. At least 24 species have been recorded from East Africa.

KEY TO THE SPECIES

1a.	Subarticular tubercles single	3
1b.	Subarticular tubercles double (Fig. 30)	2

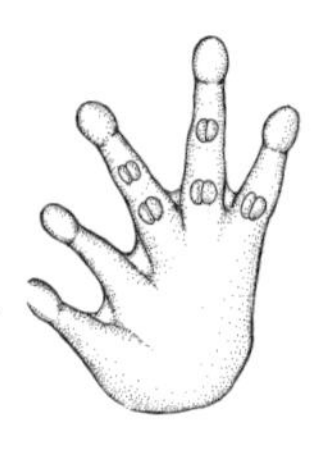

Fig. 30 Double subarticular tubercles.

2a.	Tympanum visible	15
2b.	Tympanum not visible	17
3a.	A distinct tarsal ridge present (Fig. 31)	4
3b.	Tarsal ridge absent or replaced by a short line of tubercles	13

Fig. 31 Tarsal ridge.

4a.	Tarsal ridge not serrated	5
4b.	Tarsal ridge serrated	*Bufo fuliginatus*
5a.	Spiny warts present on the side of the body	*Bufo camerunensis*
5b.	No spiny warts on the side of the body	6
6a.	A dark band running over lower edge of parotid and continuing backward for half of the body	7
6b.	No lateral dark band	8

7a. Two phalanges of third and fifth toes free of webbing (Fig. 32) *Bufo brauni*
7b. One phalanx of third and fifth toes free of webbing *Bufo reesi*

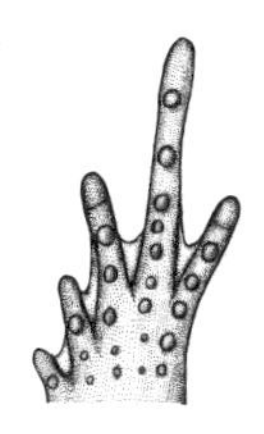

Fig. 32 Two phalanges of third and fifth toes free of web.

8a. A cross on head formed by light bands 9
8b. No light cross on head 21

9a. Parotid glands distinct, raised 10
9b. Parotid glands indistinct, flattened *Bufo maculatus*

10a. Dorsal pattern strongly contrasting 11
10b. Pattern muted, red markings on thigh *Bufo xeros*

11a. At least two phalanges of fifth toe free of webbing 12
11b. Less than two phalanges of fifth toe free of webbing *Bufo kisoloensis*

12a. Width of parotid gland equal to distance between anterior borders of eyes *Bufo kerinyagae*
12b. Width of parotid gland less than distance between anterior corners of eyes 22

13a. Three phalanges of fourth toe free of webbing 14
13b. More than three phalanges of fourth toe free of webbing *Bufo steindachneri*

14a. Tympanum as large as eye, longitudinal dorsal markings present (Fig. 33) *Bufo vittatus*
14b. Tympanum smaller than eye, no longitudinal markings *Bufo funereus*

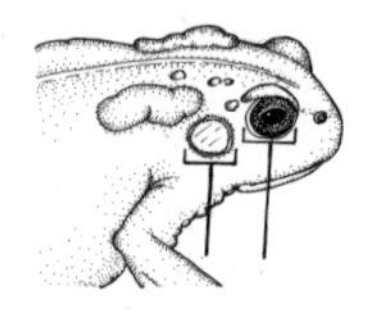

Fig. 33 Size of tympanum in relation to eye.

15a. First phalanx of third and fifth toes webbed, others with a margin of web — 16
15b. Webbing rudimentary, no toes with margins of web — *Bufo urunguensis*

16a. Tubercles of forearm and foot enlarged — *Bufo lughensis*
16b. Tubercles of forearm and foot not enlarged — *Bufo parkeri*

17a. Two enlarged carpal tubercles, one at base of first finger larger than palmar tubercles — 19
17b. One enlarged carpal tubercle (Fig. 34) — 18

Fig. 34 One enlarged carpal tubercle.

18a. Foot half of body length (Fig. 35) — *Bufo taitanus*
18b. Foot more than half of body length — *Bufo lonnbergi*

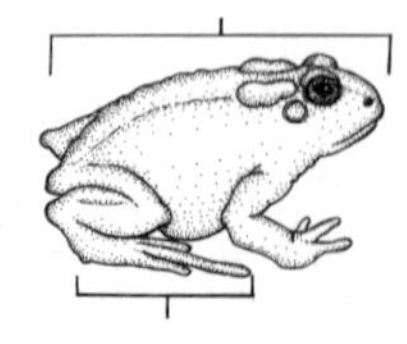

Fig. 35 Foot length half of body length.

19a. Parotid glands distinct, raised — 20
19b. Parotid glands indistinct, flattened — *Bufo lindneri*

20a. Rhomboid markings on back (Fig. 36) — *Bufo uzunguensis*
20b. Markings not strongly rhomboid — 23

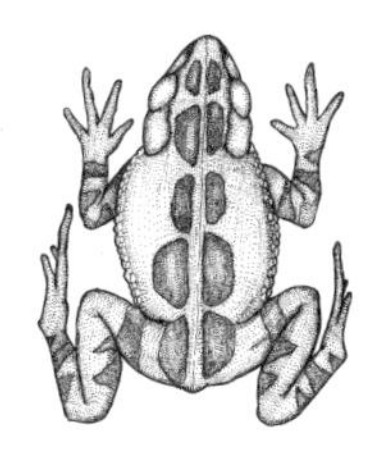

Fig. 36 Rhomboid markings on back.

21a. Paired dark markings on upper snout *Bufo turkanae*
21b. No paired dark markings on snout *Bufo garmani*

22a. No red markings on femur, male throats gray to black *Bufo regularis*
22b. Red markings on femur, male throats yellow with overlain dark speckling *Bufo gutturalis*

23a. Parotid glands adjacent to eyelid *Bufo nairobiensis*
23b. Parotid glands separated from eyelid *Bufo mocquardi*

Braun's Toad

Bufo brauni Nieden, 1911
(Plate 2.4)

Named for Dr. Braun, botanist at the Biology and Agriculture Institute of Amani.
Dead-leaf toad, *jula* in Shambara.

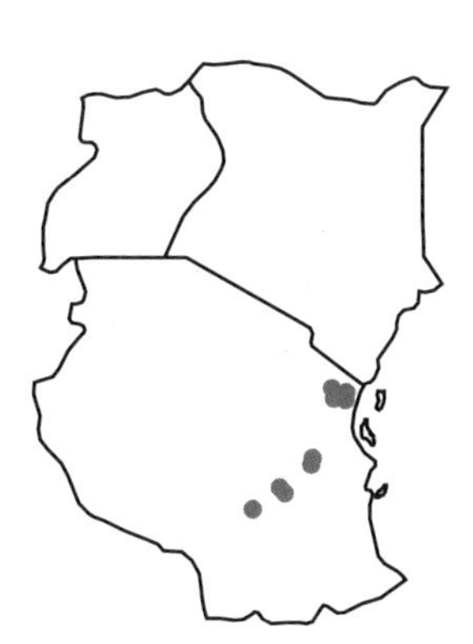

Fig. 37 *Bufo brauni.*

DESCRIPTION: This toad is large, with the females up to 110 mm in length and the males 65 mm. The snout length is equal to the horizontal diameter of the eye, with the nostrils closer to the snout tip than to the eyes. The tympanum is distinct, more than half the eye diameter. In older animals the tympanum is slightly vertically elongated. The first finger is longer than the second. Subarticular tubercles are single, and the toes are webbed to the first phalanx. Two phalanges of third and fifth toes free of webbing. The inner metatarsal tubercle is well developed, larger than the outer one. Glandular warts are prominent on the top and sides of the body. The underside is granular. The parotid glands are smooth but porous. The upper surfaces are bright gray-brown to yellowish, often with a red infusion, with the sides of the head and body darker. A dark band starts anterior to the eye and terminates just behind the forelimbs. There is a series of large, light-edged black blotches on the back. Some individuals have a hairlike yellow vertebral line. The belly is creamy off-white. This species is well camouflaged against a background of dead leaves.

HABITAT AND DISTRIBUTION (FIG. 37): This forest toad can sometimes be found in farmland and gardens on the forest edge. It is known in Tanzania from the Usambara, Udzungwa, and Uluguru mountains, at both high and low altitudes.

ADVERTISEMENT CALL: The male calls from streams in forest, usually concealed by vegetation or the stream bank. The call is a loud rattling snore, with around 72 pulses/s. Each call has a duration of 0.5–0.7 s. The emphasized harmonics are around 740 Hz and 1.5 kHz.

BREEDING: This species breeds from mid-September through to March, coinciding with the long rains. Strings of eggs are deposited in shallow streams.

TADPOLES: Unknown.

NOTES: Known food items include grasshoppers, locusts, crickets, beetles, and millipedes. The adults are common along streams but can be found solitary in forest.

KEY REFERENCES: Nieden 1911b, Barbour & Loveridge 1928a, Tandy 1972.

Cameroon Toad

Bufo camerunensis Parker, 1936

Oban toad.

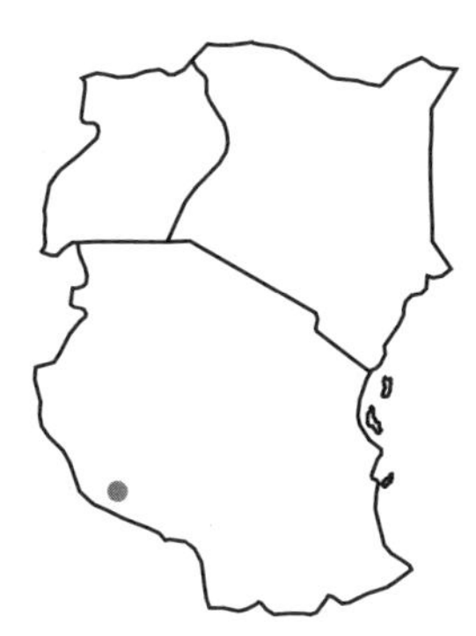

Fig. 38 *Bufo camerunensis*.

DESCRIPTION: This species is large, with the females exceeding 70mm in length. There are two pairs of black spots, one on the scapular and the other on the sacral region, with one or two dark interorbital bars. A light vertebral line is usually present. Breeding animals have prominent spines on the side of the head and behind the tympanum. In the brightest specimens, the sides of head and body are tinged with pink. This coloration serves to distinguish the toad in the field.

HABITAT AND DISTRIBUTION (FIG. 38): The Cameroon toad is confined to rain forest and outlying forest islands. It is known from Nigeria, the island of Bioko, eastern Democratic Republic of Congo, to southwestern Tanzania.

ADVERTISEMENT CALL: Males call from ground level at the edge of shallow ponds. The call is a rapid snore with a duration of 600–900ms. The emphasized harmonic is at 1600Hz, with the fundamental harmonic at 850Hz. The calls are repeated 30–40 times/min.

BREEDING: Unknown, although the condition of female toads indicates the breeding season in the Congo Basin is from May to October.

TADPOLES: Unknown.

NOTES: Known food items include a range of insects, myriapods, spiders, and snails. The Cameroon toad is recorded as the prey of Günther's green tree snake *Dipsadoboa unicolor*. It is similar in morphology and advertisement call to *Bufo togoensis* from the high forests of the Congo Basin and West Africa.

KEY REFERENCES: Noble 1924 (as *Bufo polycercus*), Márquez et al. 2000.

Sooty Toad

Bufo fuliginatus Witte, 1932

The specific name *fuliginatus* means "sooty," referring to the uniform coloration.
Shaba Province toad.

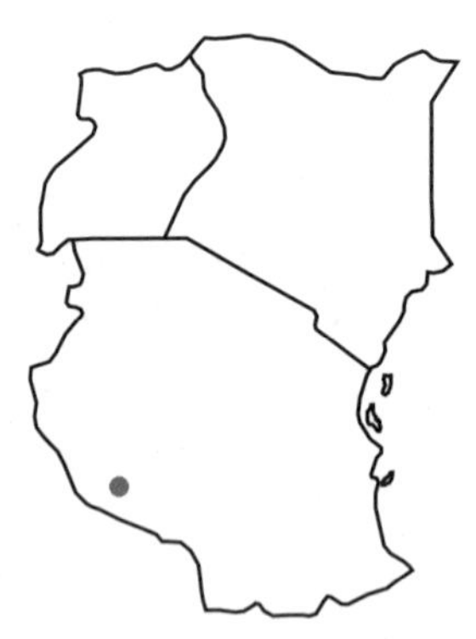

Fig. 39 *Bufo fuliginatus.*

DESCRIPTION: This is a medium-sized toad, with the male reaching 59 mm and the female up to 66 mm in length. A serrated tarsal fold and tympanum are present. Two indistinct rows of tubercles mark the position of glands under the forearm. The parotid glands are prominent. Webbing is moderate, with the third and fifth toes having at most two phalanges free of web. The back is spiny. Males lack both vocal sacs and dark throats. A dark eye-stripe extends from the eye onto the snout, but the back is nearly uniform brown, without strong contrasts.

HABITAT AND DISTRIBUTION (FIG. 39): This little-known species has been recorded from montane gallery forest of Shaba Province in the

Democratic Republic of Congo, to the highlands west of Lake Rukwa in southwestern Tanzania.

ADVERTISEMENT CALL: Each call consists of a rapid series of pulses, with dominant energy at 1.9 kHz. The call duration is 0.2–0.6 s.

BREEDING: No details of the breeding biology are known for this species. Females exhibit a cyclic breeding pattern, with a peak in the rainy season.

TADPOLES: Unknown.

KEY REFERENCE: Witte 1932.

Somber Toad

Bufo funereus Bocage, 1866

The name *funereus* means "somber."
Angola toad.

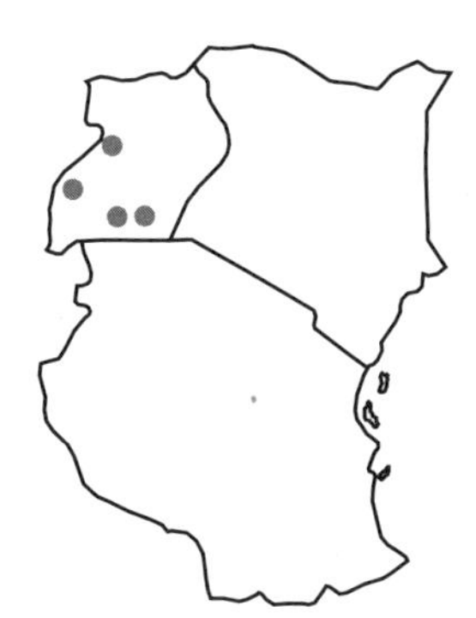

Fig. 40 *Bufo funereus.*

DESCRIPTION: The males reach 36 to 57 mm and the females, 48 to 66 mm in length. The parotid gland is slightly more than twice as long as wide, oval, and distinctly marked, separated from the eyelid. The tympanum is two-thirds to three-fourths of the eye diameter. The tarsal fold is absent, replaced by a row of small spiny tubercles. There are numerous closely set warts on the slim head. A series of separate, conical glands is present behind the angle of the jaw. Females and juveniles possess spines on the back, but these are absent in breeding males. The first finger is longer than the second. In life adults are brownish above with irregular darker markings, with an indication of a light vertebral line. The skin may be smooth or granular in the male. A distinct light band is found between the eyes, sometimes with a pale vertebral stripe. The upper jaw is marked with contrasting light and dark bands.

HABITAT AND DISTRIBUTION (FIG. 40): This frog is found in rain forest and open forest, often under fallen wood. It is known from Angola, Uganda, Burundi, Rwanda, and the northern Democratic Republic of Congo, extending northward through the Congo Basin.

ADVERTISEMENT CALL: Unknown. Males lack vocal sacs and this species may not produce an advertisement call.

BREEDING: The somber toad has a prolonged reproductive period, breeding from August to January, although females have ripe ova for most of the year.

TADPOLES: Unknown.

NOTES: Known food items include termites, beetles, ants, bugs, myriapods, spiders, caterpillars, roaches, earwigs, grasshoppers and leafhoppers, earthworms, snails, and puddle frogs. The short first finger, homogeneous warts, and lack of tarsal fold distinguish it from the common toad *Bufo regularis*.

KEY REFERENCES: Loveridge 1944, Inger 1968, Perret & Amiet 1971.

Garman's Toad

Bufo garmani Meek, 1897
(Plate 2.5)

This species was named for the ichthyologist S. W. Garman, of the Museum of Comparative Zoology at Harvard.
Light-nosed toad, olive toad, Garman's square-backed toad.

DESCRIPTION: This toad is large, with the female reaching 115 mm and the male 106 mm in length. The parotid glands are conspicuous. The back is covered by warts, each with a dark tip. The tympanum is less than half the diameter of the eye. Three or four segments of the fourth toe are free of webbing. The back varies, being reddish brown to tan, edged with black. The top of the head is characteristically free of darker

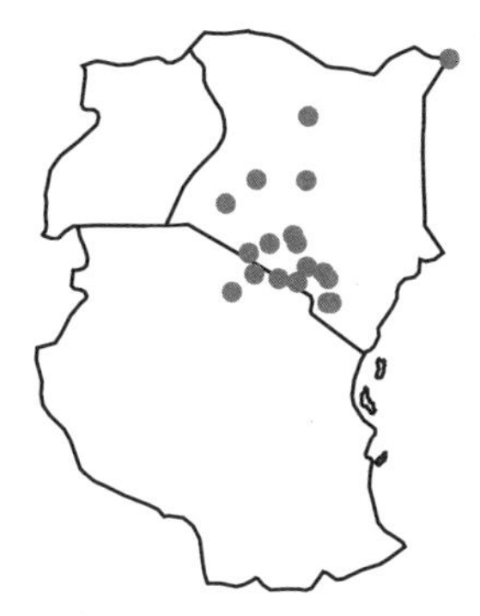

Fig. 41 *Bufo garmani.*

markings, and the bars behind the eyes do not fuse. Red patches occur on the backs of the thighs.

HABITAT AND DISTRIBUTION (FIG. 41): This savanna and grassland species is widely distributed from the northwestern parts of South Africa, Zimbabwe, northward to East Africa.

ADVERTISEMENT CALL: The males call from the edge of pools. Each call is a loud "kwaak." Call duration varies from 0.2 to 0.9 s, with pulse rates of 50–105/s. The dominant harmonic is around 1.5 kHz.

BREEDING: Eggs are laid at the start of the long rains, although these toads may breed throughout the year when ponds are available. Black eggs measuring 1.2 mm are laid in two strings. Between 12,000 and 20,000 eggs are produced per clutch. When laid, the eggs are 0.8 mm in diameter.

TADPOLES: See the chapter on tadpoles.

NOTES: During breeding, males attempt to grasp any female within reach. Females that are not ready to breed signal the male to release them by quivering and arching the back. Eggs are eaten by the serrate hinged terrapin *Pelusios sinuatus*, Müller's platanna *Xenopus muelleri*, and even Garman's toad tadpoles. The adult toads are eaten by the young of the Nile crocodile, *Crocodylus niloticus*. Displaced toads are able to return to a specific pond from up to a kilometer. The species is similar to the guttural toad *Bufo gutturalis* and the common toad *B. regularis*.

KEY REFERENCES: Taylor 1982; Channing 1991, 2001.

Guttural Toad

Bufo gutturalis Power, 1927
(Plate 2.6)

The specific name *gutturalis* refers to the loud guttural advertisement call.
Leopard toad, square-marked toad, greater cross-marked toad, common toad, *ikiyula* in Nyakusa, *chula tengu* in Nyungwe, *kikere* in Ganda and surroundings, *mtuvu* (male) and *ichcheri* (female) in Kiga, *kuvifata* in Rwanda, *ligumi* in Yao, *liumi* in Makonde and Maviha.

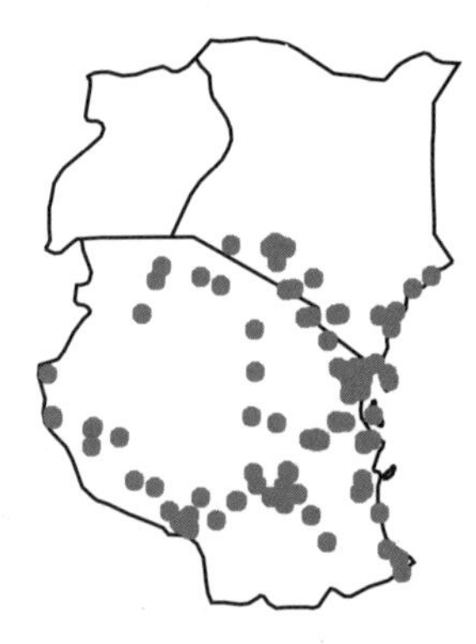

Fig. 42 *Bufo gutturalis.*

DESCRIPTION: This species is large, with the female reaching 120 mm and the male 90 mm in length. The glands under the forearm form a distinct row of pale tubercles. These are never fused into a ridge. The parotid glands are large and without warts. The back markings are usually arranged in pairs, outlined in a darker brown. A pale cross is present on the top of the head, with the transverse bar running between the eyes. Frequently a thin vertebral line is present. Red patches are found behind the thighs and in the groin. These patches are especially prominent in breeding animals. The ventral surface is white. Breeding males possess pale yellow throats with darker speckling and a single pale vocal sac.

HABITAT AND DISTRIBUTION (FIG. 42): The guttural toad is found in savanna and thornbush and seems to occur in very dry areas only where a permanent source of water is available. It is common in towns around buildings and widely distributed in eastern, central, and southern Africa, from Angola, Uganda, Kenya, and southward to South Africa.

This species is very common where it occurs, and has been introduced to the islands of Réunion and Mauritius.

ADVERTISEMENT CALL: The male calls from shallow water near the edge of a pool, and very large calling aggregations may be formed. The call is loud and can be heard from some distance. It is a slow, pulsed snore, and the individual pulses are usually clearly audible. The dominant frequency varies from 1.0 to 1.3 kHz. Each call consists of about 15–20 pulses, with a duration of 0.8–1.2 s, at a rate of 15/s. Like many other toads, pairs of males call alternately. This behavior is referred to as *antiphony*.

BREEDING: This toad breeds during the day and night. The female is approached by the male after she enters the breeding pool. The male apparently reacts to the stimulus of the female disturbing the water surface. A male will attempt to displace other males in amplexus, with the larger male usually winning. A pair will lay between 15,000 and 25,000 eggs in one clutch. They are laid in shallow water at the edge of pools, often wound in and around vegetation. The eggs are small, 1.4–1.5 mm in diameter with one black pole, and produced in two gelatinous strings, each 5 mm thick, with about 6 eggs/cm along a newly laid string.

TADPOLES: See the chapter on tadpoles.

NOTES: If handled roughly, these toads secrete a white milky fluid from the parotid glands. This substance is poisonous and very sticky and will cause the death of other species of frogs placed in a bag together with the toad. One of the compounds identified in the parotid secretion is epinephrine. Mammals up to 5 kg will vomit and die if they absorb small amounts of the secretion. The venom of the parotid glands is extremely irritating to the mucous membranes of humans, and handling a toad when you have cuts on your hand, or when wiping your eyes or lips, will result in an intensely irritating, burning sensation.

Guttural toads become quite used to humans and will happily return to the lighted veranda of a house night after night to feed on moths and beetles. There are records of toads even spending the days under wardrobes or under floorboards in older houses.

Known food includes insects, earthworms, centipedes, spiders, scorpions, lizards, and other frogs like the common squeaker *Arthroleptis stenodactylus* and tree frogs *Leptopelis*. The tadpoles are eaten by dwarf bream *Pseudocrenilabrus philander*, clawed toads *Xenopus* spp., water insects, and birds. Adults are eaten by the black-necked spitting cobra *Naja nigricollis*, the rhombic night adder *Causus rhombeatus*, the Angolan green snake *Philothamnus angolensis*, other green snakes *Philothamnus* spp., and the white lipped snake *Crotaphopeltis hotamboeia*. Other known predators include the serrated hinged terrapin *Pelusios sinuatus* and the African civet *Viverra civetta*. The toads are not defenseless, however, as the parotid secretion is a powerful poison.

Molecular evidence suggests that this species has been separated from the northern *Bufo regularis* for 5 to 6 million years.

KEY REFERENCES: Loveridge 1944, Tandy 1972, Telford & Sickle 1989.

Kerinyaga Toad

Bufo kerinyagae Keith, 1968
(Plate 2.7)

The species is named for the village of Kerinyaga in Kenya, where it was first collected.
Keith's toad.

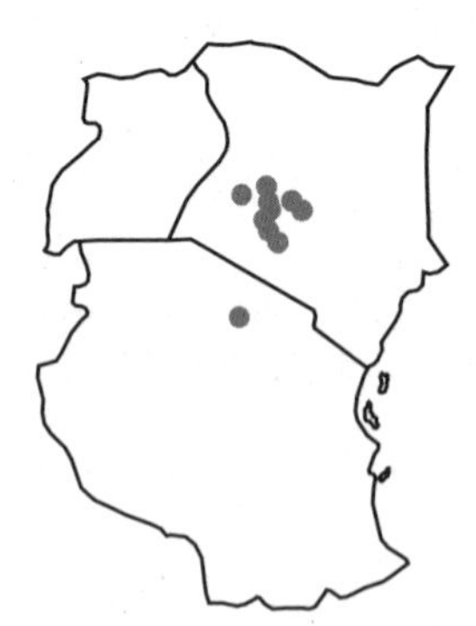

Fig. 43 *Bufo kerinyagae.*

DESCRIPTION: This species is similar in size and overall morphology to other toads in the *Bufo regularis* group, with males up to 85 mm and females up to 86 mm in length. The tympanum is distinct, vertically oval, and smaller than the eye diameter. The tympanum is larger in males than females. The first finger is longer than the second. The head width is 35% of the body length. Parotids are about half as wide as long, parallel, and separated clearly from the eye. The width of the parotid gland is equal to the distance between the anterior borders of the eyes.

The skin is covered with distinct sharp-tipped warts. The vocal sac has one or two internal openings. About half the specimens have an elongated gland on the back of the forearm; others have a series of glands. The background color is greenish tan with red-brown parotids and upper eyelids. A thin pale vertebral stripe is often present. A dark band runs from the eye along the side of the snout to the upper lip. There are about four pairs of dorsal brown blotches with dark margins. Breeding males have a dark nuptial pad, from the base of the first finger to the middle of the last phalanx, extending on to the second finger and inner metacarpal tubercle. Breeding males do not have a dark throat.

HABITAT AND DISTRIBUTION (FIG. 43): The Kerinyaga toad is found on high plateaus above 1800 m, on both sides of the rift valley, below the forests on the steep slopes. It is known from flat open grassland, where rainfall varies from 500 to 1000 mm, in Ethiopia, Kenya, Uganda, and Tanzania.

ADVERTISEMENT CALL: The male calls from shallow water in temporary pools. The call is a snore that drops in pitch. The call duration is 1.0–2.1 s, with 42–80 pulses at a rate of 38–48/s. Two frequencies are emphasized, the lower starting at 0.9 kHz and rising slightly to 1.1 kHz, and a second starting at 2.8 kHz, decreasing to 2.2 kHz. The steady drop in frequency of the upper emphasized band is characteristic of this species.

BREEDING: The toad breeds in temporary pools early in the rainy season, once there is standing water.

TADPOLES: Unknown.

NOTES: This species is often sympatric with the Nairobi dainty frog.

KEY REFERENCE: Keith 1968.

Kisolo Toad

Bufo kisoloensis Loveridge, 1932
(Plate 2.8)

This species was named for the town of Kisolo in Uganda, where it was first collected.
Montane golden toad, *Bergkröte* in German, *crapaud alticole* in French.

Fig. 44 *Bufo kisoloensis.*

DESCRIPTION: This toad is large, with the male up to 71 mm and the female 87 mm in length. It has a pointed snout, slender tapering fingers, and extensive webbing, which reaches or nearly reaches the tips of the toes, excepting the fourth one. The tympanum is distinct, but smaller than the eye. The parotid glands are two or three times longer than wide. The male has a smooth back with flattened warts, while the female is more typical of the *Bufo* genus with a granular warty back. Coloration is a uniform olive, sometimes with darker markings. A light thin vertebral line is often present. The breeding male can be easily identified as it is a bright yellow during amplexus and for a brief period afterward, although it retains the normal coloration when it starts calling.

HABITAT AND DISTRIBUTION (FIG. 44): This species is associated with rivers and is found in cool, moist montane forests up to an altitude of 1800 m. The Kisolo toad is known from the highland areas of northern Malawi and northeastern Zambia, northward to the Congo Republic, the Democratic Republic of Congo, Rwanda, Uganda, Tanzania, and Kenya.

ADVERTISEMENT CALL: The male calls from shallow streams, partly beneath vegetation. The call is a slow snore, about 1 s long, repeated rapidly. The emphasized frequency is around 600 Hz. The pulse rate is 66/s.

BREEDING: The male turns a bright yellow for the few hours before and during amplexus. Details of egg laying are unknown.

TADPOLES: See the chapter on tadpoles.

NOTES: Known food items include arthropods, worms, and snails.

KEY REFERENCES: Curry-Lindahl 1956, Fischer & Hinkel 1992, Channing & Drewes 1997.

Lindner's Toad

Bufo lindneri Mertens, 1955
(Plates 3.1, 3.2)

This frog was named for the collector Erwin Lindner, head of the entomology department at the State Museum of Natural History in Stuttgart. Lindner's dwarf toad, Dar es Salaam toad.

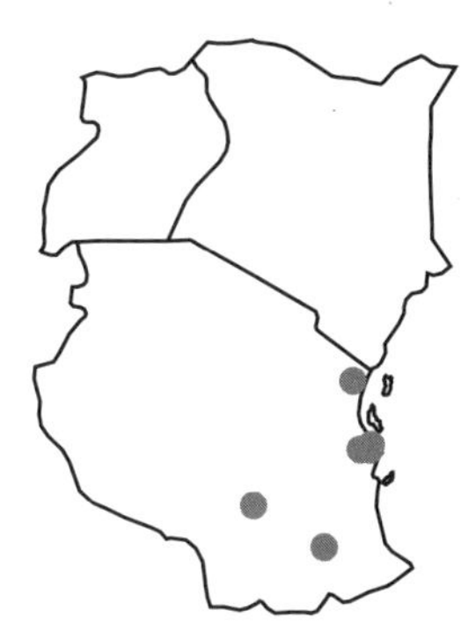

Fig. 45 *Bufo lindneri.*

DESCRIPTION: The male reaches 23 mm in length, with the female up to 34 mm long. The skin is rough and spinose, except for the terminal phalanges that are smooth. The head is as wide as long. This small toad has a characteristic pointed head. The nostrils are separated by a groove. The distance between the eyes is about equal to the width of the eyelid. The parotid glands are more than twice as long as wide, indistinct and flattened, and narrower posteriorly. Finger and toe tips are somewhat pointed. Two palmar tubercles are present. Only a little webbing is found between the toes. The throat is covered with small spines, each of which in turn is surrounded by a ring of smaller white spines. The back is gray to brown, with orange to red tubercles. The color pattern is characteristic, with a vertebral stripe originating between the shoulders. Darker inclined markings are present on the snout and upper eyelid. Below, the back of the throat to the belly is marked with a series of gray flecks, or a single blackish longitudinal line. This contrasts with the "three-pronged" ventral marking of *Bufo taitanus*. A specimen from Dar es Salaam had pink thigh patches.

HABITAT AND DISTRIBUTION (FIG. 45): This toad is found in sandy areas with open bush, and in the lowland coastal forests like Pugu, south of Dar es Salaam. It has been collected from the inland savanna and coastal lowlands of Tanzania, Malawi, and Mozambique.

ADVERTISEMENT CALL: Unknown.

BREEDING: Unknown.

TADPOLES: Unknown.

NOTES: This frog inhabits thicket in areas that were probably formerly forested. Specimens have been collected from fallow fields.

KEY REFERENCE: Clarke 1988.

Lönnberg's Toad

Bufo lonnbergi Andersson, 1911
(Plate 3.3)

This toad was named for Einar Lönnberg, a Swedish naturalist who collected in Africa during 1910–1911.

Fig. 46 *Bufo lonnbergi.*

DESCRIPTION: The female may be up to 43 mm long. The male is smaller, up to 38 mm in length. The head width/tibia ratio is about 80%. The mean head/body length ratio is 31%. There is no tarsal fold. Foot is longer than half body length. The tympanum is lacking, and there is no vocal sac in the male. The skin is fairly smooth, although covered by very small warts. A few larger warts are scattered on the back and legs. The parotid glands are elevated, narrow, and long, up to 3 times longer than wide. There is a clear gap between the eye and parotid gland. The first finger is shorter than the second. The snout is rounded in profile. Webbing is reduced, almost absent, with slight webbing present between the third and fourth toes. Characteristic markings are a light vertical stripe extending from the flank, behind the arm, to the parotid gland. There is also a dark streak between the nostril and the eye and a dark longitudinal band along the lower half of the parotid. A yellow vertebral line is present. The underside of the body and legs possesses small yellow pustulations and gray flecks. Males are bright greenish yellow in the breeding season.

HABITAT AND DISTRIBUTION (FIG. 46): This small toad is known from open grassland. It is endemic to the Kenyan highlands, at altitudes above 2000 m.

ADVERTISEMENT CALL: This species does not possess an advertisement call, although males do make a release call.

BREEDING: This species breeds from early November through late February in the southern part of its range, and probably throughout the year farther north. Spawning has been observed during the day. Eggs are laid in a string in small shallow pools. The string is 5.5 mm in diameter. Each egg is 2.5 mm in diameter and is largely black, except for a small pale area at the vegetal pole. Clutch size is 125.

TADPOLES: Unknown.

NOTES: This toad moves slowly, walking on its toes, and can walk on the bottom of a pond under water. Known predators include the gray-bellied skaapsteker *Psammophylax variabilis* and the rat *Lophuromys flavopunctatus*.

KEY REFERENCES: Loveridge 1944, Poynton 1997.

Lugh Toad

Bufo lughensis Loveridge, 1932
(Plate 3.4)

The specific name refers to Lugh in Somalia, where the species was first collected.

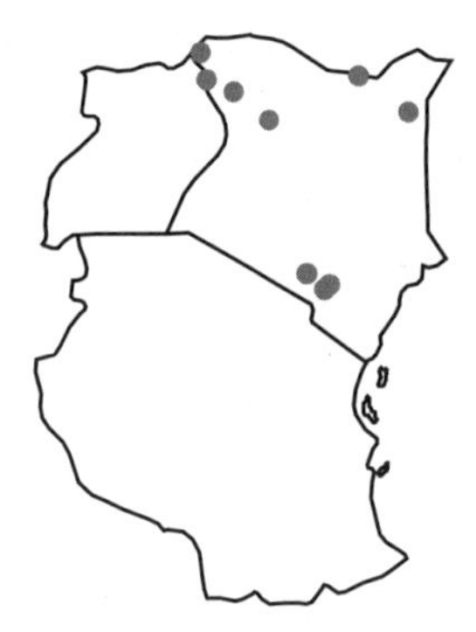

Fig. 47 *Bufo lughensis.*

DESCRIPTION: Females are known up to 47 mm long, with the smaller males reaching a length of 36 mm. A tympanum is present. The subarticular tubercles are double. Two to three phalanges of the fourth toe are free of web, and the tubercles of the forearm and foot are prominently enlarged. Two large, smooth, flat to conical metatarsal tubercles are present. The belly is speckled with black, with the throat and anterior chest are immaculate. The back is pale with dark specks, each with a pale center.

HABITAT AND DISTRIBUTION (FIG. 47): This species is found in arid areas and is known from Ethiopia, Somalia, Sudan, and Kenya.

ADVERTISEMENT CALL: The call sounds like a goat bleating. The "bleats" are repeated about once per second, and each consists of about 43 pulses at a rate of 38–70/s. The dominant frequency is between 1.7 and 2.6 kHz.

BREEDING: Unknown.

TADPOLES: Unknown.

NOTES: The toads have been found under rocks and other objects, or occupying rodent burrows.

KEY REFERENCES: Loveridge 1932b, Tandy & Tandy 1976.

Flat-Backed Toad

Bufo maculatus Hallowell, 1856
(Plate 3.5)

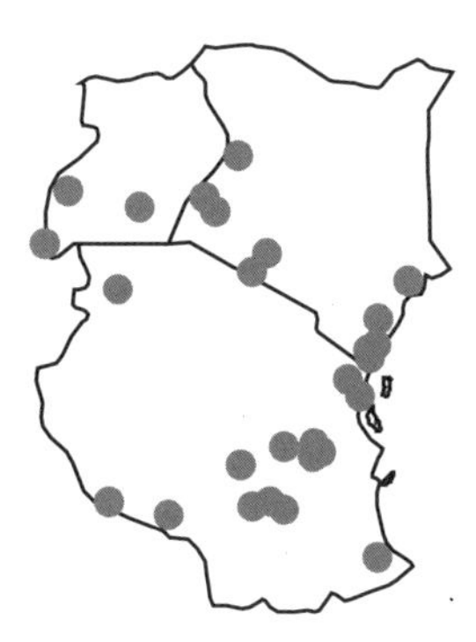

Fig. 48 *Bufo maculatus.*

The specific name *maculatus* refers to the darker spots on the back. Lesser cross-marked toad, Hallowell's toad.

DESCRIPTION: This toad is medium to large, with the female reaching 80 mm and the male 64 mm in length. A distinct row of white tubercles is present under the forearm. The back is warty, and each wart has a sharp dark tip. These warts are prominent in sexually active animals. The tympanum is less than one-half the eye diameter in males, although slightly larger in females. The parotid glands are large and obscured with smaller warts, but flattened and indistinct, giving the frog its common name—the flat-backed toad. A thin pale vertebral line is often present. A light cross is present on top of the head. The pattern of the back resembles that of the guttural toad *Bufo gutturalis*, with which it could be confused. However, the flat-backed toad does not possess red marks on the thighs. The underside is off-white, often with darker speckles.

HABITAT AND DISTRIBUTION (FIG. 48): This toad is frequently associated with river courses, although it is found away from water in open savanna and disturbed areas in forest. It occurs widely in West and East Africa southward to Angola and northeastern South Africa.

ADVERTISEMENT CALL: The male calls from the edge of pools and rivers, concealed under vegetation and well spaced. Each call has a duration of 0.5 s, with dominant harmonics between 1.8 and 2.1 kHz. The male may call antiphonally with other males.

BREEDING: This species has an explosive breeding system where hundreds of pairs may breed simultaneously in shallow water. The males and females may acquire a yellow color during breeding. The males fight over females. Eggs are laid in a string 3 mm thick. Each egg is 1.2 mm in diameter. The breeding takes place on one day only at the start of the rains, leaving the substrate in shallow water covered in strings of eggs.

TADPOLES: The tadpoles are common in quiet pools. They reduce the risk of predation by aggregating. See the chapter on tadpoles.

NOTES: Blood-sucking sandflies are known to feed on this toad. Predators include the night adder *Causus rhombeatus*. During a breeding frenzy in Luangwa in Zambia, large river crabs came out in the dozens and fed on the toads, sometimes tearing a female apart and eating her from below while the male was still clasping her.

KEY REFERENCES: Stewart 1967, Spieler & Linsenmair 1999, Rödel 2000, Channing 2001.

Mocquard's Toad

Bufo mocquardi Angel, 1924

This species is named for François Mocquard, assistant to Léon Vaillant, professor in charge of the laboratory of reptiles and fishes at the Natural History Museum in Paris.

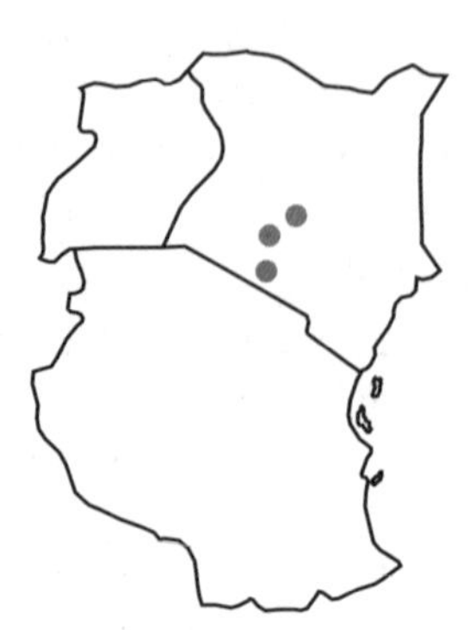

Fig. 49 *Bufo mocquardi*.

DESCRIPTION: This species is small, with females up to 34 mm long and the males shorter. The tympanum is not visible and there is no tarsal fold. The legs are short. The snout is as long as the diameter of the eye, and the head width is equal to the length of the tibia. Parotid glands do not reach the eyelids and have straight sides, not narrowing behind. The first finger is much shorter than the second. The toes are about one-third webbed, with three phalanges of the fourth toe free. Subarticular tubercles are double. Both metatarsal tubercles are of equal size. The dorsal color varies from yellow to black, with a pale vertebral line usually present. Sometimes black spots form a transverse band between the eyes and border a uniform area on the neck.

HABITAT AND DISTRIBUTION (FIG. 49): This is a high-forest species, known from Mt. Kinangop and the lower forests of Mt. Kenya, and the Nairobi highlands.

ADVERTISEMENT CALL: Unknown.

BREEDING: Unknown.

TADPOLES: Unknown.

NOTES: Fieldwork is required to confirm the status of this species.

KEY REFERENCES: Angel 1924b, Poynton 1997.

Nairobi Toad

Bufo nairobiensis Loveridge, 1932

This species takes its name from the city of Nairobi, where it was first found.

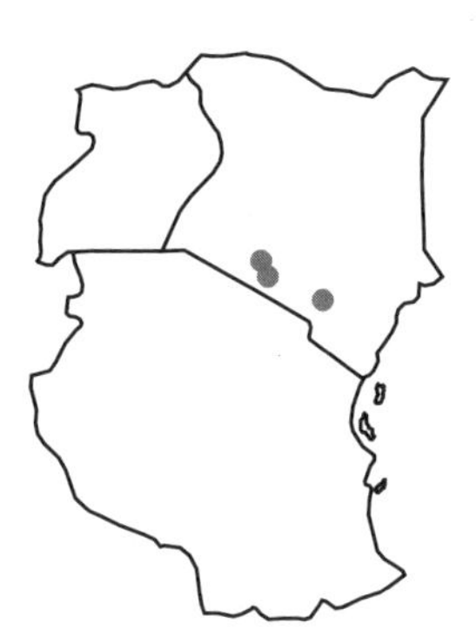

Fig. 50 *Bufo nairobiensis.*

DESCRIPTION: This species is small, with the males reaching 29 mm long and the females 39 mm. The tibia is about one-third of the body length. Adults have a background olive color with a pale vertebral line, and are white underneath. An interorbital darker marking is found in some specimens. Female markings are very distinct, with more warty skin.

HABITAT AND DISTRIBUTION (FIG. 50): The Nairobi toad is known from the wooded savanna at mid altitudes in Kenya.

ADVERTISEMENT CALL: Unknown.

BREEDING: Unknown.

TADPOLES: Unknown.

NOTES: Fieldwork is required to confirm the status of this species.

KEY REFERENCES: Loveridge 1932b, Poynton 1997.

Parker's Toad

Bufo parkeri Loveridge, 1932

This frog was named for the herpetologist H. W. Parker, a curator of the Natural History Museum in London from 1923 to 1957.
Mangasini toad.

Fig. 51 *Bufo parkeri.*

DESCRIPTION: This species is small, with the males up to 31 mm long and the females 35 mm. The tympanum is vertically elliptical, half the size of the eye. The interorbital space is equal to the length of the upper eyelid. The back is gray to black like the cotton soil, with brown or dull red warts. Male throats are yellow.

HABITAT AND DISTRIBUTION (FIG. 51): This is a savanna species, known from the southern rift valley in Kenya through the Lake Natron basin to southern Tanzania.

ADVERTISEMENT CALL: Males call from the ground near water. The call is a high-pitched buzz, repeated about once per second. Each call is just over one-tenth of a second long, at a dominant frequency of 4.5 kHz, and consists of 13 initial pulses followed by 3 slower terminal pulses. The first part of the call has a pulse rate of 110/s, with the slower terminal pulse rate being 56/s.

BREEDING: This species breeds with the onset of the long rains.

TADPOLES: Unknown.

NOTES: In the dry season this toad shares abandoned termitaria with large ants.

KEY REFERENCE: Loveridge 1932b.

Rees's Toad

Bufo reesi Poynton, 1977

This species was named for Allen Rees, who worked for the Tanzanian Wildlife Department as a principal game warden.

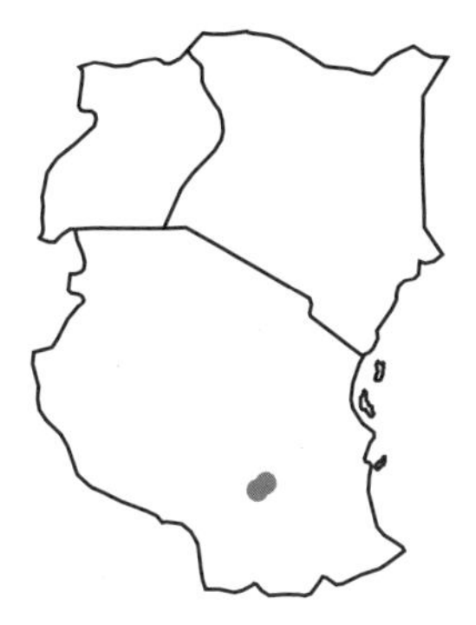

Fig. 52 *Bufo reesi.*

DESCRIPTION: Males reach a length of 57 mm and females, 63 mm. The tympanum is large, circular, with a horizontal diameter two-thirds of the diameter of the eye. The parotid glands are flattened, close to the tympanum. The webbing is extensive, with a margin of webbing to the tips of the toes. One phalanx of third and fifth toes free of webbing. The subarticular tubercles are single. The tarsal fold is prominent, extending the length of the tarsus. Males are spiny on the upper surfaces, with some larger spines surrounded by a rosette of smaller rounded warts. Tibia length is less than half the body length. The warts and parotid glands are darker than the background, there are dark paired markings over the shoulders and in the sacral area, and a dark lateral band runs from the side of the snout to the arm insertion. The belly is a deep yellow.

HABITAT AND DISTRIBUTION (FIG. 52): This species is a floodplain inhabitant, known from the Kihansi-Kilombero floodplain, in Mahenge district, southern Tanzania.

ADVERTISEMENT CALL: Unknown.

BREEDING: Unknown.

TADPOLES: Unknown.

KEY REFERENCE: Poynton 1977.

Common Toad

Bufo regularis Reuss, 1833
(Plate 3.6)

The specific name refers to the regular back pattern.
Ikiyula in Nyakusa, *chula tengu* in Nyungwe, *kikere* in Ganda and surroundings, *Mtuvu* (male) and *ichcheri* (female) in Kiga, *kuvifata* in Rwanda, *ligumi* in Yao, *liumi* in Makonde and Maviha.

DESCRIPTION: This species is large, with the female reaching 120 mm and the male 90 mm in length. The glands under the forearm form a distinct row of pale tubercles, which are never fused into a ridge. The parotid glands are large and without warts. The back markings are usually arranged in pairs, outlined in a darker brown, with small white spots on the back. A pale cross is present on the top of the head, with

Fig. 53 *Bufo regularis.*

the transverse bar running between the eyes. Frequently a thin vertebral line is present. Red patches are seen in the groin. These patches are especially prominent in breeding animals. The ventral surface is white. Breeding males possess dark throats and a single pale vocal sac.

HABITAT AND DISTRIBUTION (FIG. 53): The common toad is found in savanna and farmbush, and seems to occur in very dry areas only where a permanent source of water is available. Due to difficulties in separating it from other species like the guttural toad and the desert toad, the precise range of this species in East Africa is not known. It is found across northern Africa from Senegal eastward to Ethiopia, south to Uganda and northern Kenya, and north to the Mediterranean coast on the Nile Delta and in the oases of Djanet in Algeria and Gat in Libya.

ADVERTISEMENT CALL: The call is a very slow snore with clearly discernable pulses. The pulse rate is 29/s, with emphasized frequencies below 1 kHz.

BREEDING: This toad has two peaks of breeding, one at the beginning of the rainy season and one at the end. It may breed throughout the year in areas with permanent lakes or rivers, or breed seasonally in arid areas.

TADPOLES: Unknown.

NOTES: Molecular evidence suggests that this species has been separated from *Bufo gutturalis* for 5 to 6 million years. It eats a range of small arthropods. It is eaten by the night adder *Causus* sp. and white-lipped snake *Crotaphopeltis hotamboeia.*

KEY REFERENCES: Loveridge 1944, Tandy 1972, Tandy et al. 1976, Bachmann et al. 1980.

Steindachner's Toad

Bufo steindachneri Pfeffer, 1893
(Plate 3.7)

This species is named for Franz Steindachner, curator and later director of the Natural History Museum in Vienna from 1861 until his death in 1919.

Fig. 54 *Bufo steindachneri.*

DESCRIPTION: The male may reach 52 mm and the female 54 mm in length. The tympanum is distinct, slightly oval vertically, and more than half the diameter of the eye. The first finger is shorter than the second. A large palmar tubercle is present. There are large shovel-like subarticular tubercles under the third phalanx of the third and fourth fingers, and the penultimate phalanx of the second finger. Webbing is very slight. The subarticular tubercles under the toes are single. The parotid glands are not very distinct. The underside is very warty posteriorly, but finely speckled in the middle with larger spots anteriorly. The tarsal fold is absent, and the snout is sharp. The back is very warty. Males are yellow or brown above with dark brown markings, with a light or dark V behind the eyes. Breeding males have red on the back of the thigh, on top of the foot at the base of the fourth toe, and on the back of the upper arm. The soles of the hands and feet are dark.

HABITAT AND DISTRIBUTION (FIG. 54): This species is found in wooded savanna and coastal marshes, from Chad eastward to Ethiopia, Somalia, Kenya, Uganda, and Tanzania.

ADVERTISEMENT CALL: Males call concealed beneath vegetation at the water's edge, or at the base of vegetation. The call is a series of metallic knocks, similar in some respects to a very slow call of a guttural toad. Three or four pulses are grouped together, and these are produced at a rate of three or four groups per second. The emphasized harmonics of each pulse are at 1.1 and 2.5 kHz.

BREEDING: Unknown.

TADPOLES: Unknown.

NOTES: Known food items include ants, beetles, spiders, and woodlice. Steindachner's toad is eaten by the olive marsh snake *Natriciteres olivacea*, white-lipped snake *Crotaphopeltis hotamboeia*, and one of the night adders *Causus* sp. They run, seldom hopping.

KEY REFERENCES: Loveridge 1936, Tandy & Keith 1972.

Taita Toad

Bufo taitanus Peters, 1878
(Plate 3.8)

The specific name *taitanus* refers to the Taita Hills in Kenya.
Dwarf toad, black-chested dwarf toad, Taita dwarf toad.

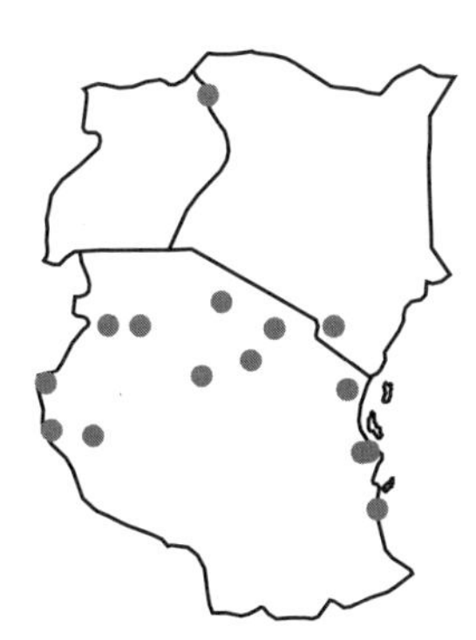

Fig. 55 *Bufo taitanus*.

DESCRIPTION: The female reaches 33 mm in length, while the male is smaller, up to 27 mm long. This toad is slender with spindly limbs.

The nostrils are nearer the snout than the eye. The internarial distance is usually two-thirds the width of the upper eyelid. The first finger is shorter than the second. Double conical subarticular tubercles are present under the fingers and toes, with supernumerary tubercles. The metatarsal tubercles are conical. A single large metacarpal tubercle is present. The palms and soles have spinose tubercles encircled by even smaller tubercles. The back is covered by small warts. The parotid glands are flattened but distinct, extending from behind the eye, continuous with the upper eyelid, to the level of the shoulder, and are more than four times longer than broad. The toes are slightly webbed, with two phalanges free of webbing on the inner side of the second toe. There is only a fringe of webbing on the basal phalanx of the fourth toe. One phalanx on the fifth toe is free of webbing. The underside is granular. The dorsal color pattern varies. Males are light gray or tan with three pairs of darker markings, while females have a uniform light brown back with darker sides. The underside is pale with a characteristic dark pattern with three anterior points. The limbs are also mottled with black marks. Males are thinner than females, with markedly more slender legs. Breeding males have dark nuptial pads that cover the entire dorsal and medial surfaces of the first two fingers.

HABITAT AND DISTRIBUTION (FIG. 55): The Taita toad is found at high elevations and on the coast. It occurs widely even in disturbed habitats. This species is found in sandy habitats in grassland and open savanna. It is recorded from Tanzania and Kenya, to southeastern Democratic Republic of Congo, Malawi, Mozambique, and Zambia. It is also recorded from Songo Songo Island.

ADVERTISEMENT CALL: This toad does not have an advertisement call. The sound production and hearing apparatus—tympanum, middle ear, and vocal sac—are absent.

BREEDING: Breeding takes place in early to mid summer. The 2-mm eggs are pale with a dark pole and are laid in strings in small pools. Clutch size is 125.

TADPOLES: Unknown.

NOTES: Unlike most other toads, this species is able to walk on toe tips. Recorded predators include the harsh-furred mouse *Lophuromys aquilus* and the southern striped skaapsteker *Psammophylax tritaeniatus*.

KEY REFERENCES: Stewart & Wilson 1966, Stewart 1967.

Turkana Toad

Bufo turkanae Tandy & Feener, 1985

This species is named for the type locality, Lake Turkana in Kenya. Lake Turkana toad.

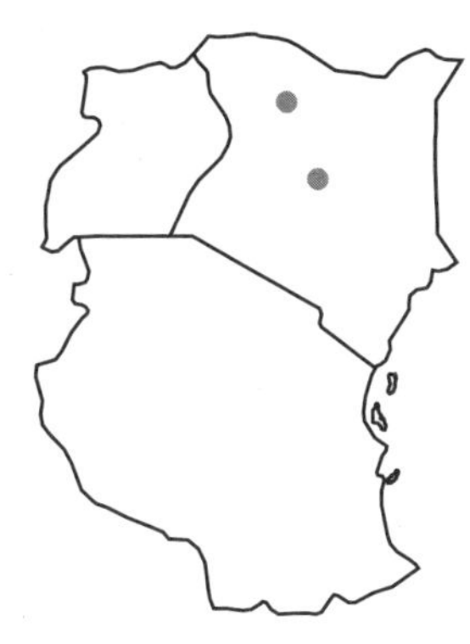

Fig. 56 *Bufo turkanae.*

DESCRIPTION: This species is small, with a body length of up to 43 mm in males and 48 mm in females. The tympanum is visible. The first finger is longer than the second. Warts on the lower surface of the forearm are not fused into a ridge. The head is wider than long, and the tympanum is smaller than half the diameter of the eye. Parotids are well separated from the eyes. The anterior border of the parotids is above the posterior border of the tympanum. Subarticular tubercles are single, except one under the first finger is double. The outer metacarpal tubercle is twice the length of the inner one. The fourth toe has three phalanges free of webbing on the inner side. Small pale spines are present in the webbing. The tarsal fold is distinct, covering about two-thirds the length of the tarsus. Tibia length is less than half the body length. Warts on the back usually have a single small dark spine, although some larger warts may have up to five spines. Spinules extend anteriorly to the eyelids. The back is a uniform gray or brown with a pair of markings on the snout between the eyes. Sometimes a pale vertebral line and other lesser markings are present. White spots are found on the back.

HABITAT AND DISTRIBUTION (FIG. 56): The Turkana toad is found in areas with permanent water in desert and arid savanna that have a mean annual rainfall of less than 500 mm, at altitudes from 350 to 1500 m. It is known from central and northern Kenya.

ADVERTISEMENT CALL: The call consists of a grating croak, with 44 pulses/s, at an emphasized frequency of 1.2 kHz. Sometimes the second harmonic is emphasized at 2.4 kHz. The call duration is about 0.5 s.

BREEDING: The species apparently breeds in the shallow water at the edge of Lake Turkana, and in the quiet edges of the rivers feeding into the lake.

TADPOLES: Unknown.

NOTES: The Turkana toad is morphologically indistinguishable from *Bufo blandfordi* of Somali and Ethiopia and *B. langanoensis* of central Ethiopia, but it is easily differentiated from them by the advertisement call.

KEY REFERENCE: Tandy & Feener 1985.

Urungu Toad

Bufo urunguensis Loveridge, 1932

This species is named for the Urungu region of southern Tanzania.

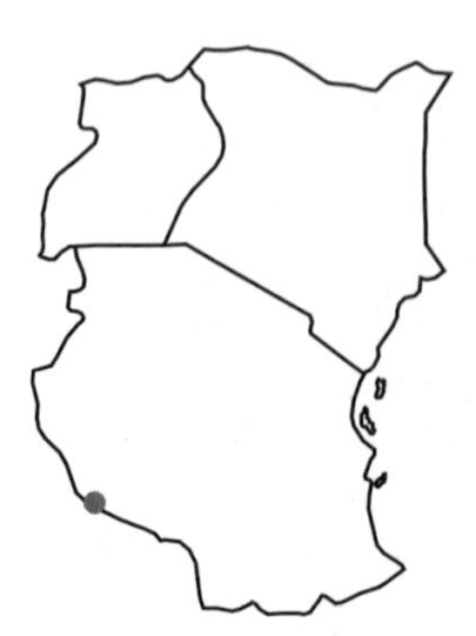

Fig. 57 *Bufo urunguensis.*

DESCRIPTION: This frog is small, with the female up to 29 mm long. The fingers are long and pointed. The tubercles under the fingers are double, and the tarsal fold is absent. Two conspicuous conical metatarsal tubercles are present. The parotid glands have a straight outer edge, which does not project below the level of the pupil, although the edge may not always be clear. The tympanum is rounded, with the vertical diameter about equal to the distance between the nostrils, and the diameter half that of the eye. The web notch may just reach the base of the inner segment of the fourth toe. The back is very rough, with spines on warts of different sizes. The female is gray, tinged with red and traces of purple. The back pattern consists of paler areas on the snout, neck, and lower back, usually with a V-shaped blotch facing backward, above the vent. The pale snout marking extends backward between the eyes onto the parotids while the underside is irregularly marked in purple or black. The soles of the hands and feet are dark with white tubercles.

HABITAT AND DISTRIBUTION (FIG. 57): The Urungu toad is known from rain-forest remnant, in Tanzania and Zambia near the southeastern corner of Lake Tanganyika.

ADVERTISEMENT CALL: Unknown.

BREEDING: Unknown.

TADPOLES: Unknown.

NOTES: Recorded food items are termites.

KEY REFERENCES: Loveridge 1932b, 1933.

Udzungwa Toad

Bufo uzunguensis Loveridge, 1932
(Plate 4.1)

This species is named for the Udzungwa Mountains of Tanzania.
Tofula in Hehe, *ikiyula* in Nyakusa.

DESCRIPTION: This toad is small, with males and females up to 30 mm in length. It lacks a tarsal fold. The parotids are distinct, three times longer than wide in males, but four to five times longer in females, contiguous with the upper eyelid. The outer margin of the parotids does

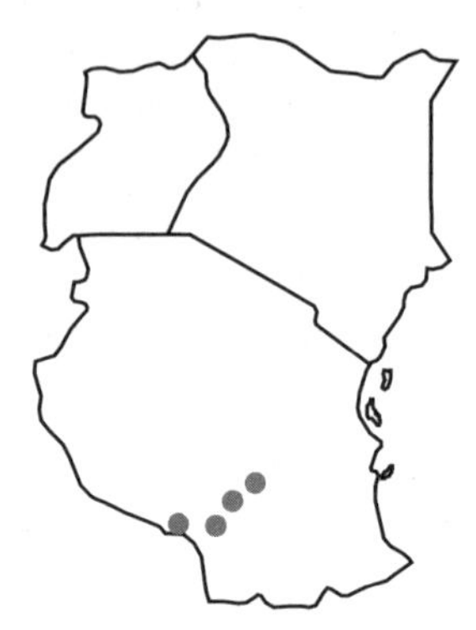

Fig. 58 *Bufo uzunguensis.*

not extend below the level of the outer rim of the eyelids. The internarial distance is two-thirds the width of the upper eyelid. The first finger is shorter than the second. Subarticular tubercles are generally double. The back and upper surfaces of the limbs have small single conical tubercles. The sacrum, flanks, and limbs have, in addition, larger tubercles ringed by smaller ones. Dorsal rhomboid markings are symmetrical. A thin yellow vertebral line is present. Usually a small median brown blotch is visible in the middle of the chest, flanked by a streak in front of each forearm. The limbs are cross marked. A V-shaped cream spot is present above the vent. Webbing occurs only as a remnant at the base of the toes. Nuptial pads are found on the dorsal surface of the first and second fingers of breeding males.

HABITAT AND DISTRIBUTION (FIG. 58): The Udzungwa toad is found in swamps and on slopes at altitudes above 1800 m. It is endemic to the Udzungwa Mountains and southern highlands of Tanzania.

ADVERTISEMENT CALL: This toad lacks a middle ear and vocal sac, and appears not to call.

BREEDING: Unknown. The eggs are pigmented.

TADPOLES: Unknown.

KEY REFERENCES: Loveridge 1932b, 1933.

Lake Victoria Toad

Bufo vittatus Boulenger, 1906

The species name means "striped," referring to the color pattern.

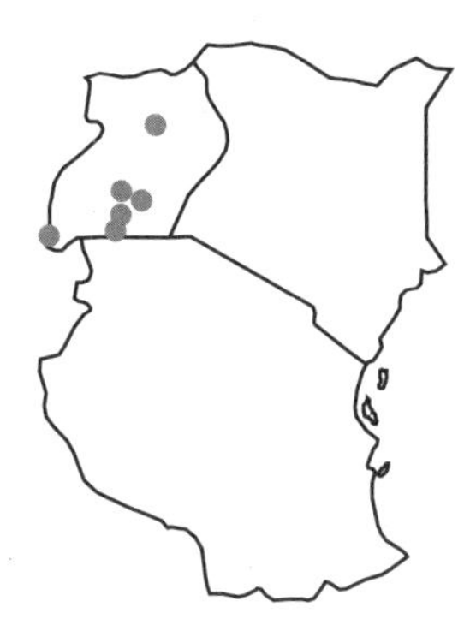

Fig. 59 *Bufo vittatus.*

DESCRIPTION: This toad is small, with the females up to 37 mm in length. The tympanum is distinct, nearly as large as the eye. The first finger is shorter than the second. The toes are one-third webbed, with three phalanges of the fourth toe free. Subarticular tubercles are single. There is no tarsal fold. Conical warts occur on the sides of the body. The parotid glands are narrow, broken up into warts. The back is reddish brown with six interrupted black longitudinal bands on the back. The belly is pale brick-red with large grayish spots.

HABITAT AND DISTRIBUTION (FIG. 59): This species is known from grasslands in low-lying areas west of Lake Victoria in Uganda.

ADVERTISEMENT CALL: Unknown.

BREEDING: Unknown.

TADPOLES: Unknown.

KEY REFERENCE: Boulenger 1906.

Desert Toad

Bufo xeros Tandy et al., 1976
(Plate 4.2)

The specific name means "dry," referring to the arid environment where this species has been found.

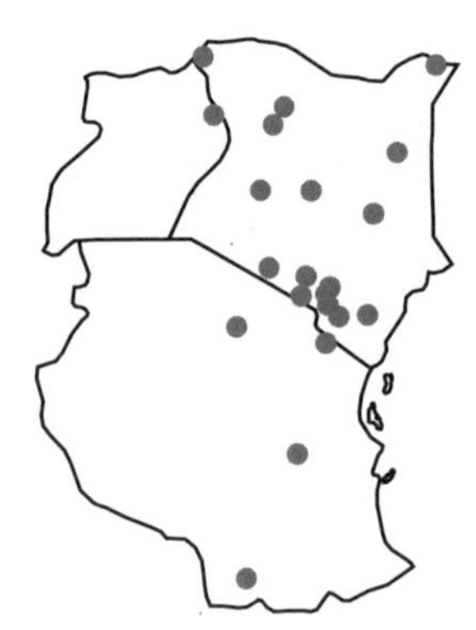

Fig. 60 *Bufo xeros.*

DESCRIPTION: This toad is large, with females up to 97 mm long and males 82 mm long. The parotid glands are narrow, diverging posteriorly, and narrowly separated from the eye. The nostrils are closer to the snout tip than to the eyes. The tympanum is vertically oval, and its diameter is 60% that of the eye. Subarticular tubercles are mostly single. The inner metacarpal tubercle is smaller than the outer one. A prominent elongated gland is present under the forelimb. The first finger is longer than the second. On the fourth toe, three and one-third phalanges are free of webbing on the inside. The tarsal fold is three-fourths the length of the tarsus. The back is pale brown with six pairs of dark brown blotches, each with darker margins. Bright red markings are present behind the thigh. There are many dark spinules on the back and sides. The gular sac in males is darkly pigmented. The nuptial pad extends onto the first three fingers.

HABITAT AND DISTRIBUTION (FIG. 60): This toad is known from oases in the Hoggar Mountains of Algeria, and savannas of sub-Saharan Africa from Western Sahara and Senegal, south to Cameroon and Tanzania.

ADVERTISEMENT CALL: The male calls from the water's edge or while sitting in the water. The call resembles the hoot of a large owl. It is rapidly pulsed (207 pulses/s), with a relatively low emphasized frequency of 0.7 kHz. There are 76–84 pulses/call with a duration of 0.4 s.

BREEDING: Unknown. Eggs are darkly pigmented.

TADPOLES: The tadpole is large for the genus, reaching 34 mm. See the chapter on tadpoles.

NOTES: This species is known to eat spiders and insects, including large numbers of beetles.

KEY REFERENCES: Tandy et al. 1976, Salvador 1996.

Genus *Churamiti*

This genus was erected to accommodate a single species of tree toad from the Ukaguru Mountains of Tanzania. It is remarkable for its bright metallic colors and large discs.

Beautiful Forest Toad

Churamiti maridadi Channing & Stanley, 2002
(Plates 4.3, 4.4)

The specific name *maridadi* is Swahili for "beautiful."

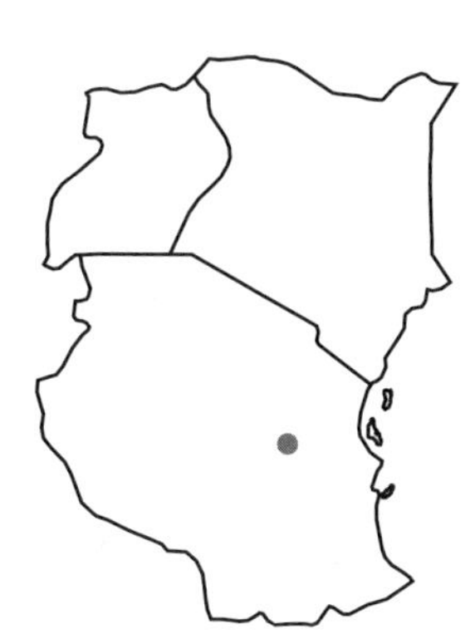

Fig. 61 *Churamiti maridadi.*

DESCRIPTION: The largest of the females known is 53 mm long. The head is wide and flattened, with a blunt snout in profile. The eyes protrude, are wider than the jaw, and have a glandular upper eyelid. The parotids are absent, but a cluster of rounded glands occurs behind the eyes, with others on the head and back. The fingers are long with large truncated terminal discs. The hind limb is long, with the tibia length 45% of the body length. Spatulate discs are present on all toes but are largest on the fourth and fifth toes. The inner metatarsal tubercle is longer than the outer one. The back is smooth with many rounded glandular warts that extend onto the limbs. The warts coalesce into large glands on the forearm, wrist, and the upper surfaces of the fingers, tibia, and foot. The back is a deep metallic yellow, silvery black, or green that extends irregularly onto the forearms and thighs and is present in small patches on the upper distal surfaces of the limbs, and on the wrist and upper foot. The pigment on the limbs is a brighter yellow, or paler green, than on the back. The small rounded warts are red-brown and stand out above the shiny back. The eyelids are reddish, as is the snout. The upper surfaces of the limbs not covered in yellow or green pigment are red to pinkish. The irregular margins of the yellow or green pigment on the limbs resemble lichen. The upper jaw is cream, and the lower eye and sides of the snout are dark brown, overlaid with fine yellow vermiculations. The eye is silvery gold with a horizontal pupil.

HABITAT AND DISTRIBUTION (FIG. 61): This forest species is presently only known from the Ukaguru Forest Reserve in Tanzania.

ADVERTISEMENT CALL: Unknown.

BREEDING: Unknown, although the females have pigmented eggs.

TADPOLES: Unknown.

NOTES: Only four specimens are known.

KEY REFERENCE: Channing & Stanley 2002.

Genus *Mertensophryne*

There is only one small species in this genus, known from lowland wooded savanna in Kenya and Tanzania. This frog is unusual for its internal fertilization. The eggs are laid in small pockets of water, and the tadpoles develop with a crown of tissue around the eyes and nostrils.

Woodland Toad

Mertensophryne micranotis (Loveridge, 1925)
(Plates 4.5, 4.6)

The specific name means "small ear" and refers to the size of the tympanum.

Fig. 62 *Mertensophryne micranotis.*

DESCRIPTION: The males reach 22 mm in length and the females, up to 24 mm. The fingers and toes are reduced with distinct tubercles, and webbing is absent. The tympanum is not visible. The parotids are present only as small raised areas. The skin is smooth. The throat on the male is white with black marbling, and the belly is black. The back is black, sometimes with a hairlike pale vertebral line.

HABITAT AND DISTRIBUTION (FIG. 62): This species is known from lowland forest and open woodland, reaching the boundary between mountain forest and miombo at an altitude of around 500 m. It has been recorded from the coastal lowlands of Kenya and Tanzania, inland as far as the lower slopes of the Udzungwa Mountains. It is also found on the islands of Songo Songo and Unguja (Zanzibar).

ADVERTISEMENT CALL: Males select water-filled containers, such as large snail shells or holes in trees, from which to call. The males produce a soft chirp, also during amplexus, that has been suggested to attract females to the water-filled containers, where breeding takes place. The male also makes a ticking sound that elicits a chirp in response from the female being clasped.

BREEDING: The breeding details are based on careful observations made by Sanda Ashe at Watamu, Kenya. Amplexus lasts from 5 to 8 hours. The male clasps the female under the arms and commences to tap the lower back of the female with his feet. The male places his cloaca close to the female's cloaca and releases a drop of fluid that is taken up by the female. This is repeated at intervals for some hours, with bouts of feeding by the female in between. Belly-to-belly amplexus has been observed. Fertilization is internal, and the female later lays between 17 and 32 eggs in two strings into one of the small water-filled snail shells or other crevices. Eggs are sometimes laid in empty coconut shells or in tree holes more than a meter above ground. Eggs may develop for about 6 hours in the oviducts before they are laid.

TADPOLES: The tadpole has a characteristic "crown," a ring of tissue surrounding the eyes and nostrils that probably serves to increase the area for oxygen uptake while the tadpole floats at the surface in anoxic water. See the chapter on tadpoles.

NOTES: These small dark toads scuttle around on the ground like little beetles. They are active through the day after rain. They are very cryptic and might be mistaken for a cricket or dark grasshopper. When disturbed they may remain motionless, or slip under a leaf on the forest floor. Known food items include beetle larvae, beetles, flies, springtails, bugs, termites, and snails.

KEY REFERENCES: Loveridge 1944, Grandison & Ashe 1983, Poynton 2000a.

Genus *Nectophrynoides*

Forest toads, spray toad.

There are 12 species described in this genus, all restricted to the mountains of Tanzania. The toads have wide heads with protruding eyes and spindly limbs. In some species the tadpoles are retained in the oviducts until the young are born alive. This is known as *ovoviviparity.* These small toads are easily overlooked, or confused with similar species. As we are preparing this book, other species await discovery and naming.

KEY TO THE SPECIES

1a. Ventral interfemoral glands surrounded by pigmented area (Fig. 63) *Nectophrynoides wendyae*
1b. No interfemoral pigmentation 2

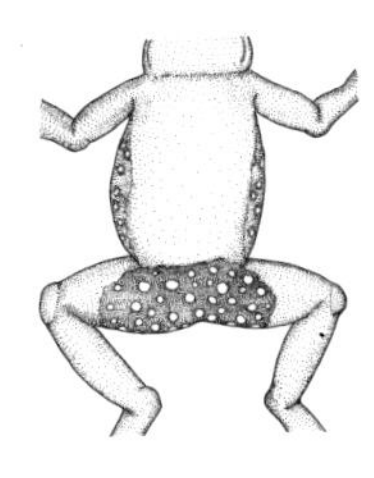

Fig. 63 Interfemoral pigmentation.

2a. Tympanum not visible 3
2b. Tympanum visible 7

3a. Finger tips expanded 4
3b. Finger tips not expanded 5

4a. Parotid gland larger than eye, no webbing on hand *Nectophrynoides laevis*
4b. Parotid gland smaller than eye, hand webbing present *Nectophrynoides pseudotornieri*

5a. Parotid glands present, even if inconspicuous *Nectophrynoides cryptus*
5b. Parotid glands absent 6

6a. Webbing present on hand, foot longer than tibia *Nectophrynoides asperginis*
6b. No webbing on hand, foot shorter than tibia *Nectophrynoides frontieri*

7a. No massive glands on limbs 8
7b. Massive glands on limbs, parotid gland continuous with eyelid (Fig. 64) *Nectophrynoides viviparus*

Fig. 64 Parotid glands extending behind tympanum.

8a. Tips of fingers rounded, sometimes slightly expanded 9
8b. Finger tips expanded and truncated (Fig. 65) *Nectophrynoides tornieri*

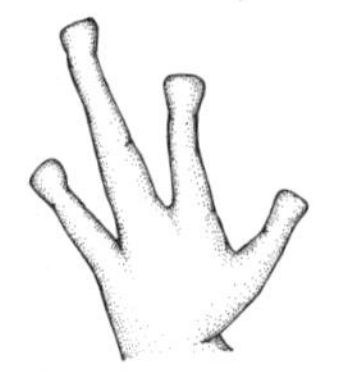

Fig. 65 Truncated finger discs.

9a. Parotid glands continuous from tympanum to scapular region, foot equal or longer to tibia *Nectophrynoides vestergaardi*
9b. Parotid glands discontinuous, foot shorter than tibia 10

10a. Parotid glands much larger than other glands on back *Nectophrynoides poyntoni*
10b. Parotid glands discontinuous, each part the size of other glands on the back *Nectophrynoides minutus*

Kihansi Spray Toad

Nectophrynoides asperginis Poynton et al., 1998
(Plates 4.7, 4.8, 5.1, 5.2)

The species name means "spray" and refers to the unique spray wetlands where this species is found.

DESCRIPTION: This is a small species, with large females reaching 29 mm in length and males up to 19 mm. The skin is smooth, with no tympanum or parotid gland visible. The eyes protrude and are visible from below. The nostrils are closer to the snout tip than to the eyes. The fingers and toes are webbed at the base. In the fourth toe, two and a half to three phalanges are free of webbing. The adults are yellow with dark

Fig. 66 *Nectophrynoides asperginis.*

brown markings, which take the form of a distinct lateral band in males but indistinct bands in females. Both sexes have a brown sacral mark. In the dry season the yellow ground color becomes more brownish, and the skin becomes more granular. Juveniles are purple with lime-green markings.

HABITAT AND DISTRIBUTION (FIG. 66): This species is endemic to the Kihansi Gorge, on the Udzungwa escarpment of Tanzania. It is only known from the small wetlands below the waterfalls that receive permanent spray, an area of about 1 hectare.

ADVERTISEMENT CALL: Two common call types are heard in the wetlands. The first is the advertisement call, which is used to attract females. The second call is a male aggression call, used to space males. Each advertisement call consists of a single note with 1–4 pulses. The emphasized frequency is 4.2–4.4 kHz, and the pulses are produced at a rate of 167/s. Unlike most frogs, it calls during the day as well as at night. The aggression call consists of a long series of notes, effectively a series of "advertisement" calls strung together. Males may respond to the aggression call by producing their own aggression call.

BREEDING: The male deposits sperm in the female's body by pressing his cloaca tightly against hers. During amplexus the male clasps the female under the armpits and places his feet on top of her thighs so that he is not touching the rock surface. The fertilized eggs are retained in her oviduct and go through the tadpole stage there. Tiny toadlets are "born" at the base of vegetation. The average clutch size is 11 and ranges from 10 to 16.

NOTES: This species is classified as Critically Endangered according to the IUCN criteria; see the conservation chapter for details. Predators include crabs *Potomonautes* sp. and driver ants *Dorylus* sp. The spray toads gather in large concentrations on rock faces exposed to spray at night; if a calling male is approached by another, calling increases in frequency until the "intruder" leaves. The territorial male may move toward the intruder in a threatening stance, with legs extended so that the body is raised high off the substrate.

KEY REFERENCE: Poynton et al. 1998.

Uluguru Forest Toad

Nectophrynoides cryptus Perret, 1971

The species name refers to the hidden nature of this species, which remained unrecognized as a distinct species for 43 years.

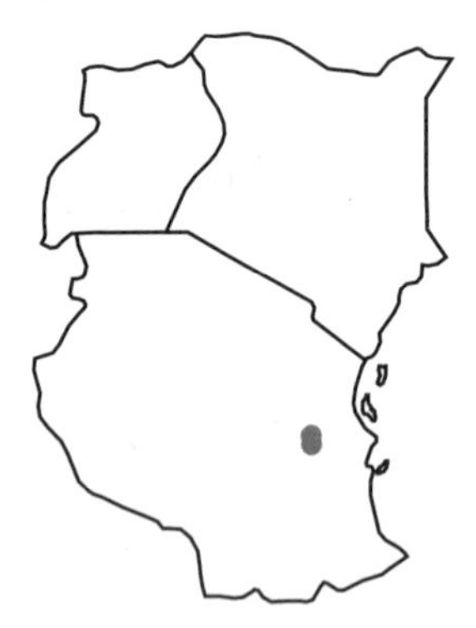

Fig. 67 *Nectophrynoides cryptus.*

DESCRIPTION: This toad is small, with the male up to 26 mm and the female up to 34 mm long. The body is slender with spindly legs. The snout is pointed and is as long as the horizontal diameter of the eye. The nostrils are closer to the snout tip than to the eyes. The tympanum is not visible, but a depression marks the tympanic position. Parotid glands are present, separated from the eyelid. The finger and toe tips are not expanded. The first finger is shorter than the second. Two metacarpal tubercles are present, with the outer being the larger one. Webbing is present between the toes and fingers and is moderate, with three phalanges of the fourth toe free. The dorsal skin is warty, while the ventral surface is coarsely granular. A dark middorsal stripe may be present, as well as white blotches on the back of some specimens.

HABITAT AND DISTRIBUTION (FIG. 67): This small toad is found on the high grasslands of the Uluguru Mountains in eastern Tanzania, at altitudes above 1800 m.

ADVERTISEMENT CALL: Unknown.

BREEDING: This species has been shown to be ovoviviparous. The oviducts of one female contained 25 juveniles, each 7 mm long.

KEY REFERENCES: Perret 1971, 1972.

Frontier Forest Toad

Nectophrynoides frontieri Menegon et al., 2004

The specific name refers to Frontier-Tanzania, whose members collected the type specimens.

DESCRIPTION: The largest male known is 18.3 mm long. No female has yet been collected. Finger and toe tips are pointed to round but not expanded. The tympanum is visible, although the hearing apparatus may not be fully functional. The eye is visible from below, and the sides of the head are vertical. Parotid glands are absent, but a ridge of glands is present on the edge of the eyelid. A small amount of webbing is present at the base of the toes. Two metatarsal tubercles are present. The frog is brown with paler blotches above, and a grey belly with white speckles.

HABITAT AND DISTRIBUTION: Found in moist forest at around 900 m. Known only from the Amani-Sigi Forest, in the Amani Nature Reserve in the East Usambaras of Tanzania.

ADVERTISEMENT CALL: Unknown.

BREEDING: Unknown.

KEY REFERENCE: Menegon et al. 2004.

Smooth Forest Toad

Nectophrynoides laevis Menegon et al., 2004

The species name derives from the skin texture, which is smooth (*laevis* in Latin).

DESCRIPTION: The only specimen known is a male, 24.8 mm long. The foot is longer than the tibia. There is no tympanum. The parotids are twice as long as wide, and longer than the eye. The eyes are just visible from below. The eyelid is edged with a raised glandular ridge. There is a little webbing on the toes, with three phalanges free of main web on the fourth toe. The finger tips are expanded, and the tip of the third toe is slightly truncated. The inner metatarsal tubercle is white and larger than the outer. The skin of the back appears smooth and is covered by very small warts. A thin pale vertebral line is present. Below, a thin dark line runs from the chin to the vent, and on the inner surfaces of the limbs.

HABITAT AND DISTRIBUTION: The specimen was collected from the Uluguru South Forest Reserve, Tanzania, from the remnant forest on the slopes of the mountain at 2000 m.

ADVERTISEMENT CALL: Unknown.

BREEDING: Unknown.

KEY REFERENCE: Menegon et al. 2004.

Dwarf Forest Toad

Nectophrynoides minutus Perret, 1972

The specific name refers to the very small size of this species.

DESCRIPTION: This is a small species, with the male being up to 19 mm long and the female up to 22 mm. The body is slender and the arms and legs are very spindly. The snout is shorter than the horizontal diameter of the eye. The first finger is shorter than the second. Subarticular tubercles are distinct. The finger and toe tips are not expanded, and webbing between the toes is rudimentary, not extending beyond the proximal subarticular tubercles. The plantar surfaces are covered with small tubercles. The tympanum is visible. Parotid glands are absent.

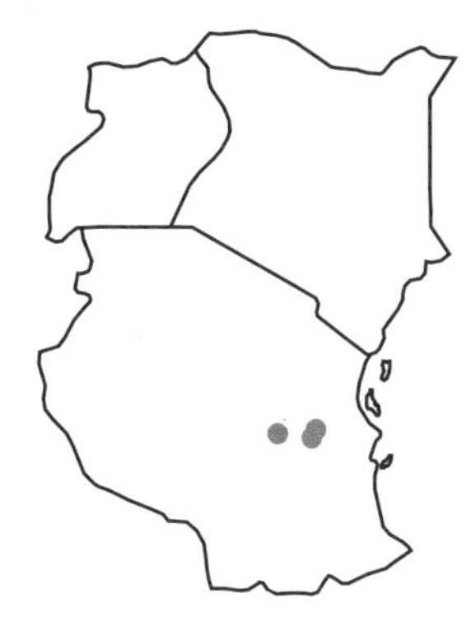

Fig. 68 *Nectophrynoides minutus.*

Webbing between the toes is well developed. A distinct palmar tubercle is present. The back is uniform brown, with darker speckles on the pale belly.

HABITAT AND DISTRIBUTION (FIG. 68): The dwarf forest toad is found in forest and grassland at high altitudes, above 2000 m, on the Uluguru and Rebeho mountains in Tanzania.

ADVERTISEMENT CALL: Unknown.

BREEDING: This ovoviviparous species produces 20–31 young.

KEY REFERENCE: Perret 1972.

Poynton's Forest Toad

Nectophrynoides poyntoni Menegon et al., 2004

This species was named for J. C. Poynton.

DESCRIPTION: This species reaches 24 mm long. The foot is shorter than the tibia. A tympanum is visible. The parotids consist of an anterior row of small glands, and a posterior part over the scapular region that is twice as long as wide. The posterior part of the gland is as long as the width of the eye. Eyes are visible from below. Webbing present at base of fourth and fifth toes. The tips of the fingers and toes are simple and rounded. The background color is brown with a black stripe running from the snout tip to the posterior edge of the parotid glands. The pale vertebral stripe is interrupted by a blotch in the middle of the back. The pattern of markings is very variable.

HABITAT AND DISTRIBUTION: Specimens were collected from forest at 1200 m in the Udzungwa Scarp Forest Reserve, Tanzania.

ADVERTISEMENT CALL: Calling can be heard from late afternoon. Males call from trees and bushes near streams. The call consists of a high-pitched buzz. Each buzz lasts about one second, at a pulse rate of 6/s and is produced at 2.9 kHz, with a second harmonic at 8.7 kHz.

BREEDING: Females have 8–10 large yolky eggs, suggesting that this species is ovoviviparous.

NOTES: Toads can be found at night on leaves up to 1.6 m above the ground.

KEY REFERENCE: Menegon et al. 2004.

Pseudo Forest Toad

Nectophrynoides pseudotornieri Menegon et al., 2004

The species name refers to its resemblance to *N. tornieri*.

DESCRIPTION: Males reach a length of 25 mm, and females 29 mm. The foot is about equal to the length of the tibia. The tympanum is absent. The parotids are weakly developed but visible in the scapular area. The eyes are visible from below. The fingers are webbed at the base, and the toes are slightly webbed, with three phalanges of the fourth toe free of web. The finger tips are expanded and truncated. The inner metatarsal tubercle is twice as large as the outer. The back is brown with various markings, while the underside is cream with speckling on the throat.

HABITAT AND DISTRIBUTION: Frogs were found in forest at 1000–1400 m, in the Uluguru North Forest Reserve, Tanzania.

ADVERTISEMENT CALL: Unknown.

BREEDING: Unknown.

KEY REFERENCE: Menegon et al. 2004.

Tornier's Forest Toad

Nectophrynoides tornieri (Roux, 1906)
(Plate 5.3)

This species was named for the German herpetologist Gustav Tornier (1859–1938).
Kijula in Shambala.

Fig. 69 *Nectophrynoides tornieri.*

DESCRIPTION: These small toads reach a length of 28 mm in males and 34 mm in females. A rudimentary sixth toe is present, visible only as a swelling on the inside of the first toe. Finger tips are expanded and truncated. The sexes are dimorphic; adult males are uniformly brown to red on the upper surfaces, with a faint light line extending from the eye along the parotid glands. The males may be patterned. The sides and undersurface are marbled with gray. The female is rusty red above, with the center of the back yellowish, a single black bar on the tibia, and another on the foot. The belly is transparent so that internal organs and eggs are visible. This species is quite variable in color, with each individual showing slight differences. In a sample of 68 specimens from the Magrotto Mountains, no two had the same color pattern.

HABITAT AND DISTRIBUTION (FIG. 69): This species is found in forest-floor growth, under leaf litter, and in bamboo stalks. They have also been found in soil cracks, in rock crevices, and under rocks in drying streambeds where some moisture remains. It is known to move into agricultural land adjacent to forest, like oil palm plantations. This Tanzanian endemic is known from the East Usambaras and the Udzungwa Mountains to Mt. Rungwe.

ADVERTISEMENT CALL: Males call in February and March at the start of the long rains. They call from the ground or up to 30 cm above the forest floor in bushes. The call is a characteristic "pink-pink," produced during the day. Each note is a brief 45-ms click at 3.5 kHz, produced at a rate of 1/s. The note is rapidly pulsed.

BREEDING: Evidence from females holding eggs or tadpoles indicates that they breed over a long period. Up to 35 young have been recorded in one female.

NOTES: These toads are found on the ground but can climb up into vegetation. Most of the time they are seen at night on herb leaves, about 25 cm above the forest floor, calling or silent. When disturbed, the male will play dead with the limbs retracted, even remaining on its back for a minute or so. Known food records include ants.

KEY REFERENCES: Barbour & Loveridge 1928, Loveridge 1944.

Vestergaard's Forest Toad

Nectophrynoides vestergaardi Menegon et al., 2004

This species was named for Martin Vestergaard who recognized these frogs as new to science.

DESCRIPTION: Females reach 24 mm in length. Length of foot subequal to tibia. Tympanum visible. The parotids are elongated. The eyes are visible from below. There is slight webbing at the base of the fingers, and the toes are webbed at the base, with four or less phalanges free of web. Finger tips are rounded. The head and back are relatively smooth, with the sides and limbs rough. The sides are darker than the back, at least in preserved animals. A thin black vertebral stripe is present. The underside is cream.

HABITAT AND DISTRIBUTION: This species occurs in montane forest, at over 1200 m in the West Usambaras, Tanzania.

ADVERTISEMENT CALL: Unknown.

BREEDING: One female contained 18 embryos, indicating that this species is ovoviviparous.

KEY REFERENCE: Menegon et al. 2004.

Robust Forest Toad

Nectophrynoides viviparus (Tornier, 1905)
(Plates 5.4, 5.5)

The specific name means "live-bearing," a reference to the breeding strategy of this group of toads.

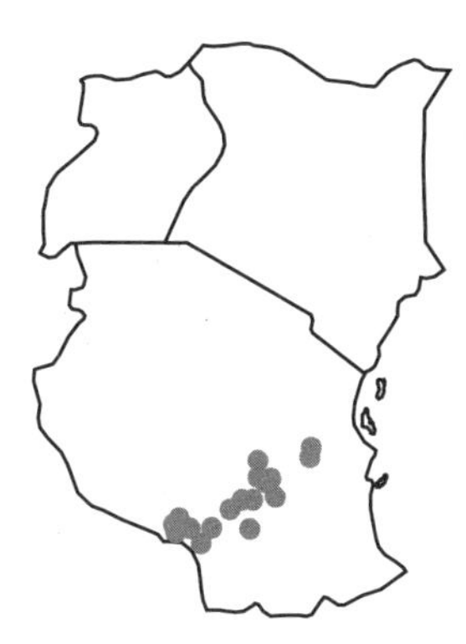

Fig. 70 *Nectophrynoides viviparus.*

DESCRIPTION: The body is stocky, with the male up to 56 mm long and the female 60 mm long. The nostrils are closer to the snout tip than to the eyes. The tympanum is very small, rounded, but sometimes difficult to see. The fingers and toes are slender. The fingers have a slight web only at the base. The toes have a clear web on the first phalanx. Webbing is well developed, with large metatarsal and subarticular tubercles. The skin is smooth, with enormous glands on the limbs and large parotid glands. This species is probably the most variable forest toad in coloration. It can be gray, reddish, pale green, or dark olive according to the background leaves. The undersurface varies from jet black to pure white and is frequently mottled. Often the limb glands contrast in color with the rest of the body.

HABITAT AND DISTRIBUTION (FIG. 70): These forest-floor dwellers can be found under rotting logs, but sometimes climbing a meter or more off the ground, particularly in the breeding season. They survive under shaded cover in agricultural land a few hundred meters from the forest edge. This Tanzanian endemic is known from the Udzungwa, Uluguru, and Ukinga mountains, including Mt. Rungwe and the southern highlands.

ADVERTISEMENT CALL: Males call during the dry winter season when temperatures may get as low as 6 °C. After sporadic rain, individuals can be found in tussock grass above the tree line at 2800 m, in bamboo, *Hagenia abysinica* forest, grasslands at the forest edge, montane forest, and reedbeds in the grasslands. They call from trees and from ground level. Preferred call sites appear to be leaf litter, bamboo stems lying at an angle on the ground, or mats of dead leaves some meters off the ground. The call is a series of slow creaks, at an emphasized frequency of 1.9 kHz. Each creak is 0.6–0.7 long, with 35 pulses at a rate of 52/s.

BREEDING: This ovoviviparous species has been recorded with up to 114 toadlets (6 mm long), or 135 tadpoles carried by the female. A big female will produce up to 70 toadlets in several bouts of birthing lasting for 24 hours.

NOTES: Food records include beetles, whip scorpions, grasshoppers, termites, spiders, snails, millipedes, woodlice, moths, caddis-fly larvae, caterpillars, and ants, the main item of diet. The advertisement call is similar to that of Barbour's tree frog *Leptopelis barbouri*, which also calls from trees.

KEY REFERENCES: Barbour & Loveridge 1928a, Menegon & Salvidio 2000.

Wendy's Forest Toad

Nectophrynoides wendyae Clarke, 1989

The species is named for Wendy Clarke, wife of the describer, Barry Clarke.

DESCRIPTION: This is a small species, with the male up to 18 mm long and the female 22 mm long. The skin is smooth, with scattered white subconical spines on the side of the head and body. The snout projects beyond the upper lip. Parotid glands are very narrow, extending posteriorly as narrow dorsolateral folds. The tympanum and middle ear are absent. The first finger is longer than the second. The toes are almost free of web, and the first toe is very small. The subarticular tubercles are single. There is a characteristic brown pectoral marking and dark chocolate brown interfemoral patch with prominent white tubercles. The soles of the feet are similarly chocolate brown, with the inner and outer

Fig. 71 *Nectophrynoides wendyae.*

metatarsal and subarticular tubercles a prominent white against the brown. There is a white line on the upper lip. The dorsal margin of the iris has a silvery white patch. Breeding males have a crimson throat and a nuptial pad on the inner surface of the thumb.

HABITAT AND DISTRIBUTION (FIG. 71): This endemic species is only known from the Udzungwa scarp above Chita, from forest at altitudes of 1500–1650 m.

ADVERTISEMENT CALL: This species has no hearing apparatus and presumably has no advertisement call.

BREEDING: Unknown.

NOTES: This species is very common where it occurs.

KEY REFERENCE: Clarke 1989.

Red Toad—Genus *Schismaderma*

There is only a single species in this genus. The red toad differs from the other toads in its skin structure and the mid-water schooling behavior of its tadpoles. Recent biochemical evidence suggests that the red toad has been separated from the other toads for 55 million years.

Red Toad

Schismaderma carens (Smith, 1849)
(Plate 6.1)

The specific epithet *carens,* meaning "head," refers to the lack of parotid glands.
Red-backed toad, African split-skin toad, *kazoli* in Lwena and Manganja, *conga* in Sena, *naliwonde* in Yao, *zonde* in Chewa.

Fig. 72 *Schismaderma carens.*

DESCRIPTION: The size of this toad is moderate to large, with the male up to 88 mm long and the female reaching 92 mm. The tarsal fold is present. The back is less warty than many similar-sized toads, and there is a distinct glandular ridge running from above the tympanum to the hind leg. The tympanum is large, round, and with about the same diameter as the eye. The eye-nostril distance is equal to the tympanum diameter. The breeding male has a vocal sac, and the first three fingers have nuptial pads. Parotids are not visible. The back pattern is characteristic of this species: a pair of small dark brown marks on the lower back, with a pair of markings on the shoulders. The ground color is pale brown, often pinkish. The outer part of the dorsolateral ridge has a darker lower edge. The flanks are pale or very dark, contrasting strongly with the reddish back. The underside is speckled with gray.

HABITAT AND DISTRIBUTION (FIG. 72): This toad is found in many different habitats and is even successful around human settlements. It is widespread in eastern, central, and southern Africa, known from southern Kenya, Tanzania, and the Democratic Republic of Congo, southward through Malawi, Zambia, Zimbabwe, Botswana, and Mozambique to South Africa.

ADVERTISEMENT CALL: Calling is restricted to midsummer, unlike most other toads that call early in the rainy season. The male calls while floating in shallow water. The call is a very loud, long whoop. Each call has a duration of 0.9–1.2 s. The dominant harmonics are between 0.1 and 0.8 kHz. Calls may be heard during the day.

BREEDING: Breeding starts after heavy rains. The toads breed by day in deep muddy water. The males gather among emerging vegetation in the deep water, separated by as little as 300 mm from each other. Calling males actively attempt to mate with other frogs, resulting in a lot of splashing as the calling males chase each other. Females enter the area of the calling males and leave after they have laid their eggs. The eggs are laid in a double string while the pair moves slowly along in the water, sometimes back and forth, producing a number of egg string rows. The eggs may be attached to vegetation. Each egg is 1.6–2.5 mm in diameter. Clutch size is 2500, but as large numbers of toads breed in the same place, tens of thousands of eggs may be deposited in the same part of the pond. Development from egg to toadlet may take 37–52 days.

TADPOLES: The tadpole is quite different from other toad tadpoles in this region on two counts: a peculiar flap of skin on the head, and its gregarious behavior. Red toad tadpoles have been found in a mixed swarm with African bullfrog *Pyxicephalus adspersus* tadpoles. See the chapter on tadpoles.

NOTES: The mass of tadpoles may be preyed on by dragonfly nymphs, the hammerkop *Scopus umbretta*, the helmeted terrapin *Pelomedusa subrufa*, and hinged terrapins *Pelusios* sp. Young toads are eaten by the savanna vine snake *Thelotornis capensis*. Adults are known to be eaten by the white-lipped snake *Crotaphopeltis hotamboeia* and Verreaux's eagle owl *Bubo lacteus*. The adult molts at about 4-day intervals.

KEY REFERENCES: Stewart 1967, Balinsky 1969, Maxson 1981.

Forest Toads—Genus *Stephopaedes*

This group of small brown toads is restricted to forests and remnant forests. They have a characteristic tadpole with a ring of tissue surrounding the eyes and nostrils. Four species are known, of which three are found in East Africa.

KEY TO THE SPECIES

1a. Parotid glands confluent with the row of glands behind the angle of the jaw (Fig. 73) *Stephopaedes loveridgei*
1b. Parotid glands separated from the glands behind the angle of the jaw 2

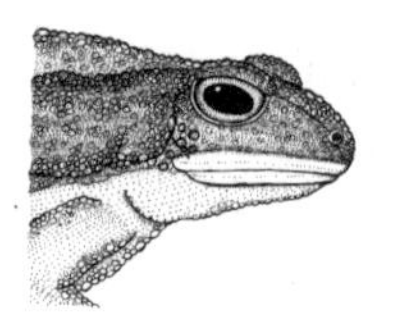

Fig. 73 Parotid glands joining glands behind jaw angle.

2a. Small spines grouped in clusters or rosettes behind the angle of the jaw (Fig. 74) *Stephopaedes howelli*
2b. No rosettes of spines behind the angle of the jaw *Stephopaedes usambarae*

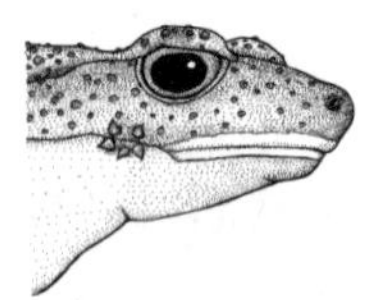

Fig. 74 Spines behind angle of jaw.

Mrora Forest Toad

Stephopaedes howelli Poynton & Clarke, 1999
(Plates 6.2, 6.3, 6.4)

This species is named for Kim Howell.
Mlola earless toad.

DESCRIPTION: This toad is small, with the female up to 45 mm long (41 mm to the end of the urostyle), while the males are about 35 mm long. The parotid glands are wider than the upper eyelid and flattened to form a platform across the shoulders. The lower border of the parotid gland is widely separated from the rounded glands behind the angle of

Fig. 75 *Stephopaedes howelli.*

the jaw, and never reaches the level of the lower margin of the eye. These glands carry small spines that are arranged in rosettes. The subarticular tubercles are double. The inner metatarsal tubercle is longer than the outer one. The upper surfaces are covered with small pale sharp spines, especially dense around the eyes. The length of the tibia is less than half of the body length.

HABITAT AND DISTRIBUTION (FIG. 75): This species is known from open coastal forest in a dolomitic region where the ground has many open cracks. It has been found on Zanzibar and Mafia islands.

ADVERTISEMENT CALL: This species has no ear and probably does not call.

BREEDING: Unknown. Females have up to 60 eggs. The eggs have a pigmented pole, typical of eggs laid in water.

TADPOLES: Unknown.

NOTES: This species is threatened by the destruction of forest for subsistence agriculture.

KEY REFERENCE: Poynton & Clarke 1999.

Loveridge's Forest Toad

Stephopaedes loveridgei Poynton, 1991
(Plate 6.5)

This species is named for Arthur Loveridge, who made important collections and published works on East African amphibians and other animals from the mid-1920s until the late 1950s.
Loveridge's earless toad.

Fig. 76 *Stephopaedes loveridgei.*

DESCRIPTION: Females are known to be up to 38 mm long, with the males up to 35 mm. Tibia length is less than 40% of body length, measured to the tip of the urostyle. The parotid glands extend ventrally to the level of the lower jaw, confluent with the row of glands behind the angle of the jaw. The skin is covered with dark-tipped spines, with over 45 spines on the eyelid. Females have rosettes of smaller spines together with large spines on the sides and lower back where the male makes contact in amplexus. Adult males do not have these spines. The outer metatarsal tubercle is three-fourths the length of the inner one. Two phalanges of the third toe are free of webbing. A margin of web extends onto, or nearly onto, the last phalanx of the fourth and fifth toes. The back is brown with an inverted darker V in the shoulder region. A pair of dark spots is sometimes present on the sacrum, as is a broken line between the eyes. A pale line runs forward from the vent, at least as far as the sacrum. The underside is white, sometimes with a single dark blotch in the pectoral area. Juveniles may have flecked bellies.

HABITAT AND DISTRIBUTION (FIG. 76): This species is known from coastal woodland and forest, from the Kilombero Valley south of the Rufiji River in Tanzania.

ADVERTISEMENT CALL: Unknown. This species has no ear and probably does not call.

BREEDING: Unknown.

TADPOLES: Unknown.

KEY REFERENCES: Poynton 1991, Poynton & Clarke 1999.

Usambara Forest Toad

Stephopaedes usambarae Poynton & Clarke, 1999

This species is named for the East Usambara Mountains in Tanzania, where it was collected.

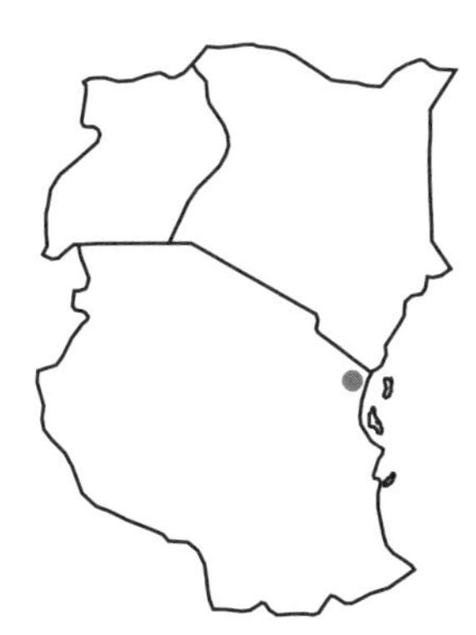

Fig. 77 *Stephopaedes usambarae.*

DESCRIPTION: This is a small toad, with the females up to 45 mm long (41 mm to the end of the urostyle), while the males are about 35 mm long. The parotid glands are wider than the upper eyelid and flattened to form a platform across the shoulders. The lower border of the parotid gland is widely separated from the rounded glands behind the angle of the jaw and never reaches the level of the lower margin of the eye. These glands carry small spines that are not arranged in rosettes. The subarticular tubercles are doubled. The inner metatarsal tubercle is longer than the outer one. The upper surfaces are covered with small pale sharp spines, especially dense around the eyes. Tibia length is less than 40% of body length. The back is brown with a pale upper lip. Markings vary, with some having a dark bar between the eyes, a pair of markings between the parotid glands, a pair on the shoulders, and a pair at the tip of the urostyle. A light vertebral line is present in some specimens. The underside is lightly freckled.

HABITAT AND DISTRIBUTION (FIG. 77): This species is known from natural forest and plantations in the East Usambara foothills, below the altitude of 410 m.

ADVERTISEMENT CALL: Unknown. This species has no ear and probably does not call.

BREEDING: Unknown. Males develop small brown rough areas on the upper and inner surfaces of the first and second fingers. The females have up to 60 eggs, each one being 2.4 mm in diameter with a dark pole.

TADPOLES: Unknown.

KEY REFERENCE: Poynton & Clarke 1999.

Snout-Burrowers—Family Hemisotidae

This family was erected for a single genus of snout-burrowing frogs that is widespread in Africa. These are also commonly known as shovel-nosed frogs.

Snout-Burrowers—Genus *Hemisus*

Snout-burrowers have globular bodies with short muscular limbs. The snout is sharp and hardened for digging. A groove runs transversely behind the eyes. Burrowing forward, as these species do, is more efficient and faster than burrowing backward, such as practiced by most other African burrowing frogs, like the sand frogs and bullfrogs. Eight species are found throughout Africa south of the Sahara, with three species present in East Africa. This genus is known as *ndizaala* in Luganda.

KEY TO THE SPECIES

1a. Small size, males not over 30 mm, females less than 60 mm long *Hemisus marmoratus*
1b. Moderate size, males 30–40 mm, females 40–60 mm long 2

2a. Fifth toe 49% to 64% of the metatarsal tubercle length, light dots on back and a vertebral line extended to each thigh
Hemisus brachydactylus

2b. Fifth toe 65% or more of metatarsal tubercle length
Hemisus guineensis

Short-Fingered Snout-Burrower

Hemisus brachydactylus Laurent, 1963

The species name refers to the short fingers.

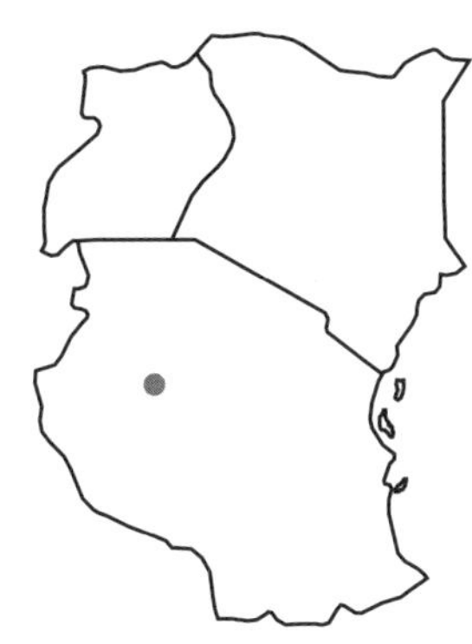

Fig. 78 *Hemisus brachydactylus.*

DESCRIPTION: This is a medium-sized species, with females known to be up to 49 mm long. The free part of the fifth toe is very short, and the inner metatarsal tubercle is pronounced. The free part of the fifth toe is 49–64% of the inner metatarsal tubercle length. The head is broader than long. The snout is slightly longer than the eye width. Eye to nostril distance is more than twice the nostril to snout tip distance. The third finger, measured from the base of the second, is shorter than the snout. Tibia length is 85% to 95% of foot length and about one-third of body length. The brown back has many light spots and a vertebral line. These small spots are more numerous on the sides of the body and the limbs. The pale vertebral line extends onto the thighs.

HABITAT AND DISTRIBUTION (FIG. 78): This species is known from two localities in open woodland in central Tanzania.

ADVERTISEMENT CALL: Unknown.

BREEDING: Unknown.

TADPOLES: Unknown.

KEY REFERENCE: Laurent 1972a.

Guinea Snout-Burrower

Hemisus guineensis Cope, 1865

The specific name refers to Guinea, where the species was first collected.

Fig. 79 *Hemisus guineensis.*

DESCRIPTION: This moderately sized species has a large metatarsal tubercle. The male reaches 39 mm in length and females, up to 53 mm. The fingers and toes are short. The first finger is longer than the second. The inner metatarsal tubercle is up to 5 mm long in large females, more than 85% of the length of the free part of the fifth toe. The tibia is shorter than the foot. The skin is smooth, and the back has a few small pale spots on a uniform dark background. Some individuals have a pale thin vertebral line than extends onto the thighs. Males have black throats. Populations in southern Tanzania, especially breeding males, tend to have a distinctly yellow background.

HABITAT AND DISTRIBUTION (FIG. 79): This species is a savanna and grassland inhabitant, known from Senegal to Angola, northeastern and southeastern Democratic Republic of Congo around the rain forest, Kenya, Uganda, Tanzania, and Mozambique.

ADVERTISEMENT CALL: Males call from the surface near the edges of temporary pools. The call is a long trill. The call has a duration of 1.8 s, consisting of 43 double notes with a dominant frequency of 3 kHz.

BREEDING: The male clasps the female in front of her legs. She digs down and prepares a small hollow in the sandy soil just below leaf litter. The eggs are fertilized while being deposited. The male leaves and the female remains with the eggs to protect them from predators such as ants.

TADPOLES: Unknown.

NOTES: These snout-burrowers have been collected from burrows up to 10 cm below the surface. There is a high degree of variation in body proportions across the range of this species, suggesting that some populations are genetically isolated and may represent distinct species.

KEY REFERENCES: Inger 1968, Laurent 1972a, Amiet 1991.

Marbled Snout-Burrower

Hemisus marmoratus (Peters, 1854)
(Plates 6.6, 6.7, 6.8)

The specific name refers to the mottled color of the back.
Mottled shovel-snouted frog, pig-nosed frog, mottled burrowing frog, marbled shovelnose frog.

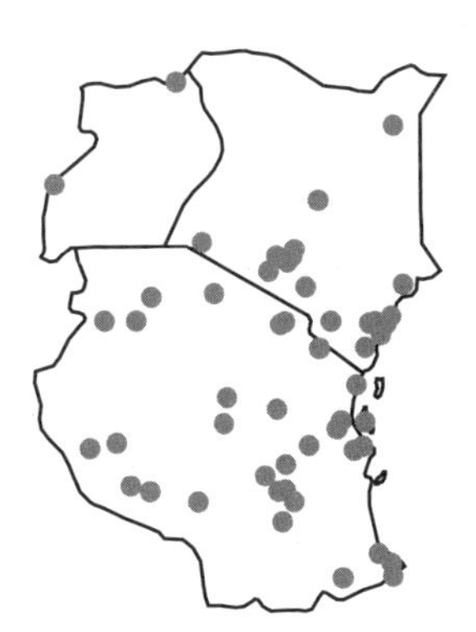

Fig. 80 *Hemisus marmoratus.*

DESCRIPTION: This small to medium frog (male 30 mm long, female less than 60 mm long) is easily recognized by the sharp snout and transverse fold between and above the small eyes. The inner metatarsal tubercle is shorter than the eye-nostril distance. The front limbs are muscular, and the snout is hardened, enabling this species to dig nose-first into soft soil. The toes are slightly webbed. Coloration is variable, with dark gray or brown marbling or spots on a paler brown background. A light vertebral line is often present. Males have black throats. Some individuals are uniformly colored.

HABITAT AND DISTRIBUTION (FIG. 80): This species is common in grasslands and open woodland. It is known from Senegal to Eritrea, western Ethiopia, and Somalia, and south into Uganda, Kenya, Tanzania, and the northern and northeastern parts of South Africa.

ADVERTISEMENT CALL: The male calls from a concealed site under vegetation at the edge of pools, usually on wet mud. The call is a long buzz, repeated frequently. Each call may last several seconds, at a pulse rate of 70–90/s, with most energy at 4 kHz. The pulses are double.

BREEDING: Breeding takes place during the short rains. The males clasp the females and then are dragged into a burrow, with the female digging. The females lay the eggs in a burrow or under a log or stone and remain with them. About 150–200 eggs are laid in a compact mass 25 mm × 13 mm, each egg being 2.0–2.5 mm in diameter within a 3–4-mm capsule. Clutches with as few as 30–35 eggs have been found. Many empty capsules protect the top of the clutch. Burrows are usually found in wet soil under shade of vegetation or leaf litter, or in the roots of wild bananas, a little back from the water. Continuing rains cause the water level to rise to the level of the tadpoles, which liberates them. Tadpoles appear to be able to develop either when left in a moist mass or when in water. When the water does not rise high enough to allow the tadpoles to swim out into the pond, the female may carry tadpoles to the water on her back, or they may follow her by swimming on the wet mud.

TADPOLES: See the chapter on tadpoles.

NOTES: Known food items include termites and ants. Recorded predators include fiscal shrikes and the sharp-beaked snake *Rhamphiophis acutus*, the dimorphic egret *Egretta dimorpha*, and the house snake *Lamprophis lineatus*.

KEY REFERENCES: Loveridge 1936, 1942c; Rödel et al. 1995; Kaminsky et al. 1999.

Tree Frogs—Family Hyperoliidae

This family consists of a large number of mostly climbing frogs, many of which are brightly colored. The group is centered on the tropics, although some species reach the temperate region in southern Africa. In East Africa there are five genera: the reed frogs *Hyperolius*, the spiny reed frogs *Afrixalus*, the tree frogs *Leptopelis*, the running frogs *Kassina*, and the wot-wots *Phlyctimantis*. All tree frogs are known as *akawulula* in Luganda.

KEY TO THE GENERA

1a. Pupil horizontal to round — *Hyperolius*
1b. Pupil vertical (Fig. 81) — 2

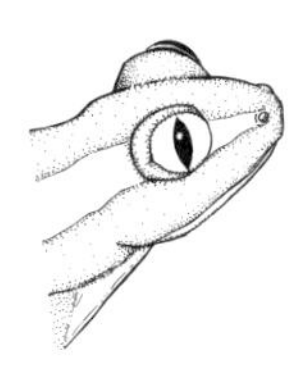

Fig. 81 Vertical pupil.

2a. Vomerine teeth present — 3
2b. Vomerine teeth absent — *Afrixalus*

3a. Opening of vent directed ventrally — 4
3b. Opening of vent not directed ventrally — *Leptopelis*

4a. Male gular gland straplike, attached anteriorly and posteriorly (Fig. 82) — *Kassina*
4b. Male gular gland free posteriorly and laterally — *Phlyctimantis*

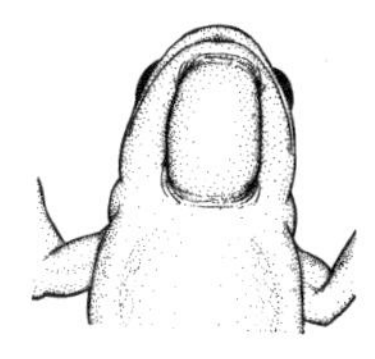

Fig. 82 Straplike gular gland.

Spiny Reed Frogs—Genus *Afrixalus*

These small climbing frogs are characterized by minute sharp hardened warts or spines on the skin. These can be observed as a roughness or sometimes felt by touching the back lightly. This feature gives the group the name of spiny reed frogs. Many species lay eggs on leaves above water, and the eggs are often protected by covering them with the leaf and gluing the edges of the leaf together. This may serve as a defense against egg predators. The spiny reed frogs are found in sub-Saharan Africa in forest, woodland, and savanna. In East Africa there are 12 species.

KEY TO THE SPECIES

1a. Tibia with a transverse or oblique band (Fig. 83) — 2
1b. No transverse band on tibia — 5

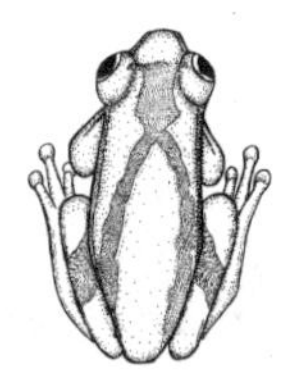

Fig. 83 Transverse band on tibia.

2a. A dark transverse lumbar band present (Fig. 84) — *Afrixalus sylvaticus*
2b. No transverse lumbar marking — 3

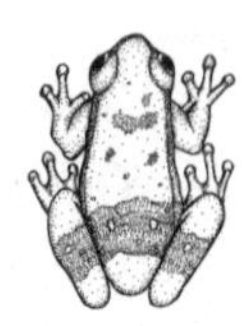

Fig. 84 Transverse lumbar band.

3a. A dark patch between eyes that diverges posteriorly into two short stripes (Fig. 85) — *Afrixalus brachycnemis*
3b. Various longitudinal stripes present, no patch between eyes — 4

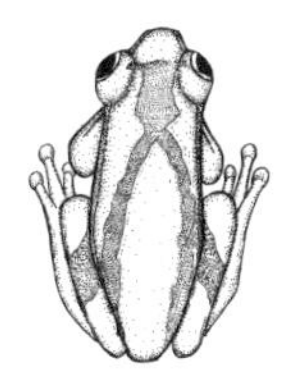

Fig. 85 Patch behind eyes that diverges posteriorly.

4a. A dark middorsal stripe present (Fig. 86) *Afrixalus septentrionalis*
4b. No middorsal dark stripe *Afrixalus morerei*

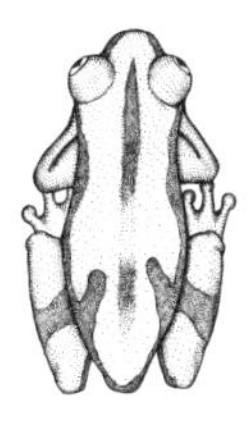

Fig. 86 Middorsal stripe present.

5a. A thin pale marking from the lower back that widens around a dark rectangular patch anteriorly (Fig. 87) *Afrixalus osorioi*
5b. Not as above 6

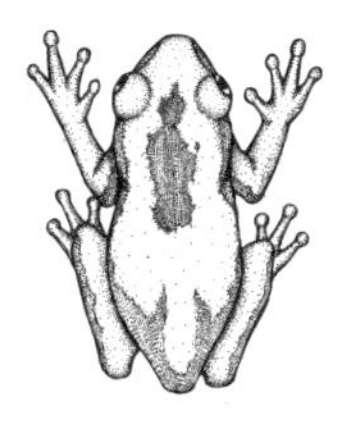

Fig. 87 Dark rectangular patch.

6a. Three light longitudinal stripes 7
6b. Not as above 8

7a. Three light stripes that converge on the head *Afrixalus fulvovittatus?*
7b. Middle pale stripe not converging on lateral stripes (Fig. 88) *Afrixalus wittei*

Fig. 88 Middle pale stripe that does not converge with lateral stripes.

8a. Two broad dorsolateral pale stripes that meet on the head *Afrixalus fornasini*
8b. Not as above 9

9a. Back without lines, except the dark dorsolateral lines running back from eye 10
9b. Two dark dorsal stripes 11

10a. Back yellow with brown markings *Afrixalus laevis*
10b. Back white with irregular spots *Afrixalus uluguruensis*

11a. Thin dark dorsolateral lines that converge on the head (Fig. 89), body up to 27 mm long *Afrixalus orophilus*
11b. Two dark dorsal stripes that may merge into one, body up to 40 mm long *Afrixalus stuhlmanni*

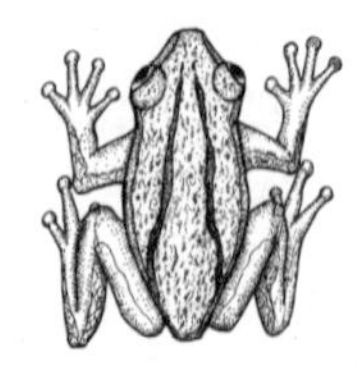

Fig. 89 Thin dark stripes that converge on the head.

Short-Legged Spiny Reed Frog

Afrixalus brachycnemis (Boulenger, 1896)
(Plate 7.1)

The specific name refers to the short lower leg.
Golden afrixalus, short-legged banana frog, short-limbed banana frog, golden leaf-folding frog, *kachula kachena* in Nyungwe.

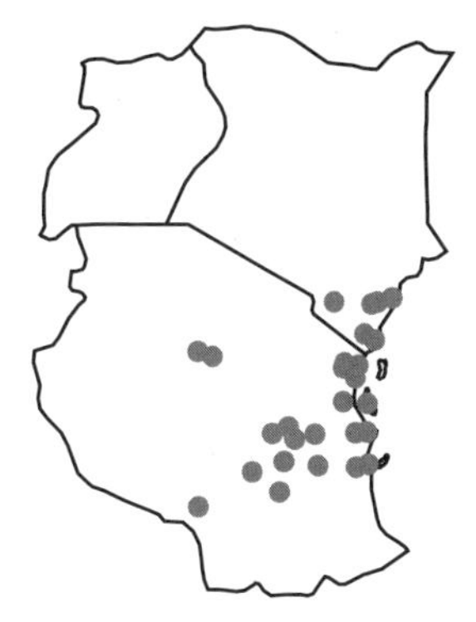

Fig. 90 *Afrixalus brachycnemis.*

DESCRIPTION: Males reach 25 mm in length, while the female may be slightly longer, at 27 mm. The back of the male is yellow to translucent silver with small black spots. There is a dark lateral line from the nostril to the groin, with small white spots. A dark blotch between the eyes forms a patch that becomes two backward-diverging stripes. A dark oblique band over the tibia extends onto the lower back when the limbs are tucked in. The throat and digits are pale yellow. The females are white below. Males have small spines all over, while the females have spines only on the head. The throat is yellow in males and white in females. The markings are variable.

HABITAT AND DISTRIBUTION (FIG. 90): This species occurs in grassland and coastal forest, where it is found in shallow swamps and adjacent rice fields. The species is known from Malawi to coastal Kenya.

ADVERTISEMENT CALL: Calling males gather on flooded grass. The call is a long buzz, with a dominant frequency at 4.4–4.6 kHz. The pulses are produced at about 27/s.

BREEDING: The eggs are laid on blades of grass. Unlike most other species the sides of the leaf are not glued together.

TADPOLES: Unknown.

NOTES: Predators include the spotted bush snake *Philothamnus semivariegatus.*

KEY REFERENCES: Schiøtz 1999, Channing 2001.

Fornasini's Spiny Reed Frog

Afrixalus fornasini (Bianconi, 1849)
(Plates 7.2, 7.3)

This frog was named for Carlo Fornasini.
Fornasini's leaf-folding frog, greater leaf-folding frog, Mozambique banana-frog, silver-banded banana frog, *pasa* in Nyakusa, *chitowa* in Yao, *kitowa* in Makonde.

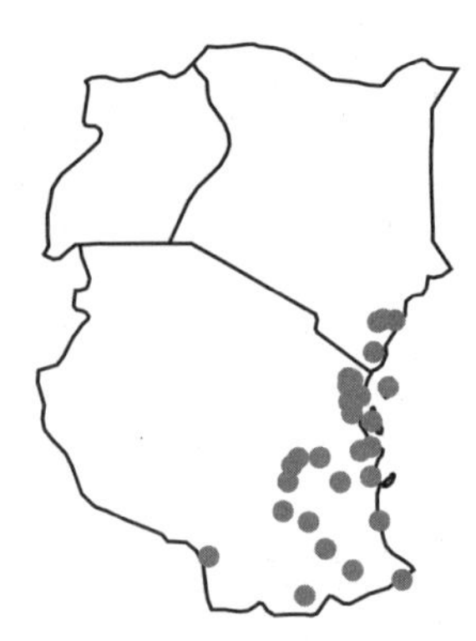

Fig. 91 *Afrixalus fornasini.*

DESCRIPTION: The male may reach 38 mm long and the female, 40 mm. The back is covered with small black spines, each on a pale wart. The typical color pattern of this frog is distinctive: a darker brown vertebral band, with wide white or light brown bands on each side. The middle band begins between the eyes and continues over the vent. In some animals the back may be uniform brown or, in the northern areas into Tanzania, uniform white. The upper halves of the femur and tibia are white, and the lower halves dark. The contrast between the bands may be very slight at some times, as in other species, so caution is advised when using pattern to identify them. The fingers are webbed at the base, and the toes are webbed to the discs, except the fourth toe, which has one phalanx free of web.

HABITAT AND DISTRIBUTION (FIG. 91): This species is associated with ponds where reeds and sedges are growing. It has been found in Kenya southward through eastern and southern Tanzania, including the low slopes of the Usambara, Uluguru, and Udzungwa mountains, south to Malawi, Mozambique, and northeastern South Africa.

ADVERTISEMENT CALL: The male calls from vegetation hanging low over water. The call is a very loud series of "clacks," about 7–12/s with a dominant frequency at 2.3–2.6 kHz, preceded by a short, quiet buzz. One male stimulates the next to call, resulting in a considerable overlap in the calls. There is some variation in the length of the two components of the call.

BREEDING: Breeding occurs during early to mid summer. The eggs are laid in vegetation up to 1 m above water. About 80 eggs are laid on a reed leaf, starting at the tip, the edges of which are then folded toward one another and glued together, presumably as a defense against predators. Sometimes the female deposits the eggs on more than one leaf. The eggs are white, 1.6 mm in diameter, and enclosed in a 3.5-mm jelly capsule.

TADPOLES: See the chapter on tadpoles.

NOTES: The frog sits out on vegetation, and when disturbed, it drops into the water below. Apart from insects, the adults prey on the eggs and developing larvae of the gray tree frog *Chiromantis xerampelina*, the tinker reed frog *Hyperolius tuberilinguis*, the spiny-throated reed frog *Hyperolius spinigularis*, and the eggs of other *Afrixalus fornasini* individuals. They get to the foam-nest frog eggs by sticking their heads deep into the nest. Fornasini's spiny reed frog arrives late at the breeding sites in order to have eggs of other species available. The skin contains small amounts of substances called tachykinins, which are responsible for upsetting the heartbeat and serve to detract mammal predators. This species is eaten by the spotted bush snake *Philothamnus semivariegatus* and the marbled tree snake *Dipsadoboa aulica*.

KEY REFERENCES: Roseghini et al. 1988, Schneichel & Schneider 1988, Drewes & Altig 1996, Channing 2001.

Four-Lined Spiny Reed Frog

Afrixalus fulvovittatus? (Cope, 1861)
(Plate 7.4)

The species name means "tawny ribbon," referring to the color of the dorsal stripes.
Four-lined leaf-folding frog.

Fig. 92 *Afrixalus fulvovittatus?*

DESCRIPTION: Males and females reach 28 mm in length. The three light bands over the length of the back are never confluent. The light bands are reddish brown, and the dark stripes reddish or purple.

HABITAT AND DISTRIBUTION (FIG. 92): This frog is found in savanna and grassland regions of tropical Africa from western and southeastern areas of the Democratic Republic of Congo, to Uganda, western Kenya, and Tanzania. The distribution is not well known due to confusion with similar species.

ADVERTISEMENT CALL: The call consists of a short buzz followed by a series of notes. The emphasized frequency is 3.3 kHz, with a duration of 1.2 s. The initial buzz is 0.2 long, each note has 3 pulses, and the note repetition rate is 11/s.

BREEDING: The eggs are attached in a small clump between leaves above or under water, with 10–80 to a clutch.

TADPOLES: Unknown, although the form from West Africa has been described.

NOTES: The taxonomy of *Afrixalus fulvovittatus* is confused, and this taxon includes a number of species. The species in East Africa has a different advertisement call from that of *A. fulvovittatus* in West Africa. This is an interesting puzzle waiting for someone to solve. Known food items include beetles, bugs, grasshoppers, and spiders.

KEY REFERENCE: Schiøtz 1999.

Smooth Spiny Reed Frog

Afrixalus laevis (Ahl, 1930)
(Plate 7.5)

The species name means "smooth," referring to the lack of prominent dorsal spines.
Forest leaf-gluing frog.

Fig. 93 *Afrixalus laevis.*

DESCRIPTION: This species is small, with males and females up to 28 mm long. The gular flap is small, about one-third of the throat width. Adults have an orange-yellow back and darker markings. Often the dense yellow pigment is concentrated anteriorly and translucent behind. A darker stripe from the nostrils passes backward through the eyes and continues posteriorly. The pattern is very variable.

HABITAT AND DISTRIBUTION (FIG. 93): This species is found in rain forest, from Cameroon to the eastern Democratic Republic of the Congo and southwestern Uganda.

ADVERTISEMENT CALL: Males call from within vegetation. The call consists of a single brief note, with an emphasized frequency of 2.7 kHz.

BREEDING: The eggs are laid on vegetation up to 2 m above water. The eggs overhang quiet tributaries where the water may be only 100 mm deep. The leaves on which the eggs are deposited are not glued together.

TADPOLES: See the chapter on tadpoles.

NOTES: This species has been found in secondary vegetation.

KEY REFERENCE: Schiøtz 1999.

Morere's Spiny Reed Frog

Afrixalus morerei Dubois, 1985
(Plate 7.6)

The species is named for Jean-Jacques Morere, who collected in West Africa.
Dabaga leaf-folding frog.

Fig. 94 *Afrixalus morerei.*

DESCRIPTION: The males reach 21 mm in length and the females, 23 mm. A parallel pair of darker dorsolateral lines runs from the eyes to the hind legs. These merge with a broader pair of lateral bands at the back of the body. The dorsolateral bands may be broken anteriorly, with a isolated pair of markings between the eyes.

HABITAT AND DISTRIBUTION (FIG. 94): This species has been found in open grasslands on the southern Udzungwa highlands in Tanzania.

ADVERTISEMENT CALL: The males produce a long buzz, with an emphasized frequency of 3.5 kHz, at a pulse rate of 26/s.

BREEDING: Unknown.

TADPOLES: Unknown.

KEY REFERENCE: Schiøtz 1999.

Montane Spiny Reed Frog

Afrixalus orophilus (Laurent, 1947)

The species name *orophilus* means "mountain loving," referring to the habitat where this species is found.
Two-lined leaf-gluing frog.

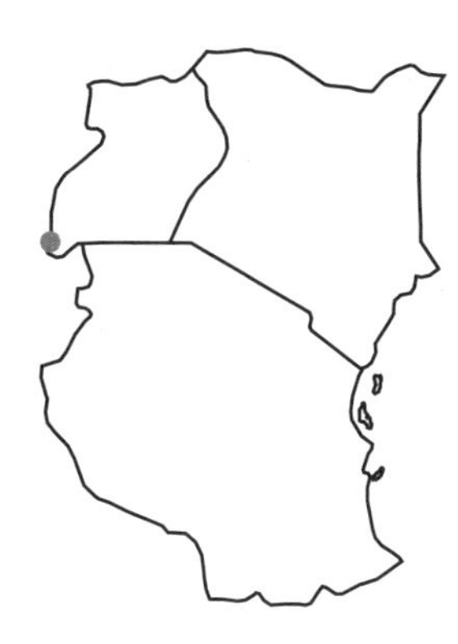

Fig. 95 *Afrixalus orophilus.*

DESCRIPTION: Both males and females are small, reaching 27 mm in length. The back is pale brown with dark speckles, and a pair of characteristic fine dark-brown dorsolateral lines diverge posteriorly from the top of the head, reaching a maximum separation over the sacral region.

HABITAT AND DISTRIBUTION (FIG. 95): This montane species is known from high savannas, and in the bamboo of the high mountains of Rwanda, Burundi, southwestern Uganda, and the extreme eastern region of the Democratic Republic of Congo.

ADVERTISEMENT CALL: Unknown.

BREEDING: Unknown.

TADPOLES: Unknown.

KEY REFERENCE: Schiøtz 1999.

Congo Spiny Reed Frog

Afrixalus osorioi (Ferreira, 1906)
(Plate 7.7)

The species was named for Balthazar Osorio.
Congo Basin leaf-gluing frog.

Fig. 96 *Afrixalus osorioi.*

DESCRIPTION: The males reach 30 mm in length while the females grow to 35 mm. The body proportions are similar to those of other small spiny reed frogs. The males have small spines on the legs and back. The dorsal pattern consists of dark chocolate markings on a silvery background. The dark pattern is an elongated blotch from between the eyes backward to the middle of the back. Dark blotches run forward from the groin parallel to the midline, separated by a thin pale rearward extension of the background color.

HABITAT AND DISTRIBUTION (FIG. 96): This species has been found in forest proper and grassland islands within the rain forest. It is known from the eastern Congo basin through Uganda, to western Kenya.

ADVERTISEMENT CALL: The call is a number of clicks, about 20/s with an emphasized frequency of 2.6 kHz.

BREEDING: Unknown.

TADPOLES: Unknown.

KEY REFERENCE: Schiøtz 1999.

Northern Spiny Reed Frog

Afrixalus septentrionalis Schiøtz, 1974

The species name derives from Latin and means "northern."

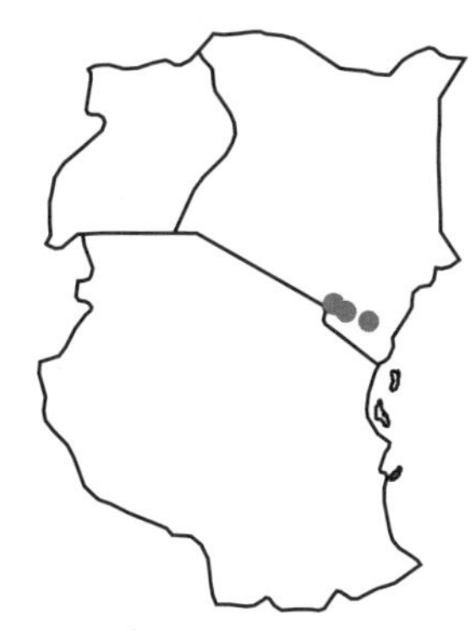

Fig. 97 *Afrixalus septentrionalis.*

DESCRIPTION: This frog is small, with males up to 21 mm long and females slightly longer. The back is smooth, with a dark middorsal stripe. Oblique dark bands on the thigh join oblique markings on the posterior of the back when the legs are folded in.

HABITAT AND DISTRIBUTION (FIG. 97): This frog has been found in the dry savanna of Kenya and is expected in northern Tanzania.

ADVERTISEMENT CALL: The male produces a deep buzz, with a pulse rate of 30/s. The emphasized frequency is 2.8 kHz.

BREEDING: Unknown.

TADPOLES: Unknown.

KEY REFERENCE: Schiøtz 1999.

Stuhlmann's Spiny Reed Frog

Afrixalus stuhlmanni (Pfeffer, 1893)
(Plate 7.8)

This species was named for F. Stuhlmann, who collected in East Africa in 1888–1889.
Stuhlmann's leaf-folding frog.

Fig. 98 *Afrixalus stuhlmanni.*

DESCRIPTION: This species is superficially similar in build to Fornasini's spiny reed frog. It is small, with males up to 25 mm long, and females up to 40 mm. The tongue is heart-shaped, with a deep split. The tympanum is not visible, and the toes are half webbed. The skin is remarkably smooth, as the dorsal warts are very small, only visible under a strong lens. Color is variable, with a light brown background covered with darker fine speckles. A dark band runs from the tip of the snout, through the eyes, and along the side of the body. In addition, there is a pair of brown dorsal stripes. In some animals these merge into a single band, while in others this becomes a series of spots.

HABITAT AND DISTRIBUTION (FIG. 98): This species is known from grassland swamps on Zanzibar and is expected to occur in coastal Tanzania.

ADVERTISEMENT CALL: Unknown.

BREEDING: Unknown.

TADPOLES: Unknown.

NOTES: This species has gone unreported for a number of years, but is plentiful during the breeding season.

KEY REFERENCE: Pfeffer 1893.

Forest Spiny Reed Frog

Afrixalus sylvaticus Schiøtz, 1974
(Plate 8.1)

The species name means "of the forest."

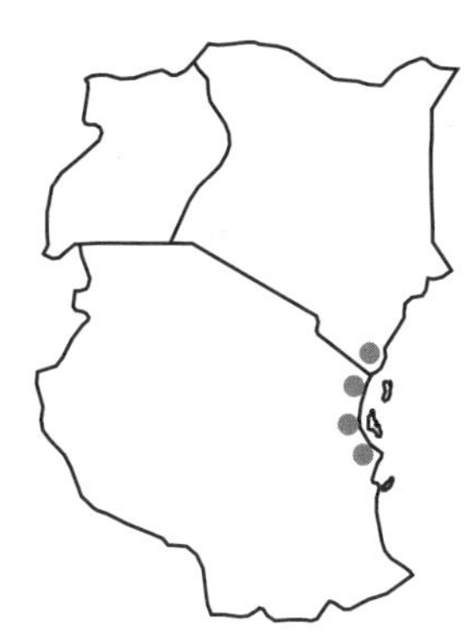

Fig. 99 *Afrixalus sylvaticus.*

DESCRIPTION: This is a small species, with females up to 24 mm long and the smaller males up to 21 mm long. The background is a pale brown with a dark transverse band at the level of the sacrum. Often there is a dark marking on the midline between the arms. Each tibia has an irregular cross-band. The skin is smooth in males.

HABITAT AND DISTRIBUTION (FIG. 99): This species is common on bushes around pools in coastal forest. It is known from northern coastal Tanzania and southern coastal Kenya.

ADVERTISEMENT CALL: The male calls from vegetation. The call is a series of 5 or 6 clicks in half a second, with an emphasized frequency of 4 kHz.

BREEDING: Unknown.

TADPOLES: Unknown.

KEY REFERENCE: Schiøtz 1999.

Uluguru Spiny Reed Frog

Afrixalus uluguruensis (Barbour & Loveridge, 1928)
(Plates 8.2, 8.3)

This species is named for the Uluguru Mountains in Tanzania.

Fig. 100 *Afrixalus uluguruensis.*

DESCRIPTION: The male reaches 25 mm in length and the larger female, up to 28 mm. The head is slightly broader than long. The tympanum is not visible. The fingers and toes are dilated at the tips, with the toes webbed to the base of the discs. The skin on the females is smooth, but the males have small pale spines. The upper surfaces are white, with reddish flecks and black speckles on the anterior part of the back and a dark pair of markings on the sacral region. A speckled reddish line runs from each nostril through the eye to the flank. The underside is yellow.
The hands and feet are yellow with orange discs.

HABITAT AND DISTRIBUTION (FIG. 100): This species is known from high-altitude ponds in the Usambara, Uluguru, and Udzungwa mountain forests, and coastal Tanzania.

ADVERTISEMENT CALL: The males call from grass and other vegetation around forest pools. The call is a buzz, lasting up to 2 s. Each note consists of 3 or 4 pulses. There are 15 notes/s, with an emphasized frequency of 2.8 kHz.

BREEDING: Unknown.

TADPOLES: See the chapter on tadpoles.

NOTES: Known food items include flies, bugs, beetles, grasshoppers, ants, earwigs, and lacewings.

KEY REFERENCES: Barbour & Loveridge 1928a, Schiøtz 1999.

De Witte's Spiny Reed Frog

Afrixalus wittei (Laurent, 1941)
(Plate 8.4)

The specific name refers to G.-F. De Witte, herpetologist at the Institut Royal des Sciences Naturelles in Brussels until 1951.
Witte's afrixalus, De Witte's leaping frog, Witte's banana frog.

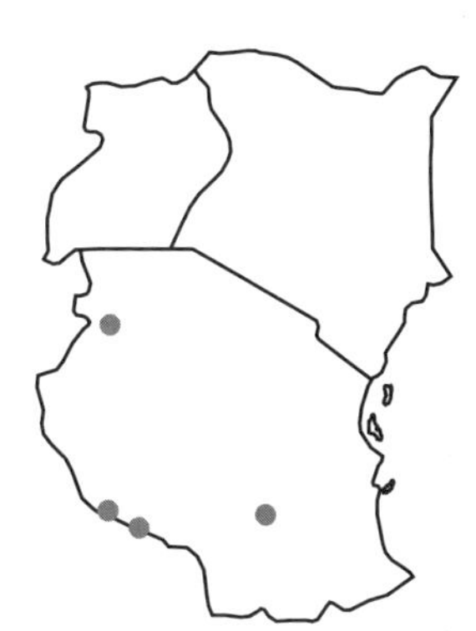

Fig. 101 *Afrixalus wittei.*

DESCRIPTION: This is one of the larger species in the genus, with females known to be up to 35 mm long. Small spines are present on the upper surface. Although similar in build and color to many other species in this genus, the dorsal pattern is characteristic. It consists of two dark dorsolateral stripes on a pale background that merge on the head and have a common origin at the back of the body. Background color is golden yellow to brown. A dark band is present along the side of the snout.

HABITAT AND DISTRIBUTION (FIG. 101): This frog is found in tropical lowland savanna, known from northern Angola, Zambia, and southwestern Democratic Republic of Congo, to Tanzania.

ADVERTISEMENT CALL: The males call from flooded grass. The call consists of a short creak that runs into a brief series of about 8 rapidly produced clicks. The call duration is up to 1 s, with an emphasized frequency of 2.8 kHz. The clicks are produced at a rate of around 12/s.

BREEDING: Unknown.

TADPOLES: Unknown.

KEY REFERENCES: Schiøtz 1999, Channing 2001.

Reed Frogs—Genus *Hyperolius*

This genus consists of small to medium-sized (20–35-mm) climbing frogs that are often brightly colored. Living specimens, especially calling males, are surprisingly easy to identify; preserved museum specimens are very difficult. The calls differ greatly between species. The pupil is horizontal, the tympanum is indistinct, and the skin is smooth. There are no cornified spines on the female, but breeding males often have small spines. The tips of the fingers and toes are expanded. Most species show a number of color patterns, and these often differ between the male and female. Juveniles often have a color phase that becomes one of many adult phases. Eggs are laid on the surface or just above water level in vegetation. The eggs are small, 0.8–1.5 mm in diameter, with 100–500 or more per clutch. The lack of field information for many of these frogs results in uncertainty of the validity of many species names.

The genus is found in the savannas and forests of Africa south of the Sahara. There are 32 species in East Africa.

The extreme color-pattern polymorphism shown by this genus prevents the construction of an identification key. Knowledge of the call, habitat, and locality data, along with the photograph and description, will enable specimens to be identified.

The Hyperolius viridiflavus Complex

This complex consists of a number of highly variable and very colorful reed frogs that have confused herpetologists for decades. Part of the confusion derives from the overlapping of color patterns in different localities. Additionally, each juvenile possesses one of a number of color patterns. As it matures, it may retain the juvenile pattern or acquire an adult pattern. About 40 different color patterns are recognized in this complex. Genes control color patterns, and various populations appear to possess various combinations of those genes. This results in particular patterns being present in localized populations, for example, the uniform brown individuals from the western shores of Lake Victoria. Various scientists have referred to these color forms as subspecies or color variants within one of three species: *Hyperolius parallelus, H. viridiflavus*, or *H. marmoratus*.

The male advertisement call, which is unique to each species, and hence useful to distinguish species, is a very short whistle. The brevity of this call made it difficult to analyze before the recent availability of suitable computer programs. Preliminary examination of *Hyperolius* calls recorded in southern and eastern Africa shows that different call types are present. Extensive fieldwork will be required to record natural calls at different temperatures and from as many different individuals in each population as possible. Detailed call analysis is required as part of the approach to understand the relationships in this species complex.

The introduction of molecular biochemical techniques in recent years has enabled particular genes to be traced in populations. This has led to an understanding of the evolution and relationships of many species, some of which were difficult to identify using only external features. Preliminary biochemical studies of some populations of this group of reed frogs indicate that some color morphs are genetically identical and can confidently be placed in one species, while other color morphs are genetically isolated from neighboring populations and should be recognized as separate species. In East Africa the following species have been separated on the basis of DNA studies: *H. argentovittis, H. glandicolor, H. mariae*, and *H. viridiflavus*. The names of the included taxa are listed in the notes under each species.

Much is known about the physiology and behavior of many populations that have been referred to as "*Hyperolius viridiflavus*." The following is a synthesis that may apply to some of the four species just listed.

ADVERTISEMENT CALL: The male calls from vegetation around pans and other water bodies. Two males may call simultaneously but alternate with a third one. The call is used to maintain spacing between males at a minimum of about 50 cm. The female selects an isolated male rather than a male calling close to another calling male. The female prefers a male with lower-frequency calls, implying that larger males are selected, as larger frogs make lower-pitched calls. The female also selects the male with the loudest call, providing that there is a reasonable difference between the calls. The male will return exactly to a previously used calling site on successive evenings.

BREEDING: A clutch consists of 150–650 small eggs, each 1.3–1.5 mm in diameter within 2.5-mm capsules. Eggs are attached to submerged roots in small clumps of about 20. The eggs are pale yellow with a darker brown pole. Some eggs have also been reported as blue-green in color. Metamorphosis takes 64–100 days in captivity. A sexually mature female can produce eggs (72–694 per clutch) every 2 or 3 weeks for up to 14 months and longer in captivity. In the field the clutch size is also variable; a female may lay a second clutch 15–60 days later, depending on the weather.

DEVELOPMENT OF COLOR POLYMORPHISM: Juveniles possess certain patterns, and adults others. The female and male color patterns develop differently. Some males retain the pattern they had as juveniles, yet others develop different ones. The female pattern is usually completely different from the juvenile pattern. The adult color phase is developed when the animal reaches sexual maturity. Bathing metamorphosing froglets in estrogen or testosterone (hormones normally only present at sexual maturity) for 1 min a day resulted in juveniles with precocious adult patterns. The natural potential for recombining the genes controlling color pattern is enormous. The amount of variation possible from a limited number of parents was illustrated by breeding six generations from one pair. The offspring showed an amazing variety of color patterns. If two individuals possess the genetic resources to create all these patterns, why do most populations in the field consist of one dominant pattern? Studies of the population genetics of these frogs in the field would be very useful to address this question.

PROTECTION AGAINST DESICCATION: These frogs spend time out in the bright sun and are at risk from drying out. Overall, however, they have a very low rate of water loss through the skin, compared to other frogs. They conserve water in various ways. Juveniles possess a skin that is more resistant to water loss than that of adults. During the dry season, water loss is reduced by inactivity, as these frogs can remain motionless on one leaf all day. Inactivity also decreases the loss of moisture by reducing secretions from the skin glands. In addition, the frog can go into aestivation, a special state of inactivity that slows the metabolism down to 50% of the normal resting rate and further reduces water loss. Above very high temperatures, like 42 °C, the frog makes use of body water stored in the bladder for cooling, by releasing it through the skin to permit evaporation. During the dry season the frog is also white, especially at high temperatures, and can reflect sunlight. This reflective color derives from the food it consumes. Food wastes are converted to substances called *purines*. The pigment cells (iridiophores) of the skin possess small purine platelets that are arranged parallel to the surface. These crystals probably act as quarter-wavelength interference reflectors, reducing the heat load and therefore water loss.

The red skin on the inside of the thighs and the side of the belly is widely reported as functioning as a startle mechanism, as it is only visible when the animal jumps. However, this red skin is only half as thick as the belly skin and is well supplied with blood vessels, making it specialized for rapid water uptake. The frog would be able to make use of drops of dew or rain to replace water lost for cooling through this red skin.

NOTES: Frogs in this group have been recorded as prey of the Angolan green snake *Philothamnus angolensis*, water snakes *Lycodonomorphus* sp., fishing spiders, the young of the Nile crocodile *Crocodilus niloticus*, birds like the woodland kingfisher *Halcyon senegalensis*, tree frogs *Leptopelis* spp., and the red-legged kassina *Kassina maculata*. It feeds on a range of small insects. Its longevity is at least 4 years 9 months in captivity.

KEY REFERENCES: Zimmerman 1975, 1979; Richards 1982; Kobelt & Linsenmair 1986; Dawson & Bishop 1987; Schiøtz 1999.

Sharp-Nosed Reed Frog

Hyperolius acuticeps Ahl, 1931
(Plate 8.5)

The species name means "sharp nosed."
Gunther's sharp-nosed reed frog.

Fig. 102 *Hyperolius acuticeps.*

DESCRIPTION: This is a small frog, with the male reaching a length of 24 mm and the female 26 mm. The frogs are slender with a distinct long snout. The tympanum is very small and is often not clearly visible. Males have small warts at the corners of the mouth. The dorsal skin is smooth. Less than two phalanges of the fourth toe are free of web. The back is green, with a white band from the eyes to the groin. The limbs and throat are green, while the gular disc is yellow. The belly is white. Some individuals have pale dorsolateral stripes. The frogs are often translucent, with the eggs visible through the belly skin of females.

HABITAT AND DISTRIBUTION (FIG. 102): This species is found in vegetation around small temporary pools in savanna. It occurs from West Africa, across to Ethiopia, and southward to eastern South Africa.

ADVERTISEMENT CALL: The male calls from reeds and other vegetation overhanging deep water. Males call in small choruses, sometimes located only 50 mm apart. The relatively long call consists of a series of pulses that usually slows down during the last half of the call. There are between 8 and 31 pulses/call. The pulse rate starts out at 130/s and then slows to 50/s, in some cases showing a clear division between the initial fast phase and the slower second phase. In some of the calls analyzed, the pulse rate is constant. The pulses are always visible as discrete elements. The emphasized frequency varies from 4.3 to 4.5 kHz. The mean call duration is 0.2 s, with some calls up to 0.5 s long.

BREEDING: Clutch size varies from 60 to 292. The eggs are small, only 0.7–1.2 mm in diameter, with a dark brown hemisphere. They can be found attached singly to both sides of submerged leaves.

TADPOLES: See the chapter on tadpoles. Tadpoles have been found developing in a small pocket of water trapped on top of a rock in a river.

NOTES: Food records include ants. This species is similar to both the green reed frog *Hyperolius viridis* and the long reed frog *H. nasutus* from the western side of the continent.

KEY REFERENCES: Schiøtz 1999 (part of *H. nasutus*), Rödel 2000 (as *H. nasutus*), Channing et al. 2002a.

Ahl's Reed Frog

Hyperolius argentovittis Ahl, 1931
(Plates 8.6, 8.7)

The species name *argentovittis* means "silvery ribbon."
Korfe in Nyakusa.

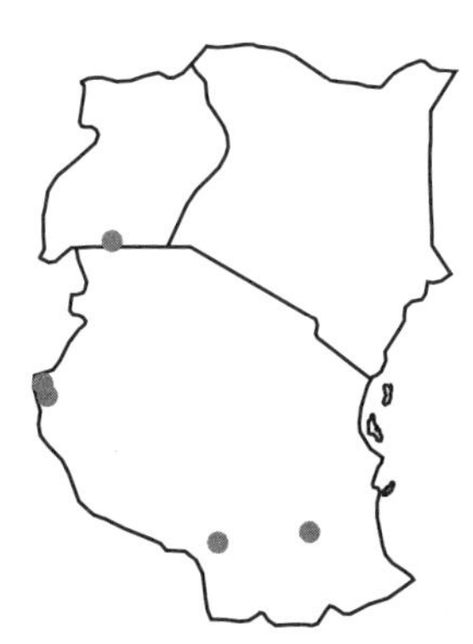

Fig. 103 *Hyperolius argentovittis.*

DESCRIPTION: Both males and females reach 25 mm in length. The tympanum is hidden. The toes are webbed to the base of the discs. Coloration is variable: a dark subdermal lateral streak is present in many populations. The back may be dark brown with a pale vertebral stripe and fine vermiculations; pale with red-brown vermiculations; dark with a pale vertebral stripe and oblique lateral stripes, each with a thin red central line; or yellow to green with a dark lateral edge with red spots on the flank. The gular disc is bright yellow behind. The limbs and undersides are red. Other patterns may be present.

HABITAT AND DISTRIBUTION (FIG. 103): This species is found near water where vegetation is lush. It is known from the highlands of southern Malawi, central Zambia, through central and western Tanzania, Rwanda, Burundi, Uganda, and eastern Democratic Republic of Congo.

ADVERTISEMENT CALL: Males call from vegetation in or near water. The call is a brief click at a dominant frequency of 3.9–4.1 kHz.

BREEDING: Unknown.

TADPOLES: Unknown.

NOTES: This frog was previously included as part of the common reed frog *Hyperolius viridiflavus*, as *H. v. albofasciatus*, *H. v. melanoleucus*, *H. v. marginatus*, or *H. reesi*.

KEY REFERENCES: Schiøtz 1999, Wieczorek et al. 2000.

Argus Reed Frog

Hyperolius argus Peters, 1854
(Plate 8.8)

This species is named for Argus the guardian of Io, the being whose 100 eyes were given to the peacock in Greek mythology.
Argus sedge frog, argus-eyed frog, argus-spotted sedge-frog.

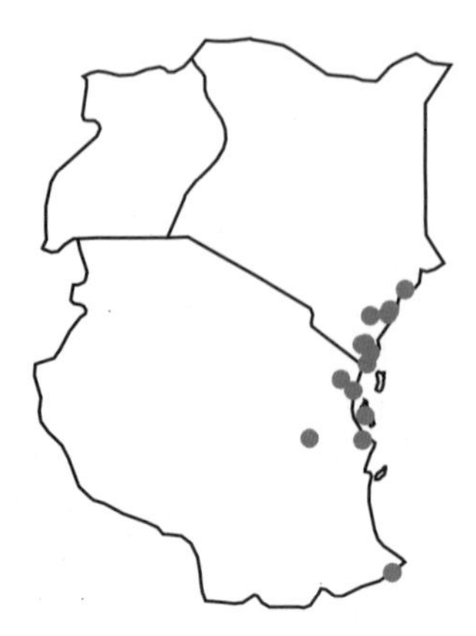

Fig. 104 *Hyperolius argus.*

DESCRIPTION: This reed frog is large, with the male and female up to 37 mm long. It has a head width/body length proportion of 36%–42%. The young are green after they metamorphose. Adult males have a green back with a dark canthal line and dark eyes, often with small brown dots on the back, and thin light lateral bands. Below they are pale with blue-green around the arm and bluish vocal sacs. The females are reddish brown with a horseshoe-shaped pale band, with a darker outline, over the snout and eyes. The bands on the back may be absent in some individuals or a series of wide dots in others.

HABITAT AND DISTRIBUTION (FIG. 104): The argus reed frog is associated with temporary and permanent pools where water lilies are growing. The species is widespread, known from the coastal savanna of eastern Africa from southern Somalia to South Africa, up to an altitude of 1000 m in the East Usambara Mountains of Tanzania.

ADVERTISEMENT CALL: The male calls from floating vegetation, such as lily pads, in deep water. The call is a brief nasal "oink" (0.4 s), with harmonics at 1.5, 2.0, and 2.5 kHz. The call is frequency-modulated, rising a little in pitch. Males make a high-pitched chirp that serves as an aggression call when other males are too close.

BREEDING: Breeding takes place around the edges and in the middle of shallow pans. Temporary pools formed in flat depressions are a favorite habitat for this species. It breeds from spring to summer after the rains have started. The eggs are 1 mm in diameter within 4-mm capsules. Clutch size is about 200. The eggs are laid in clusters of 30, each cluster being attached to vegetation hanging in water, up to 50 mm below the surface.

TADPOLES: See the chapter on tadpoles.

NOTES: The male and female color patterns develop at maturity under the influence of the hormones estradiol and testosterone.

KEY REFERENCES: Hayes & Mendez 1999, Schiøtz 1999, Channing 2001.

Balfour's Reed Frog

Hyperolius balfouri (Werner, 1908)

This species was named for J. W. Balfour, a missionary in Uganda.

Fig. 105 *Hyperolius balfouri.*

DESCRIPTION: This large species has a long snout. Females are as long as 42 mm, males as long as 34 mm. The back is yellowish gray to dark brown, often with a slate blue dorsolateral stripe. The ventral surfaces are uniformly yellowish. The discs, underside of thighs, and outer side of lower legs are bright pink. Both males and females have the same color pattern.

HABITAT AND DISTRIBUTION (FIG. 105): This species is associated with pools in savanna, from Uganda, western Kenya to Cameroon and southwestern Ethiopia.

ADVERTISEMENT CALL: The male advertisement call is a low-pitched slow creak. The call duration is 0.1 s, with an emphasized frequency of 1.4 kHz. Each creak consists of an irregular series of pulses.

BREEDING: Unknown.

TADPOLES: Unknown.

NOTES: Known food items include caterpillars, moths, and beetles.

KEY REFERENCES: Noble 1924, Schiøtz 1999.

Bocage's Reed Frog

Hyperolius bocagei Steindachner, 1867
(Plates 9.1, 9.2)

This species was named for the herpetologist J. V. Barboza du Bocage, who worked extensively in Angola.

Fig. 106 *Hyperolius bocagei.*

DESCRIPTION: This frog is small, with the male up to 26 mm long. The head is relatively broad, about one-third of the body length. The main part of the webbing is level with the middle tubercle below the fourth toe, but forming a visible margin almost to the tip of the toe. The back is brown to green or yellow, with a darker dorsolateral line or series of spots in males. The back and upper limbs often have scattered dark spots. A dark line is often present on the side of the snout. Females in breeding condition are tomato red, but brown otherwise. This species is similar in overall morphology to the common reed frog *Hyperolius viridiflavus*, but is distinguishable by its reduced webbing.

HABITAT AND DISTRIBUTION (FIG. 106): This species is associated with grassy pans and is known from Angola, southern Democratic Republic of Congo, and western Zambia through to southwestern Tanzania.

ADVERTISEMENT CALL: The male calls from elevated positions in vegetation near water. Two calls may be heard: a harsh brief buzz at 3.5 kHz with a duration of 0.03 s, and a soft slower series of 4 or 5 clicks. The latter may be the male aggression call.

BREEDING: Unknown.

TADPOLES: Unknown.

NOTES: There is some uncertainty about the correct scientific name for this frog, as more than one species might be confused.

KEY REFERENCE: Schiøtz 1999.

Brown Reed Frog

Hyperolius castaneus Ahl, 1931
(Plate 9.3)

The specific name refers to the brown color of chestnuts.
Montane reed frog, *Kastanienbrauner Riedfrosch* in German, *rainette marron* in French.

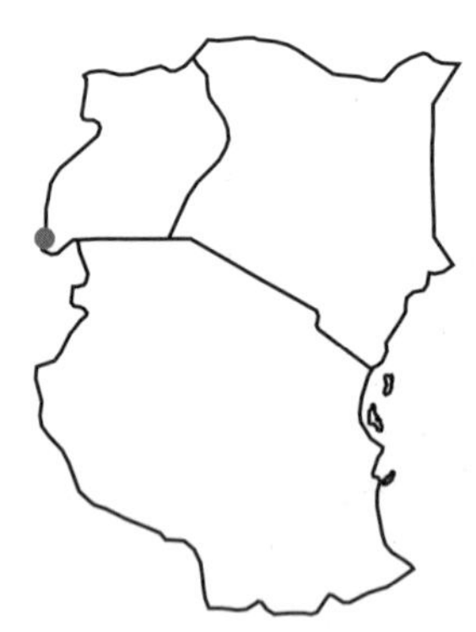

Fig. 107 *Hyperolius castaneus.*

DESCRIPTION: The brown reed frog reaches a length of 36 mm in females and 27 mm in males. The dorsal pattern is in greens and browns, like so many other reed frogs. There are many color patterns; some of the common ones include a yellowish brown back with a pinkish side, separated by an irregular thin dark-brown line that runs from the nostril, through the middle of the eye, posteriorly to the vent; a brown back with green spots, each with a black center, and a dark line from the nostril to the eye; a green back with dark irregular spots, often with a pale lateral line. The underside and feet are yellow to red. Males have a yellow throat and gular sac.

HABITAT AND DISTRIBUTION (FIG. 107): This species is found in high-altitude swamps in the forests of the mountains of western Uganda, Rwanda, and the Democratic Republic of Congo.

ADVERTISEMENT CALL: The call is brief note at an emphasized frequency of 2.4 kHz. A second harmonic at 4.8 kHz is present.

BREEDING: The unpigmented eggs are laid in a small batch attached to the upper surface of a leaf overhanging water. The tadpoles wriggle into the water to continue development. Clutch size is 100–150.

TADPOLES: Unknown.

NOTES: This species can easily be confused with *Hyperolius lateralis*, which is similar in call and color pattern. Further work is required to investigate if these two taxa represent extremes of a polymorphic species.

KEY REFERENCES: Fischer & Hinkel 1992, Schiøtz 1999.

Cinnamon-Bellied Reed Frog

Hyperolius cinnamomeoventris Bocage, 1866
(Plates 9.4, 9.5)

The specific name *cinnamomeoventris* means "cinnamon-colored belly."
Dimorphic reed frog.

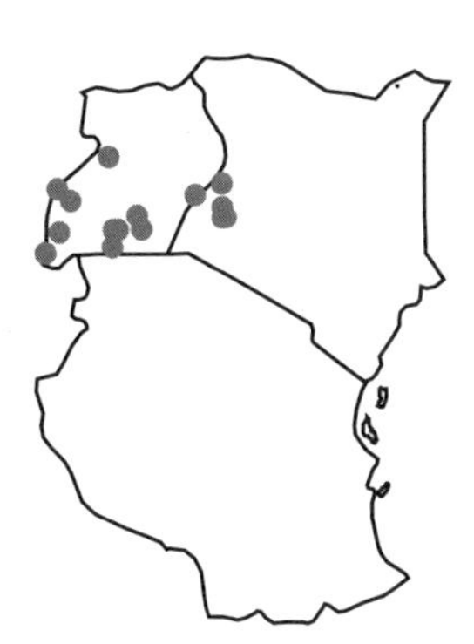

Fig. 108 *Hyperolius cinnamomeoventris.*

DESCRIPTION: The male reaches 28 mm in length, with the female up to 33 mm. The skin of the back of both sexes is rather coarse. The sexes have different color patterns, with the male brown or green with a light dorsolateral line, and the female green above and pale yellow below, with an irregular dark line separating the two colors. The male may have darker spots on the back, with a yellow throat and gular sac. The female does not have the pale dorsolateral lines, but a short dark line is present on the side of the snout. Anterior and posterior thighs are brilliant red in breeding animals.

HABITAT AND DISTRIBUTION (FIG. 108): This distinctive species is found in both forest and savanna. It is known from Cameroon, the Democratic Republic of Congo, Angola, Uganda, and Kenya.

ADVERTISEMENT CALL: The call consists of short clicks, each about 0.05 s long, with a dominant frequency of 3 kHz.

BREEDING: Clutch size is 146, and development takes about 15 days.

TADPOLES: Unknown.

NOTES: This species is unusual in that it has been collected both in forest and in savanna, and it would be very interesting to discover more about its biology. Known food items include flies, ants, roaches, crickets, and leaf hoppers. Frogflies are major egg predators. At Kibale in Uganda nearly 50% of the egg masses were infested with frogflies, which lay their eggs on the frog eggs. The fly larvae feed on the eggs and significantly reduce the breeding success of the frog.

KEY REFERENCES: Perret 1966, Inger 1968, Schiøtz 1975, Vonesh 2000, Channing 2001.

Bladder Reed Frog

Hyperolius cystocandicans Richards & Schiøtz, 1977

The specific name means "white bladder" and refers to the silvery bladder that can be seen through the ventral skin.

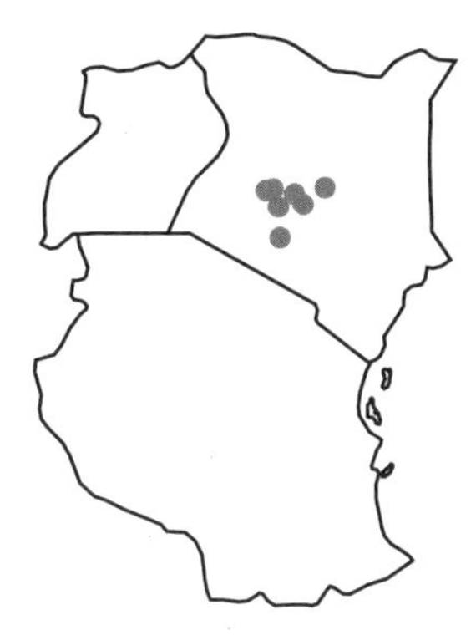

Fig. 109 *Hyperolius cystocandicans.*

DESCRIPTION: This is a large species, with the females reaching 36 mm in length, while the males are smaller, at 28 mm. In both sexes the silvery bladder can be seen through the transparent belly skin. Adults are uniform yellow-brown to green-gold, although sometimes the back is mottled. The sides and limbs are unpigmented and translucent. Males have pinkish undersides and limbs.

HABITAT AND DISTRIBUTION (FIG. 109): This species is known from high-altitude grasslands in central Kenya.

ADVERTISEMENT CALL: The males call in vegetation away from the water. The call is a series of quiet clicks. Each click is very brief with an emphasized frequency around 2.6 kHz.

BREEDING: The eggs have a dark upper hemisphere and a cream lower pole. Clutch size is around 100. This species develops from egg through tadpole to adult in 3 months.

TADPOLES: Unknown.

NOTES: This species might be confused with the montane reed frog *Hyperolius montanus*, but is distinguishable by the transparent belly skin. The males are easily distinguished from *H. montanus* males by the presence of rough glandular skin on the thumb, along the underside of the arm, and onto the chest.

KEY REFERENCES: Richards & Schiøtz 1977, Schiøtz 1999.

Highland Reed Frog

Hyperolius discodactylus Ahl, 1931
(Plate 9.6)

The species name refers to the distinct discs on the fingers.
Albertine Rift reed frog.

Fig. 110 *Hyperolius discodactylus.*

DESCRIPTION: This species is large, with the female reaching a length of 37 mm. The female is a dead-leaf brown, often with black speckles. A dark line from the nostril through the eye continues posteriorly as a series of spots. The back is flecked with yellow. The throat and belly are yellow-orange, and the undersides of the hands and feet are pink. Reproductive males have a bright green throat.

HABITAT AND DISTRIBUTION (FIG. 110): This species is found in high mountain forests along streams. Specimens have been recorded from the mountains bordering the Democratic Republic of Congo, Uganda, Burundi, and Rwanda.

ADVERTISEMENT CALL: Males call from trees near streams, up to 5 m above ground. The call is a buzz lasting 0.3–0.4 s, with an emphasized frequency of 2.0 kHz.

BREEDING: Unknown.

TADPOLES: Unknown.

KEY REFERENCE: Schiøtz 1999.

White-Snouted Reed Frog

Hyperolius frontalis Laurent, 1972
(Plate 9.7)

The specific name refers to the pale markings on the snout.

Fig. 111 *Hyperolius frontalis.*

DESCRIPTION: This is a large species, with the females reaching 35 mm and the males 29 mm in length. The pattern is characteristic: a translucent green back with a yellow or golden triangle on the snout. The snout marking is sometimes irregular. The pale markings may continue posteriorly as a short pale dorsolateral line or series of markings. Numerous fine black spots cover the upper surfaces of the back and limbs. The male vocal sac is green.

HABITAT AND DISTRIBUTION (FIG. 111): This species is known from montane forests on the border between Uganda, the Democratic Republic of Congo, and Rwanda.

ADVERTISEMENT CALL: The males call from dense vegetation overhanging small streams and from elevated positions in trees or other vegetation. The call is a brief buzz with an emphasized frequency of 2.5 kHz.

BREEDING: The eggs are opaque white, large (5 mm in diameter), and placed up to 3 m above water. Sometimes the eggs are laid above quiet tributaries, some only 100 mm deep. Clutch size is around 24. The tadpoles drop into the water to continue development.

TADPOLES: Unknown.

KEY REFERENCE: Schiøtz 1999.

Peters' Reed Frog

Hyperolius glandicolor Peters, 1878
(Plate 9.8)

The specific name *glandicolor* means "the color of acorn."

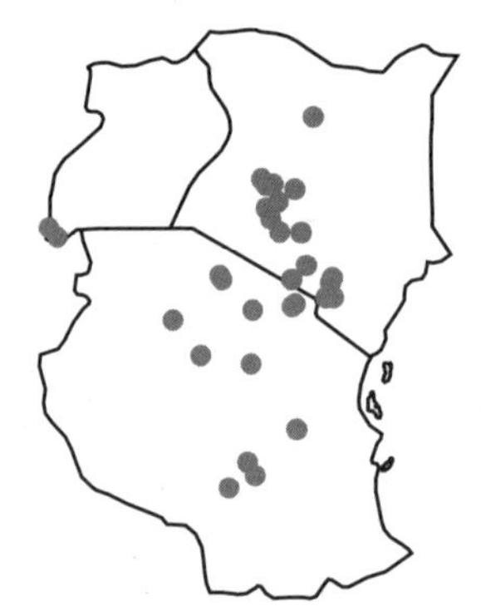

Fig. 112 *Hyperolius glandicolor.*

DESCRIPTION: Both males and females reach 25 mm in length. The tympanum is hidden. The toes are webbed to the base of the discs. Coloration is variable. The back is brown with various markings: small black spots; white spots and markings; pale marbling; yellow spots; or small white rings. Other patterns may be present.

HABITAT AND DISTRIBUTION (FIG. 112): The species is found near water where vegetation is lush. It is known from southern and central Kenya, central Tanzania including the Serengeti, to southwestern Uganda.

ADVERTISEMENT CALL: Males call from vegetation in or near water. They may suspend themselves between grass stems. The call is a brief click at a dominant frequency of 3.9–4.1 kHz.

BREEDING: Breeding takes place during the short rains, from October to January, with some additional activity during the later long rains where these are heavy. No details of eggs or breeding sites have been reported.

TADPOLES: Unknown.

NOTES: This frog was previously included as part of the common reed frog, as *H. v. ferniquei*, *H. v. pantherinus*, *H. v. goetzei*, *H. v. ngorogoroensis*, *H. v. pitmani*, and *H. v. ommatostictus*. *H. orkarkarri* from the Serengeti is regarded as a synonym, based on molecular evidence.

KEY REFERENCES: Schiøtz 1999, Wieczorek et al. 2000.

Kihanga Reed Frog

Hyperolius kihangensis Schiøtz & Westergaard, 1999
(Plate 10.1)

This species was named for the Kihanga region of the Udzungwa Mountains in Tanzania.

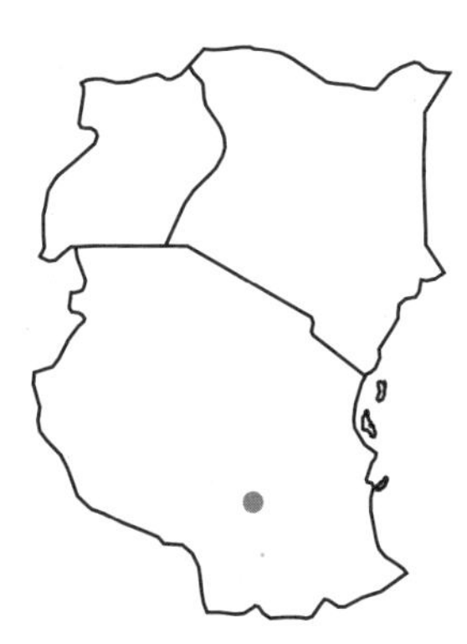

Fig. 113 *Hyperolius kihangensis.*

DESCRIPTION: This species is small, with the males reaching 19 mm in length and the females 26 mm. The back is brownish with broad pale cross-bands. Some individuals have an hourglass pattern. Breeding males have rough skin on the back. The fingers, toes, and discs are yellow to red, with a yellow tinge to the belly in males but red in females. A characteristic small white spot is present on the heel.

HABITAT AND DISTRIBUTION (FIG. 113): This species is associated with swamps and is presently only known from forest east of Iringa, but is probably widespread in the Udzungwa Mountains and the southern highlands of Tanzania.

ADVERTISEMENT CALL: This species appears to be voiceless.

BREEDING: Unknown.

TADPOLES: Unknown.

KEY REFERENCES: Schiøtz 1999, Schiøtz & Westergaard 2000.

Kivu Reed Frog

Hyperolius kivuensis Ahl, 1931
(Plate 10.2)

The specific name refers to Lake Kivu on the border between the Democratic Republic of Congo and Rwanda.
Lake Kivu reed frog.

Fig. 114 *Hyperolius kivuensis.*

DESCRIPTION: The males reach a length of 30 mm and females, up to 38 mm. This species has a slender body with a long snout. The back is brown, silvery gray, or bright green. A dark stripe on the side of the snout passes posteriorly through the eye to become a dark stripe on the side of the body. Sometimes a lighter line is present above the dark one. The underside is pale to gray. The parts of the legs and feet that are concealed when the animal is at rest are yellow to red.

HABITAT AND DISTRIBUTION (FIG. 114): This widely distributed reed frog is found in open savanna. It is known from Angola, Zambia, the Democratic Republic of Congo, Burundi, Uganda, Tanzania, and Kenya.

ADVERTISEMENT CALL: The male calls from concealed positions high in vegetation. The call is a brief, harsh chirp, lasting 30 ms, with the dominant harmonic at 2.5–3.0 kHz. This is sometimes preceded by a single click. The males also produce a call believed to serve as a male-male spacing vocalization, which consists of six evenly spaced clicks, with a duration of 0.4 s. The dominant harmonic is at 2.5–3.0 kHz.

BREEDING: The eggs are white with a dark pole and have a diameter of 1.5 mm. Clutch size is around 190. The eggs are laid on vegetation and require only 9 days to hatching.

TADPOLES: Unknown.

NOTES: Frogflies infest up to 20% of the egg masses at Kibale in Uganda. The flies deposit eggs, which hatch into larvae that feed on the frog eggs.

KEY REFERENCES: Schiøtz 1999, Vonesh 2000, Channing 2001.

Lang's Reed Frog

Hyperolius langi Noble, 1924
(Plates 10.3, 10.4)

This species was named for James Paul Lang (1889–1964), an assistant at the American Museum of Natural History who collected this species during an expedition to the Belgian Congo, 1909–1915.
Flat-headed reed frog.

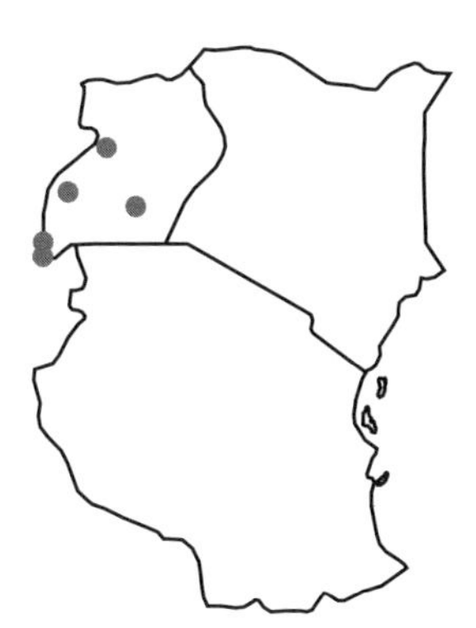

Fig. 115 *Hyperolius langi.*

DESCRIPTION: The females reach a length of 31 mm and the males, 24 mm. The head is broader than the body, and the tympanum is hidden. The back is smooth or finely granular, with the belly coarsely granular. The back pattern is variable, but a pale area above the upper lip and a short pale dorsolateral stripe are common. An alternative pattern consists of a pale snout with a dark hourglass pattern from the eyes posteriorly. The back is greenish yellow with brown markings. The lower surfaces are greenish and translucent. The legs show cross-banding.

HABITAT AND DISTRIBUTION (FIG. 115): Lang's reed frog inhabits high montane forest. It is known from Uganda and the Democratic Republic of Congo.

ADVERTISEMENT CALL: Unknown.

BREEDING: Clutch size is 56, with 1.6-mm-diameter eggs. In Uganda eggs were discovered laid between leaves. The time to hatching is 2 weeks.

TADPOLES: Unknown.

NOTES: Frogflies infest up to 50% of egg masses. The fly larvae feed on the frog eggs and significantly reduce the hatching success.

KEY REFERENCES: Schiøtz 1999, Vonesh 2000.

Side-Striped Reed Frog

Hyperolius lateralis Laurent, 1940
(Plates 10.5, 10.6, 10.7)

The species name refers to the stripe on the side of the body.
Mottle-sided reed frog.

DESCRIPTION: Females reach 30 mm in length and the males, 26 mm. The color patterns are variable across the range but tend to be consistent within any one population. Juveniles and some males are a translucent pale brown to green, with a dark stripe between the nostril and the eye. A pale dorsolateral stripe is found in many specimens. The underside is reddish, and the feet are red. Many females are a dark green or brown with a broad pale, often yellow, lateral stripe. The stripe is edged

Fig. 116 *Hyperolius lateralis.*

with black and may be broken up into a series of spots, or be completely absent. The male vocal sac is yellow.

HABITAT AND DISTRIBUTION (FIG. 116): This species is found in wooded grassland and is known from western Kenya, Uganda, eastern Democratic Republic of Congo, through Rwanda and Burundi, and western Tanzania south to the border with Zambia.

ADVERTISEMENT CALL: Males call while concealed in vegetation. Each call is a harsh brief creak, with an emphasized frequency of 2.6–3.0 kHz. The duration of each call is 0.05–1.00 s. The rasping aggression call is longer and slower pulsed.

BREEDING: The mean clutch size is 109, and the egg size 1.8 mm in diameter. Development takes 14 days.

TADPOLES: Unknown.

NOTES: Males attempt to dislodge another male in amplexus by kicking with the hind feet and by getting the snout under the male to lift him off the female. Frogflies lay eggs on the frog egg masses. The fly larvae feed on the frog eggs. Over 60% of egg masses studied in Uganda were infested. This frog has a color pattern similar to that of *Hyperolius cinnamomeoventris* and *H. castaneus.*

KEY REFERENCES: Schiøtz 1999, Vonesh 2000.

Mary's Reed Frog

Hyperolius mariae Barbour & Loveridge, 1928
(Plate 10.8)

The species name refers to Mary Loveridge, wife of Arthur Loveridge.

Fig. 117 *Hyperolius mariae.*

DESCRIPTION: Both males and females reach 25 mm in length. The tympanum is hidden. The toes are webbed to the base of the discs. The back is uniformly yellowish brown, with a characteristic dark side to the body. This dark pigmentation appears to be in the deeper layers of the skin and is often referred to as a *subdermal streak*. A dark stripe is also present from the nostril to the eye and sometimes a little farther back. The gular disc is bright yellow. The limbs and undersides are red.

HABITAT AND DISTRIBUTION (FIG. 117): This species is associated with grass and other plants near water. It is found on the East Usambara Mountains, the central and coastal lowlands of Kenya and Tanzania, and adjacent islands such as Pemba and Zanzibar.

ADVERTISEMENT CALL: Males call from vegetation in or near water. The call is a brief whistle at a dominant frequency of 3.9–4.1 kHz.

BREEDING: Unknown.

TADPOLES: Unknown.

NOTES: This frog, including *Hyperolius v. rubripes*, was previously included as part of the common reed frog. Males make a creaking aggression call when approached too closely by other calling males.

KEY REFERENCES: Loveridge 1944, Schiøtz 1999.

Dwarf Reed Frog

Hyperolius minutissimus Schiøtz, 1975
(Plate 11.1)

The specific name *minutissimus* refers to the extremely small size of this species.

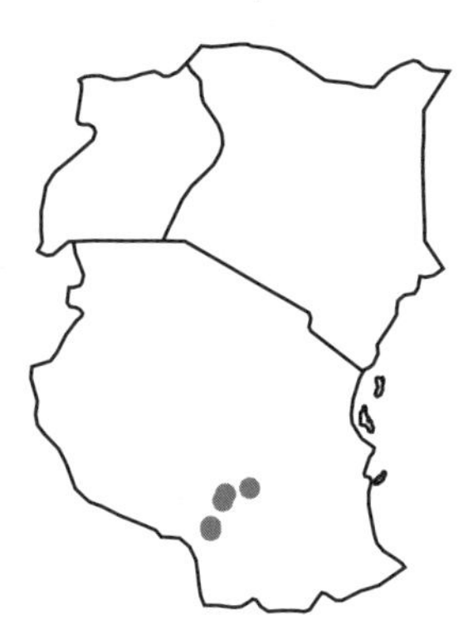

Fig. 118 *Hyperolius minutissimus.*

DESCRIPTION: This frog is very small, with exceptional males up to 17 mm long and the females 24 mm long. Males are mature at about 12 mm. The back is brown with a pale greenish or yellow stripe from each nostril to the eye, and a pair of dorsolateral stripes. Some animals have additional pale spots. Secondary sexual characters in males include small spines on the front of the lower jaw, and a relatively large pale vocal flap.

HABITAT AND DISTRIBUTION (FIG. 118): This species is known from high-altitude grasslands in southeastern Tanzania.

ADVERTISEMENT CALL: Males call from vegetation growing in water. The call is very quiet and might be easily missed. It consists of a rapid series of clicks, about 5/s, at an emphasized frequency of 4.0 kHz.

BREEDING: Unknown.

TADPOLES: Unknown.

KEY REFERENCES: Schiøtz 1975, 1999.

Mitchell's Reed Frog

Hyperolius mitchelli Loveridge, 1953
(Plates 11.2, 11.3)

The species was named for B. L. Mitchell, who was in the Department of Game, Fish, and Tsetse in Nyasaland (Malawi).

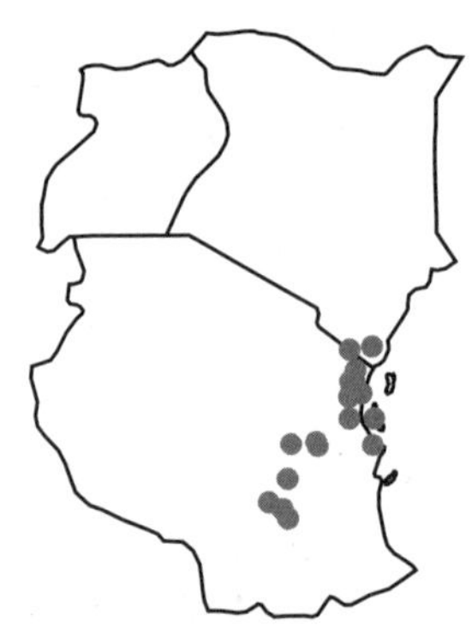

Fig. 119 *Hyperolius mitchelli.*

DESCRIPTION: Males may reach 30 mm in length and females, up to 32 mm. Head width is about one-third of body length. The main part of the webbing reaches the middle tubercle beneath the fourth toe. A light band on the side of the snout continues over the upper eyelid as a pale, wider dorsolateral band. In many animals this white band has a black border. A light heel spot is usually present. Dark spots are often found on the limbs and back. The undersurfaces are yellow to orange in the male but only orange in the female.

HABITAT AND DISTRIBUTION (FIG. 119): This species is associated with water bodies in forest. It is known from southeastern Kenya, northeastern Tanzania, through Malawi to Mozambique. They have also been found on the offshore islands.

ADVERTISEMENT CALL: The male calls from vegetation, often from the top of sedges. The call is a peculiar short chirp, about 0.1 s long, with a series of harmonics that start around 2.6, 3.1, 3.6, and 4.0 kHz and end closely spaced, with the lowest harmonic rising to 3.0 and the highest dropping to 3.6 kHz. The dominant harmonic starts at 3.6 kHz and drops.

BREEDING: Pale yellow eggs are laid on a leaf overhanging water, in clutches of 50–100. After 5 or 6 days the pale tadpoles wriggle off the leaf during rain into the pond below.

TADPOLES: See the chapter on tadpoles.

NOTES: One of the color morphs was previously known as *Hyperolius rubrovermiculatus*. This species may be confused with the spotted reed frog *H. puncticulatus,* which has a clacking call. Many individuals are plain brown and might be confused with the sympatric *H. puncticulatus*, but the white heels and shorter body are diagnostic. The morph with a broad white lateral stripe edged with black is one of the most colorful in the East African region.

KEY REFERENCES: Schiøtz 1975, 1999; De Fonesca & Mertens 1979; Channing & Crapon de Caprona 1987; Channing 2001.

Montane Reed Frog

Hyperolius montanus (Angel, 1924)
(Plate 11.4)

The specific name refers to the Aberdare Mountains, where this species was first collected, on 19 February 1912.

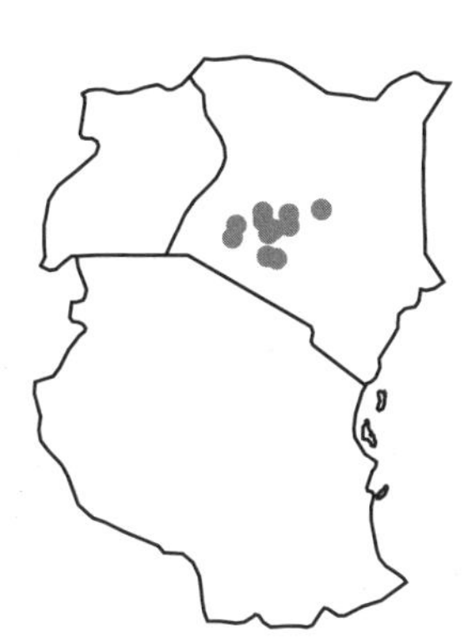

Fig. 120 *Hyperolius montanus.*

DESCRIPTION: Males reach 26 mm in length and females, up to 34 mm. The tympanum is hidden. The throat is smooth, with a group of conical granules at the angle of the mouth. The back is brown, yellow, or even greenish. A dark or pale yellow stripe from the nostril runs through the eye to the groin, but it may end behind the eye. The throat is yellow to greenish, sometimes speckled.

HABITAT AND DISTRIBUTION (FIG. 120): Montane reed frogs are known from the montane grasslands of central Kenya. They have been found on *Hagenia* trees at an altitude of 3000 m.

ADVERTISEMENT CALL: The call consists of a series of short chirp-like notes. Each chirp consists of about 10 pulses in 0.08 s. The notes are produced at a rate of around 3/s, at an emphasized frequency of 2.2 kHz.

BREEDING: Unknown.

TADPOLES: Unknown.

KEY REFERENCE: Schiøtz 1999.

Golden-Eyed Reed Frog

Hyperolius ocellatus Günther, 1859

The specific name *ocellatus* refers to the distinctive eyelike spots on the back of the females.
Ocellated reed frog.

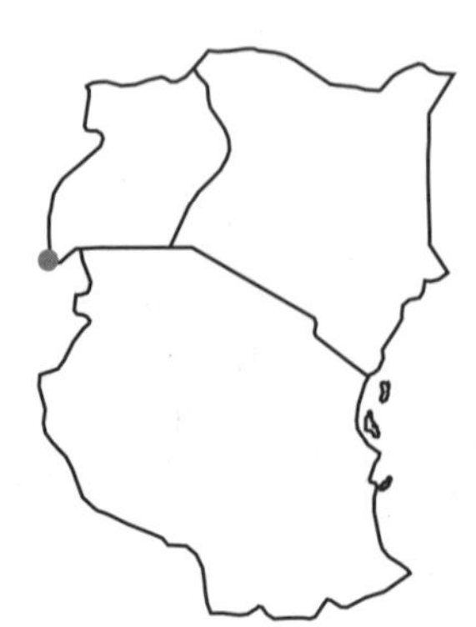

Fig. 121 *Hyperolius ocellatus.*

DESCRIPTION: The male reaches a length of 27 mm and the female, 34 mm. The sexes have different color patterns. The male is green with a pale silvery green triangle in front of the eyes that extends behind the eyes as dorsolateral stripes. The underside is white, but the limbs and toes are green. The iris is golden. The female is silvery gray with small black spots on the front half of the body. The belly is orange, the throat is yellow, and the lower surfaces of the limbs and toes are pink.

HABITAT AND DISTRIBUTION (FIG. 121): It is known from forest in Nigeria, Uganda, and the Democratic Republic of Congo.

ADVERTISEMENT CALL: The male calls from open vegetation at the edge of the forest. The call is a fine twittering, consisting of 4 or 5 pulses, at a dominant frequency of 3.8–4.0 kHz, with a duration of 0.15 s.

BREEDING: Unknown. The eggs are pale green with a small dark gray-green pole. This species breeds in forest ponds and along small streams.

TADPOLES: Unknown.

NOTES: Western populations from Cameroon have slightly different color patterns.

KEY REFERENCE: Schiøtz 1999.

Parker's Reed Frog

Hyperolius parkeri Loveridge, 1933
(Plate 11.5)

This frog was named for H. W. Parker, herpetologist of the Natural History Museum in London, 1923–1957.
Brown-or-green sedge-frog.

DESCRIPTION: Male frogs reach 27 mm in length and females, 28 mm. The snout projects beyond the mouth. Head width is one-third of body length. The male vocal sac is small. The breeding male has dark spines underneath the feet and body. Webbing is extensive, with only one joint of the fourth toe free of web. The back of the male is green, yellow, or brown, with a light band on the side of the snout and a dorsolateral band that has a dark border. The female is always green with scattered dark spots on the back. The throat is blue in reproductive animals.

Fig. 122 *Hyperolius parkeri.*

HABITAT AND DISTRIBUTION (FIG. 122): Parker's reed frog is associated with ponds in savanna and forest. This species is known from the coastal lowlands of Kenya and Tanzania to Mozambique, and inland up to an altitude of about 1000 m.

ADVERTISEMENT CALL: The male calls alongside streams in forest, or from grass in pools with standing water. Unlike other reed frogs, the male calls from a resting position, without raising the head. The call is a long loud trill, with an initial sound followed by up to 28 pulses at a dominant frequency of 4.2–5.5 kHz. Pulse rate is 17/s. The duration of the call is about 1.5 s.

BREEDING: The white eggs are attached to sedges or the underside of reed leaves just above water level. Clutch size is 36–110 eggs.

TADPOLES: Unknown.

KEY REFERENCES: Loveridge 1933; Schiøtz 1975, 1999.

Variable Reed Frog

Hyperolius pictus Ahl, 1931
(Plate 11.6)

The specific name *pictus* means "painted," referring to the back pattern. Variable montane sedge frog, Ahl's painted reed frog, *kolamwilwe* in Hehe, *tufi* in Kinga.

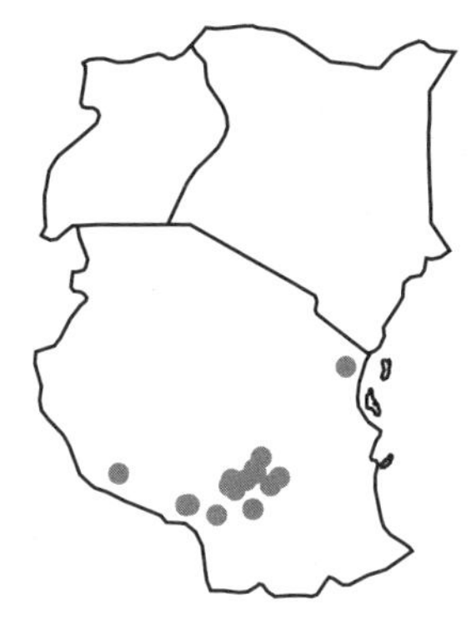

Fig. 123 *Hyperolius pictus.*

DESCRIPTION: The male reaches a length of 29 mm and the female, 38 mm. The snout has curved sides and a blunt tip. The head width is about one-third of the body length. This is a short-legged frog, with variable markings in brown and green. The back may be uniform, or with light longitudinal bands, or with irregular longitudinal dark brown or green bands. The underside is yellow to orange. The limbs are orange and the discs reddish. Several different adult patterns are known.

HABITAT AND DISTRIBUTION (FIG. 123): This species is found around water bodies in grassland and forest and is known from the East Usambaras, the highlands of southwestern Tanzania, northern Malawi, and adjoining Zambia.

ADVERTISEMENT CALL: The call is a two-part croak. The two parts may be alternated. One part consists of 15 pulses at a rate of 30/s with a duration of 0.5 s. The second part consists of about 12 pulses at a rate of 100/s. The croaks may be slower or may speed up. The dominant frequency is 2.7 kHz.

BREEDING: Breeding takes place during the summer rainy season. Eggs are laid in gelatinous clusters on vegetation above water level at the water's edge. Each 75-mm cluster contains 60–100 eggs, and each egg is white with a gray upper half and 2 mm in diameter within a 5-mm capsule. Tadpoles hatch after 18 days and may be flooded into the pools by rain. Time to metamorphosis is 3–4 months.

TADPOLES: Unknown.

NOTES: Recorded food items include bugs, beetles, wasps, and spiders. Some color patterns may resemble those of the spotted reed frog *Hyperolius puncticulatus*.

KEY REFERENCES: Loveridge 1933, Schiøtz 1999.

Mette's Reed Frog

Hyperolius pseudargus Schiøtz & Westergaard, 1999
(Plate 11.7)

The specific name refers to the similarity between this species and *Hyperolius argus*. The common name refers to the collector of this species, Mette Westergaard.

Fig. 124 *Hyperolius pseudargus.*

DESCRIPTION: This is a large reed frog, with males and females up to 35 mm long. The adults are green with a pale thin stripe running from the nostrils to the rear of the body. The upper eyelid is golden, and there are many, very small golden spots on the back. The vocal pouch is greenish when inflated. The undersurfaces are yellowish green.

HABITAT AND DISTRIBUTION (FIG. 124): This species is found on high grasslands in the Udzungwa Mountains and southern highlands of Tanzania.

ADVERTISEMENT CALL: The male calls from sedges around water or floating vegetation. The call is a long series of clicks, produced at 4/s, with a dominant frequency of 3.0 kHz. The clicking may continue for a few seconds, or may be terminated with a short buzz.

BREEDING: Unknown.

TADPOLES: Unknown.

KEY REFERENCES: Schiøtz 1999, Schiøtz & Westergaard 2000.

Spotted Reed Frog

Hyperolius puncticulatus (Pfeffer, 1893)
(Plates 11.8, 12.1)

The specific name *puncticulatus* means "spotted."
Golden sedge frog, southern broad-striped sedge-frog, coastal reed frog, *lunkelewa* in Shambala, *koti* in Sukwa in southern Tanzania.

Fig. 125 *Hyperolius puncticulatus.*

DESCRIPTION: The male is up to 33 mm long and the female, 43 mm. The snout is blunt with straight sides, and the head width is about one-third of the body length. The feet are moderately webbed, with one joint on the fourth toe free of web. The color pattern is variable, although a yellow back with a pale stripe from each nostril through the eye and continuing dorsolaterally along the body is common. The dorsolateral band may only continue as far as the arm, or may continue as a series of spots. A second series of light middorsal spots or a continuous band may also be present. Other patterns include the following: (1) Reddish brown back with a broad yellow stripe on the side of the snout and body, a black line surrounding the light stripe, and an orange undersurface; and (2) Black back with light yellow spots, yellow undersurface, and orange throat. A visit to any pond where this species is breeding will show a remarkable array of color patterns.

HABITAT AND DISTRIBUTION (FIG. 125): This forest species has been collected from southern Malawi through northern Mozambique, eastern Tanzania, to coastal Kenya. It has been found on the coastal plain, including the offshore islands, and on the Eastern Arc Mountains to Mt. Rungwe.

ADVERTISEMENT CALL: The male calls from above ground in vegetation around the edge of forest pools. The call is a series of brief clacks. Each clack is short, 0.02 s, with a dominant frequency of 2.0 kHz. The clacks are spaced half a second apart when the males are excited. The call is very loud.

BREEDING: Breeding takes place in midsummer. Clutch size is unknown, although in captivity a female deposited 19 eggs, each 2.5 mm in diameter within 5-mm capsules. The eggs are white and laid in a 25–50-mm cluster on submerged vegetation. Predators include the white-lipped snake *Crotaphopeltis hotamboeia*. The eggs of this species are yolk-rich, with nearly twice as much yolk as the eggs of other similar frogs. This results in a different pattern of egg development.

TADPOLES: Unknown.

NOTES: Known food items include beetles, earwigs, grasshoppers, and bugs.

KEY REFERENCES: Chipman et al. 1999, Schiøtz 1999, Channing 2001.

Water Lily Reed Frog

Hyperolius pusillus (Cope, 1862)
(Plate 12.2)

The specific name *pusillus* means "very small."
Translucent reed frog, water lily frog, transparent pigmy sedge-frog, dwarf reed frog.

DESCRIPTION: The male reaches a length of 21 mm and the female, 23 mm. Both head and body are broad and flat. The snout is blunt and curved, and the head width is slightly more than one-third of the body length. Webbing is reduced, with one to slightly more joints of the fourth toe free. The frog is translucent green, often with a darker line on the side of the snout. Dark spots may be present on the back and may vary

Fig. 126 *Hyperolius pusillus.*

in size, although they are absent in many populations. The male throat is white. The feet and discs are sometimes yellow; otherwise they are green. Many specimens are translucent, and the internal organs can be seen through the belly skin. The eyes are golden, in contrast to the body, with black horizontal pupils.

HABITAT AND DISTRIBUTION (FIG. 126): This frog is often found on floating vegetation such as water lily leaves. It occurs in open grassy pans, from southern Somalia to Kenya, Uganda, Tanzania, Malawi, Zimbabwe, and northern South Africa.

ADVERTISEMENT CALL: The male calls from floating vegetation. The call is a series of soft flat clicks, 0.02 s long, with a dominant frequency between 4.5 and 5.4 kHz.

BREEDING: Breeding takes place during summer. Mating occurs on the water surface, and batches of 20–120 eggs are laid in a single layer between overlapping lily leaves. Clutch size varies from 216 to 313. The eggs are white or light green and small, 0.8 mm in diameter in 2-mm capsules.

TADPOLES: See the chapter on tadpoles.

NOTES: Water lily reed frogs are known to eat ants and flying termites. Predators include the intermediate egret *Mesophoyx intermedia*. The tissue fluid of this species is green.

KEY REFERENCES: Power 1935, Schiøtz 1999, Channing 2001.

Five-Striped Reed Frog

Hyperolius quinquevittatus Bocage, 1866
(Plate 12.3)

The specific name means "five stripes" and refers to the common color pattern.
Black-striped sedge frog, tropical reed frog.

Fig. 127 *Hyperolius quinquevittatus.*

DESCRIPTION: The male reaches 28 mm in length, with females up to 35 mm long. The side of the snout is straight, and head width is 29%–33% of body length. The legs are short, with the tibia length being less than half the length of the body. The adults have two color patterns: (1) Five light stripes, with the vertebral and side stripes green, and the dorsolateral stripes golden brown, separated by darker brown areas; white belly; and red or pink limbs and toes. (2) Green back and sides with two dark brown dorsolateral stripes, white belly, and red limbs and toes.

HABITAT AND DISTRIBUTION (FIG. 127): This species is found in open savanna at high altitudes.
It is known from a broad belt through Angola, southern Democratic Republic of Congo, southern Tanzania, northern Zambia, and Malawi.

ADVERTISEMENT CALL: The male calls from low vegetation, frequently from flooded grass. The call is a short chirp, with a duration of 0.1–0.3 s. Each chirp consists of about 8–10 pulses, at a frequency of 3.9–4.0 kHz. The pulses are initially slow, but speed up at the end of the call. The call resembles that of the sharp-nosed reed frog *Hyperolius acuticeps*.

BREEDING: Unknown.

TADPOLES: Unknown.

KEY REFERENCES: Schiøtz 1999, Channing 2001.

Sheldrick's Reed Frog

Hyperolius sheldricki Duff-MacKay & Schiøtz, 1971

This species is named for David Sheldrick, the warden of Tsavo National Park (East).

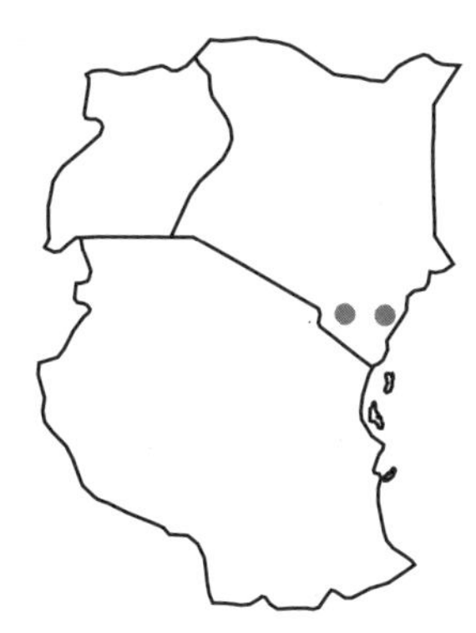

Fig. 128 *Hyperolius sheldricki.*

DESCRIPTION: This small species is similar in overall appearance to *Hyperolius viridiflavus*. Males reach a length of 24 mm, with females up to 35 mm long. The gular sac and protective flap are large. The upper eyelid has a characteristic conical protuberance. The feet are extensively webbed. The dorsal pattern consists of dark brown stripes on a light brown background, with scattered dark spots. These include a middorsal stripe, a stripe running posteriorly from the eye, and short dorsolateral stripes. The insides of the limbs and underside of the hands and feet are dark maroon.

HABITAT AND DISTRIBUTION (FIG. 128): This species is known from temporary pools in the extremely arid southeastern area of Kenya.

ADVERTISEMENT CALL: Males call from grass in temporary pools. The call is a brief whistle.

BREEDING: Unknown.

TADPOLES: Unknown.

KEY REFERENCE: Duff-MacKay & Schiøtz 1971.

Spiny-Throated Reed Frog

Hyperolius spinigularis Stevens, 1971
(Plates 12.4, 12.5, 12.6)

The species gets its name from the black spines on the male throat gland. The specific name *spinigularis* means "spiny throat."
Spiny reed frog, Mulanje reed frog.

Fig. 129 *Hyperolius spinigularis.*

DESCRIPTION: The male reaches a length of 24 mm and the female, 28 mm. The snout is blunt with straight sides. The eyes are noticeably protruding. Webbing is well developed, with the main part level with the middle tubercle beneath the fourth toe. There are spines on the lower surface and the underside of the leg in the male. Both sexes have small spines on the back. The back is green or brown, with snout stripes and dorsolateral stripes edged with brown. The throat is greenish blue. An alternative pattern includes a pale triangle on the snout with a broken dorsolateral band.

HABITAT AND DISTRIBUTION (FIG. 129): The frogs are found in vegetation above ground level. They prefer overgrown marshes, where they select broad-leaved perches. This species is known from Mulanje in Malawi and the Udzungwa, Uluguru, and Usambara mountains in Tanzania, as well as at a low altitude in coastal forest.

ADVERTISEMENT CALL: In captivity the males make a weak, rasping, high-pitched "tcheek-tcheek" call, believed to serve a territorial function. The advertisement call is unknown; there may not be one. Spiny throated reed frogs breed in ponds together with many other calling species.

BREEDING: This species breeds from the end of December to the middle of March. A single mass of 150–200 unpigmented eggs is laid among grass stems or in a single layer on leaves above water. The eggs are 1.8–2.0 mm in diameter within 3.5-mm capsules. Egg masses have been found up to 5 m above water level. The female has been reported to return on two successive nights to wet the eggs. After a few days the tadpoles drop into the water below.

TADPOLES: See the chapter on tadpoles.

NOTES: The male differs from all other reed frogs by the prominent black spines that cover the throat disc, chest, abdomen, and undersurfaces of the hind limbs. Fornasini's spiny reed frog *Afrixalus fornasini* preys on the egg masses, in some cases reducing the number of eggs that develop into tadpoles by more than 50%.

KEY REFERENCES: Schiøtz 1999, Vonesh 2000, Channing 2001.

Tanners' Reed Frog

Hyperolius tannerorum Schiøtz, 1982

This species was named for Mr. and Mrs. John Tanner, who donated Mazumbai Natural Forest Reserve to the University of Dar es Salaam.

DESCRIPTION: Females reach 34 mm in length, while the males are much smaller, at 23 mm. The back is green with a broad pale stripe that starts at the nostrils and runs through the eyes to the groin. The male gular flap is large and smooth.

HABITAT AND DISTRIBUTION (FIG. 130): Frogs were collected in a small swamp in undisturbed montane forest. At the time of writing, this

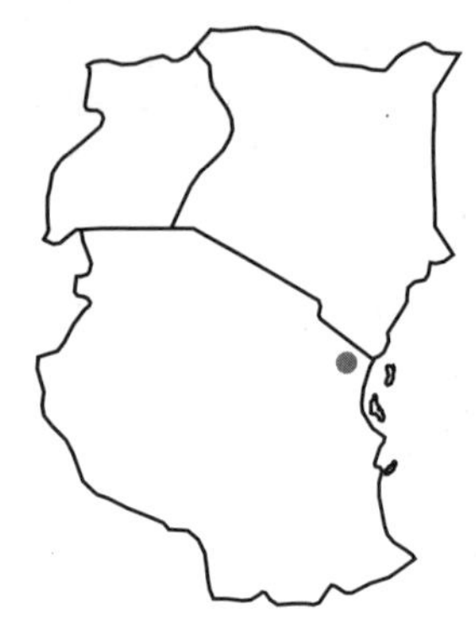

Fig. 130 *Hyperolius tannerorum.*

species is only known from Mazumbai forest, at an altitude of 1410 m, in the western Usambaras, Tanzania.

ADVERTISEMENT CALL: This species appears to have no advertisement call, although males made a soft clicking call in captivity.

BREEDING: The unpigmented eggs are laid above water in vegetation.

TADPOLES: Unknown.

KEY REFERENCES: Schiøtz 1982,1999.

Tinker Reed Frog

Hyperolius tuberilinguis Smith, 1849
(Plate 12.7)

The specific name *tuberilinguis* means "bump on tongue," referring to a papilla on the tongue.
Yellow-green reed frog, green reed frog, straw-or-green sedge-frog, Smith's reed frog.

DESCRIPTION: This is a large species. The males reach a length of 34 mm and the females, 38 mm. The snout is pointed with straight sides. Head width is more than a third of body length. The adult is often a uniform yellow or green, but sometimes brown with a backward-pointing light triangle between the eyes. The males have a bright yellow throat. Females are white below but tinged with yellow. The hidden parts of the limbs are red.

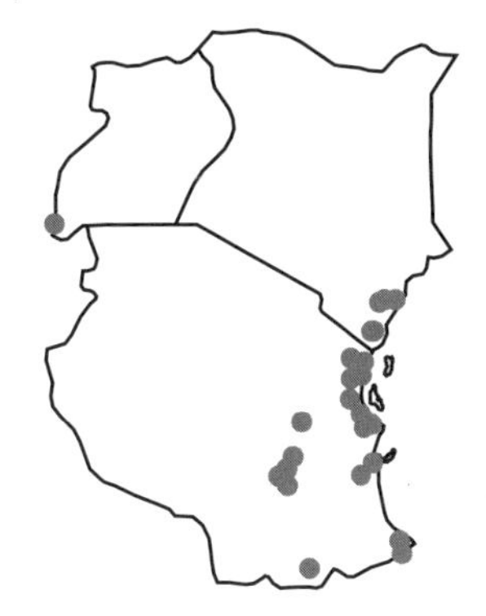

Fig. 131 *Hyperolius tuberilinguis.*

HABITAT AND DISTRIBUTION (FIG. 131): This distinctive species is found in lowland savanna and forest, often in temporary pools with dense vegetation. It is known from Uganda, Kenya, and Tanzania to the eastern coast of South Africa.

ADVERTISEMENT CALL: The male calls from above ground level in vegetation. The call is a sharp click or tap, at an emphasized frequency of 3 kHz. The male uses calls as spacing cues; vocalizations and combat also serve to maintain the distance between males. There are up to 6 clicks in each call, with the male producing more in the presence of a chorus of other males, although the female prefers calls with 2 or 3 clicks. The male also produces a creaking aggression call.

BREEDING: The eggs are laid in a mass a short distance above the surface of the water, attached to vegetation. The egg mass is very sticky. Each egg is white or yellow, 1.5 mm in diameter within a 4-mm capsule. Clutch size is 236–400.

TADPOLES: See the chapter on tadpoles.

NOTES: Predators include the marbled tree snake *Dipsadoboa aulica* and the edible bullfrog *Pyxicephalus edulis.* Tinker reed frogs have been found trapped in a spider web.

KEY REFERENCES: Loveridge 1944, Schiøtz 1999, Channing 2001.

Common Reed Frog

Hyperolius viridiflavus (Duméril & Bibron, 1841)
(Plate 12.8)

The name *viridiflavus* refers to the yellow spots common in some patterns.

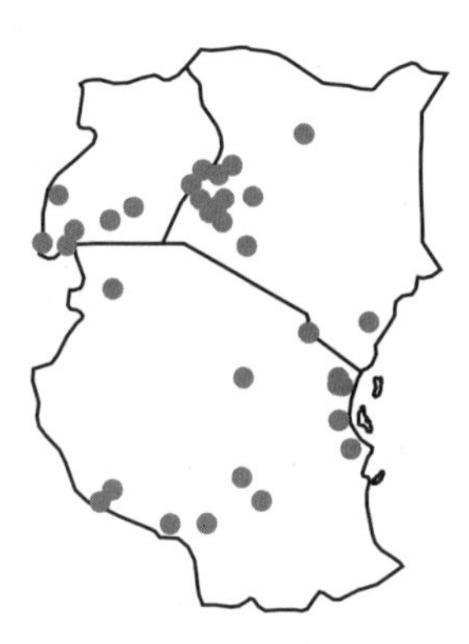

Fig. 132 *Hyperolius viridiflavus*.

DESCRIPTION: Both males and females reach 25 mm in length. The tympanum is hidden. Toes are webbed to the base of the discs. The pattern is variable, a green or brown back with small red or yellow spots. The gular disc is bright yellow behind. The limbs and undersides are red.

HABITAT AND DISTRIBUTION (FIG. 132): The common reed frog is found in savanna. It is known from Ethiopia through Uganda, eastern Democratic Republic of Congo, Tanzania and Kenya.

ADVERTISEMENT CALL: Males call from vegetation in or near water. The call is a brief click at a dominant frequency of 3.9–4.1 kHz.

BREEDING: Unknown.

TADPOLES: Unknown.

NOTES: This frog was previously known as *H. v. viridiflavus* and *H. v. bayoni*.

KEY REFERENCES: Schiøtz 1999, Wieczorek et al. 2000.

Green Reed Frog

Hyperolius viridis Schiøtz, 1975
(Plate 13.1)

The name *viridis* means green.

Fig. 133 *Hyperolius viridis.*

DESCRIPTION: This is a small slender species, with the females reaching 23 mm in length and the males being smaller. The back is green, sometimes with small black spots. A pale dorsolateral line runs from snout tip to groin. Breeding males have a yellow throat.

HABITAT AND DISTRIBUTION (FIG. 133): The green reed frog is found in grasslands at high altitudes. This species is known from the highlands linking the eastern and western rift valleys in northern Zambia and southern Tanzania.

ADVERTISEMENT CALL: The males call from vegetation. Each call consists of a short buzz at 3.7 kHz, followed by a series of 4 or 5 clicks evenly spaced. The call is made up of a very brief, rapidly pulsed initial phase, followed by a series of slower pulses. The two phases are clearly separated and resemble the calls of many of the smaller species in the genus *Afrixalus*. Some males in the populations produce a longer call, similar to the second phase of the other males in the chorus. We regard this as the male aggression call.

BREEDING: Unknown.

TADPOLES: Unknown.

NOTES: In southwestern Tanzania *Hyperolius viridis* and *H. acuticeps* are morphologically identical, and may breed in the same pond.

KEY REFERENCES: Schiøtz 1975, 1999; Channing et al. 2002.

Kassinas—Genus *Kassina*

This genus consists of mostly ground-living frogs with a vertical pupil and long legs. Some species have discs. The tadpoles develop in deep quiet pools and have high fins and characteristic accessory plates at the angle of the jaw, between the keratinized jaw sheaths. The genus is found throughout sub-Saharan Africa, with three species in East Africa.

KEY TO THE SPECIES

1a. Toes with distinct enlarged discs	2
1b. Toes with small swellings at tips	*Kassina senegalensis*
2a. Hidden parts of limbs with red patches	*Kassina maculata*
2b. Hidden parts of limbs not red	*Kassina maculifer*

Red-Legged Kassina

Kassina maculata (Duméril, 1853)
(Plate 13.2)

The specific name *maculata* means "spotted," referring to the color pattern of the back.
Red-legged pan frog, red-blotched black frog, spotted running frog, *nanhango* in Makonde.

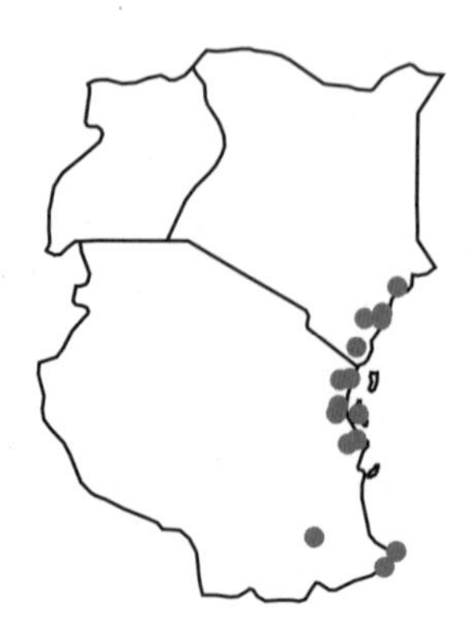

Fig. 134 *Kassina maculata.*

DESCRIPTION: The male reaches a length of 72 mm and the female, 71 mm. The head and body are stocky, and the eyes are large and protruding. The tympanum diameter is half of the eye diameter. Discs are present on the fingers and toes. The male throat flap is circular, and the female vent is directed downward and surrounded by simple lobes covered with small spines. The back is silvery gray with large black spots. A thin pale line surrounds each spot. All concealed parts of the limbs are red with black spots. The groin and the posterior face of the thigh are deep red. The underside is white with small gray specks.

HABITAT AND DISTRIBUTION (FIG. 134): The red-legged Kassina is associated with savanna, found in deep temporary pools and swamps. It occurs from Kenya along the coastal lowlands south to the tropical east coast of South Africa, including the offshore islands like Zanzibar.

ADVERTISEMENT CALL: The males call from vegetation at the surface of deep water, and from permanent water for much of the year. The call is a loud "kwack" and resembles the noise of bursting bubbles. It is a very short, rising note. Call length is less than 0.03 s, with the dominant frequency rising rapidly from 0.3 to 1.3 kHz.

BREEDING: The eggs are attached to submerged plants in small groups of up to 100. Each egg is 1.5 mm in diameter within a 2.5-mm capsule.

TADPOLES: See the chapter on tadpoles.

NOTES: The skin contains at least three kinds of defensive chemicals. The first is a peptide that stimulates the large intestine of some animals. The second is a group of peptides called *tachykinins*. At least two different tachykinins are present, and both are responsible for lowering the blood pressure and increasing the heart rate.
The third kind of defensive chemical is a peptide that has a potent action on the gallbladder and stimulates the pancreatic juice, which is rich in enzymes but poor in bicarbonate. The effect of these chemicals is to make any mammal that eats the frog, or even tastes the skin, violently ill, hopefully protecting the frog against any future attacks. These defenses are very effective, as only a few millionths of a gram of secretion per kilogram of predator will cause these effects.

Known food items include the common reed frog *Hyperolius viridiflavus* and the golden-backed spiny reed frog *Afrixalus aureus*, as well as grasshoppers. Despite the skin toxins, red-legged kassinas are eaten by the intermediate egret *Mesophoyx intermedia* and the savanna vine snake *Thelotornis capensis*. They are able to climb well and will seek shelter in leaf axils.

KEY REFERENCES: Nakajima 1981, Roseghini et al. 1988, Schiøtz 1999, Channing 2001.

Spotted Kassina

Kassina maculifer (Ahl, 1924)
(Plate 13.3)

The name *maculifer* means "spotted."

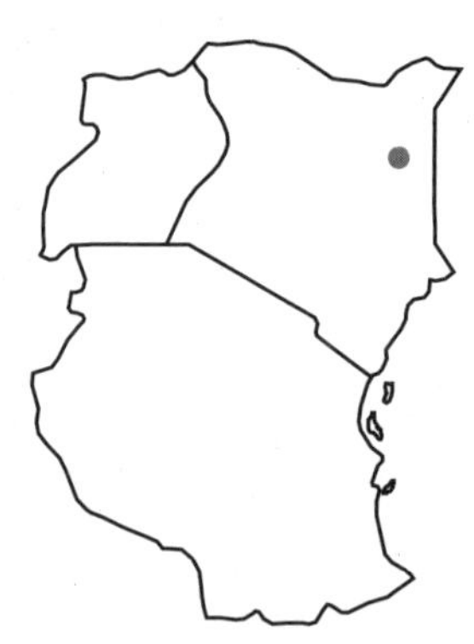

Fig. 135 *Kassina maculifer.*

DESCRIPTION: This is a large frog, with females growing up to 44 mm long and the males 39 mm. It has large digital discs. The back is gray with dark reddish brown spots. In the shoulder region a typical, somewhat X-shaped blotch is found.

HABITAT AND DISTRIBUTION (FIG. 135): This species is known from dry savanna in northeastern Africa, from Somalia, eastern Ethiopia, to northeastern Kenya.

ADVERTISEMENT CALL: The call is a very brief rising note. Duration is 0.05 s, with the emphasized frequency rising from 0.7 to 2.0 kHz.

BREEDING: Unknown.

TADPOLES: Unknown.

KEY REFERENCES: Tandy & Drewes 1985, Schiøtz 1999.

Senegal Kassina

Kassina senegalensis (Duméril & Bibron, 1841)
(Plate 13.4)

The specific name refers to Senegal, where the species was first discovered. Bubbling kassina, Senegal frog, running frog, Senegal running frog, *dorya* in Nyakusa, *kabunda* in Lega, *nanhengo* in Makonde, *Senegal-Streifenfrosch* or *Senegal-Rennfrosch* in German.

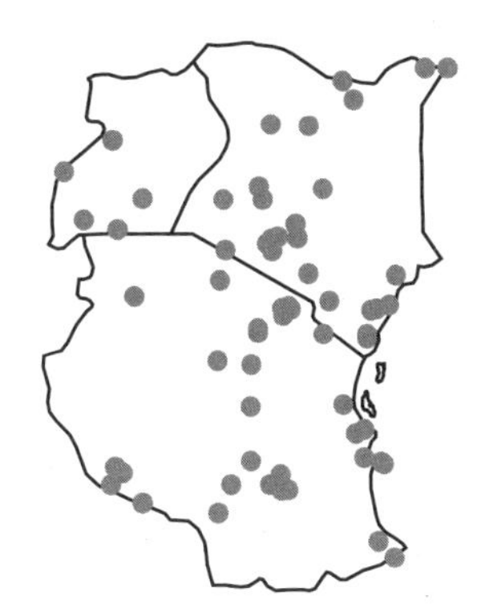

Fig. 136 *Kassina senegalensis.*

DESCRIPTION: Both sexes are about the same size, up to 49 mm long. Small swellings are present at the tips of the fingers and toes. The toes are very slightly webbed. Inner metatarsal tubercle length is less than the distance between the nostrils. The male throat gland is elongated without a free posterior edge, paler than the vocal sacs. The vent is directed downward and in females is surrounded by two pairs of lobes without spines. The back pattern is a dark vertebral band with a pair of paravertebral bands. This is usually distinct on a yellowish brown to gray background.

HABITAT AND DISTRIBUTION (FIG. 136): This common frog occurs in moist and arid savanna throughout sub-Saharan Africa, from sea level to an altitude of 2500 m.

ADVERTISEMENT CALL: The male calls under vegetation away from water, but sometimes in shallow water providing that enough cover is available. The call consists of a rising note that sounds like a bursting bubble. Males form calling aggregations, with 3–10 males in a group. They call in single bouts, with a chorus leader setting the pace and one or more frogs answering. Different males may serve as chorus leaders. Individuals in a chorus may call very rapidly in succession, producing an effect as if the calls were moving across the breeding site. Each call is short, 0.1–0.2 s. The dominant frequency increases from 0.4 to 1.5 or 2.0 kHz. This species possesses a territorial call, which is trilled with a duration of 0.2 s and dominant energy of 0.4–1.6 kHz. This call is often heard early in the evening when the males are closely spaced.

BREEDING: Breeding takes place during April and May in East Africa. The female is attracted to the male at the calling site away from the water. The male clasps the female, and then the female leads the way to shallow water. Then the female dives to grasp an underwater object and touches her vent to it while depositing an egg. The eggs sink and are soon difficult to see. The eggs are pale blue-green, 1.4–1.8 mm in diameter, in jelly capsules 3 mm in diameter. They are laid singly or in groups of 1–20, every 30 cm. Clutch size is 260–400. Development to metamorphosis takes 52–90 days.

TADPOLES: See the chapter on tadpoles.

NOTES: Known food items include flies, ants, caterpillars, and other small arthropods. During the day Senegal kassinas hide in burrows dug by other animals, such as a male kingfisher's resting burrow in a sandy bank, and in termite mounds. Maximum recorded longevity in captivity is 5 years 5 months. The animals can move over 500 m between dry-season refugia and seasonal breeding pools.

KEY REFERENCES: Loveridge 1936, Schiøtz 1999, Channing 2001, Razzetti & Msuya 2002.

Tree Frogs—Genus *Leptopelis*

The free frogs are medium to large in size, with broad heads, large discs usually, and extensive margins of webbing along the digits. Ground-dwelling species may have smaller discs and long fingers, or the discs may be absent. Most of these frogs live above ground in vegetation, but some live in burrows, and eggs are laid in shallow nests near water. The tadpoles have elongated tails with which they wriggle to the water. There are 14 species in East Africa.

KEY TO THE SPECIES—MALES

Females of different species may be very similar and are difficult to identify. The key below will only be reliable for males.

1a. Iris with red or pink — 2
1b. Iris golden or brown — 3

2a. Blue throat, pink eyes, small golden spots on back — *Leptopelis barbouri*
2b. White throat, pale transverse marks on back; red-eyed females — *Leptopelis parkeri*

3a. White spots on heels (Fig. 137) — 4
3b. No white spots on heels — 5

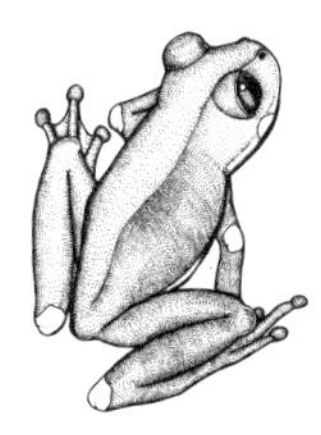

Fig. 137 White spots on heels.

4a. Dark canthal line and interorbital bar, one and a half phalanges free of webbing on the fourth toe — *Leptopelis vermiculatus*
4b. No dark canthal and interorbital bar, two phalanges free of webbing on the fourth toe — *Leptopelis flavomaculatus*

5a. Reduced webbing, three and a half phalanges or more free of webbing on fourth toe — 6
5b. Two and a half phalanges or less free of webbing on fourth toe — 8

6a. Pectoral glands present (Fig. 138), n-shaped dorsal pattern — 14
6b. No pectoral glands, no n-shaped dorsal markings — 7

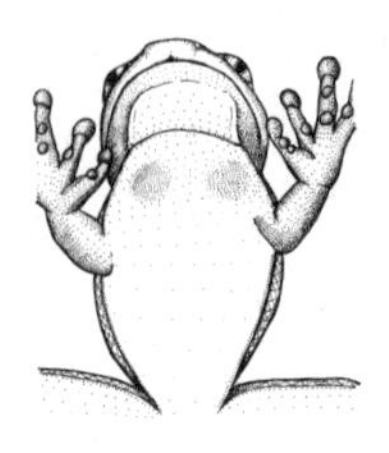

Fig. 138 Pectoral glands.

7a. Three longitudinal dorsal stripes, plus dark lateral stripes *Leptopelis oryi*
7b. Reversed Y-dorsal marking, dark interorbital bar *Leptopelis argenteus*

8a. Heels black with thin white border (Fig. 139) *Leptopelis modestus*
8b. No black heels 9

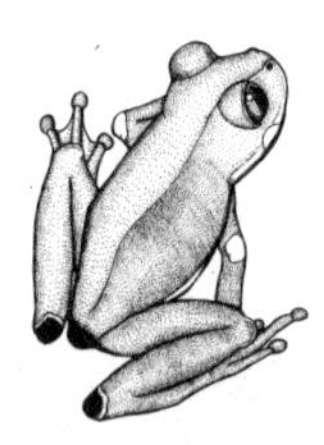

Fig. 139 Black heels with white border.

9a. Throat blue 10
9b. Throat white or gray 11

10a. Tympanum half the eye size, irregular dark markings on back *Leptopelis karissimbensis*
10b. Tympanum less than half the eye size, uniform brown or green or white blobs on back *Leptopelis uluguruensis*

11a. Two dorsolateral stripes, discs very small *Leptopelis argenteus*
11b. No dorsolateral stripes, discs moderate size 12

12a. Tympanum half the eye size, triangular marking on back *Leptopelis christyi*
12b. Tympanum more than half the eye size, no triangular marking on back 13

13a. Throat gray, white spot below the eye *Leptopelis fiziensis*
13b. Throat white, no white spot below the eye *Leptopelis kivuensis*

14a. Interorbital distance 36% or more of the distance from the nostril to the tympanum *Leptopelis parbocagii*
14b. Interorbital distance less than 36% of the distance from the nostril to the tympanum *Leptopelis bocagii*

Silvery Tree Frog

Leptopelis argenteus (Pfeffer, 1893)
(Plate 13.5)

The specific name *argenteus* means "silvery."
Broadley's tree frog, Bagamoyo forest treefrog, triad treefrog.

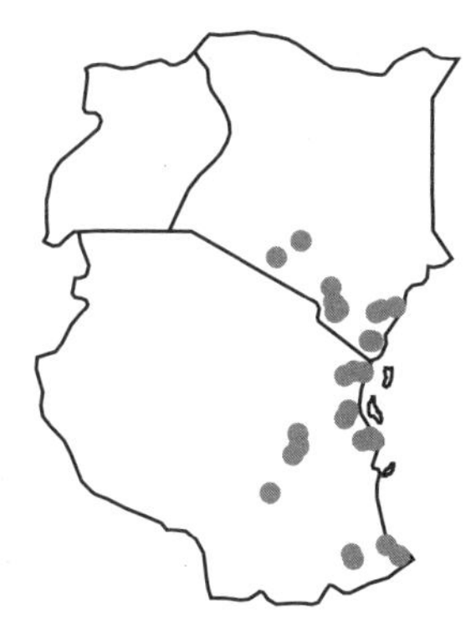

Fig. 140 *Leptopelis argenteus.*

DESCRIPTION: This is a medium-sized frog, with the male up to 45 mm long and the female reaching 52 mm. The width of the disc of the fourth toe is greater than the width of the outer tubercle beneath the fourth toe. The length of the inner metatarsal tubercle is less than the length of the inner toe, which is spadelike. The male does not possess pectoral glands. Vomerine teeth are present in two small groups between the widely spaced choanae. The tympanum is small, with a diameter less than half that of the eye. The toes have slight webbing. The back is light brown, with a darker triangle between the eyes, which points posteriorly. A pair of darker bands may separate over the lower back. A dark line along the snout often continues behind the eye. Male and female patterns are similar. Some populations have a reversed Y-shaped dorsal marking, while others have a plain back.

HABITAT AND DISTRIBUTION (FIG. 140): This species has been found on trees and grass in lightly wooded moist savanna and is common on the coastal lowlands. It is known from Kenya, Tanzania, Malawi, and Mozambique, to the eastern highlands of Zimbabwe.

ADVERTISEMENT CALL: The male calls from dense bushes or high grass or sedge, up to 3 m above ground level. Three different calls are produced: two short croaks and a longer whine. The shorter croak may precede or follow two or three of the longer whines, or either may be uttered alone. Often a croak is followed immediately by two whines, sounding as if two different species are calling. The shorter croaks are less than 0.1 s long, with dominant frequencies between 1.5 and 2.0 kHz, while the other, longer, croak is 0.2 s long, with dominant energy at 1.8 kHz.

BREEDING: Unknown.

TADPOLES: See the chapter on tadpoles.

NOTES: Although this variable species is widespread, nothing is known of its biology. The northern populations have a reversed Y pattern on the back and have been called *Leptopelis concolor*, while the southern form has been called *L. broadleyi*. All the populations have the same advertisement call and represent different color morphs, similar to the situation in many populations of reed frogs. Recorded food items include moths and caterpillars. This frog is eaten by the spotted wood snake *Philothamnus semivariegatus*.

KEY REFERENCES: Loveridge 1944, Howell 1981, Schiøtz 1999, Channing 2001.

Barbour's Tree Frog

Leptopelis barbouri Ahl, 1929
(Plates 13.6, 13.7, 13.8)

This species is named for Thomas Barbour of the Museum of Comparative Zoology, Harvard.

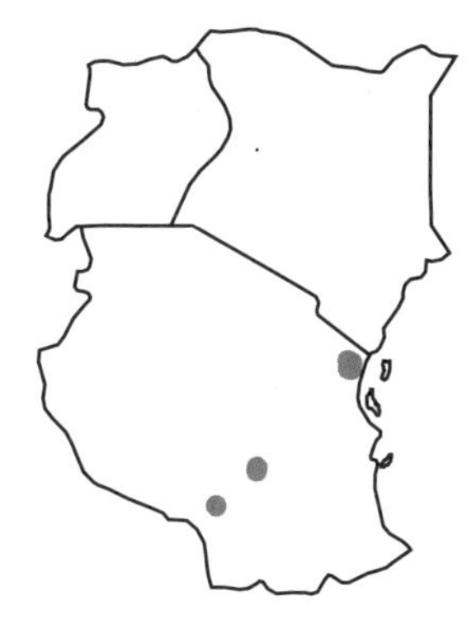

Fig. 141 *Leptopelis barbouri.*

DESCRIPTION: This tree frog is a beautiful translucent green. Males reach 39 mm long and females, 58 mm. The back has golden or silver spots. The throat of the male is blue. The males have pectoral glands. The tympanum is small and easily overlooked. The eye is creamy white with pink or red edging.

HABITAT AND DISTRIBUTION (FIG. 141): This species has been found in montane forest and is known from the Usambara and Udzungwa mountains and Mt. Rungwe, Tanzania.

ADVERTISEMENT CALL: The male calls from elevated positions on trees or other vegetation near water. These call sites may be 3 m or higher. The call is a brief buzz with an emphasized frequency of 1.5 kHz, at a pulse rate of 100/s.

BREEDING: The eggs are laid in an underground nest. In one instance, a pair was found with only the heads above ground, about 10 m from a stream. The clutch consisted of 32 eggs, each 6 mm in diameter.

TADPOLES: The egg develops into a bright yellow, green, and blue tadpole.

NOTES: Recorded food items include woodlice, crickets, grasshoppers, and bugs.

KEY REFERENCES: Barbour & Loveridge 1928a (as *L. aubryi*), Schiøtz 1999, Schiøtz & Westergaard 2000.

Bocage's Tree Frog

Leptopelis bocagii (Gunther, 1844)
(Plate 14.1)

This species is named for the herpetologist J. V. Barbosa du Bocage, director of the National Museum in Lisbon, known as the Museu Bocage. Bocage's burrowing frog, Bocage's frog, horseshoe forest tree frog, *Bocages-Waldsteigerfrosch* in German, *Grenouille grimpeuse de Bocage* in French.

Fig. 142 *Leptopelis bocagii.*

DESCRIPTION: The male is up to 52 mm long and the female, 58 mm. The width of the disc of the fourth toe is 80%–100% the width of the outer tubercle below the toe. The metatarsal tubercle is longer than the inner toe, and is spadelike to assist with digging. The male has pectoral glands. The tympanum is large, with a horizontal diameter at least half that of the eye. The discs are small, not exceeding the width of the digits. The markings are a dark n-shaped area on a lighter brown background, sometimes with a vertebral band. Irregular lines are present on the snout and sides of the body. Various markings occur between the eyes. A population in Kakamega, Kenya, has a green-backed form, with typical markings and call. It can be distinguished from the similar *Leptopelis parbocagii* by the distance between the eyes, which is less than 36% of the distance from the nostril to the tympanum.

HABITAT AND DISTRIBUTION (FIG. 142): This largely ground-dwelling treefrog occurs in open savanna. It is known along the eastern side of Africa, from Ethiopia through Kenya and Tanzania, across to Angola, northern Namibia, Botswana, and northern South Africa.

ADVERTISEMENT CALL: The male calls from ground level, but also from within mammal burrows and even a meter or two above ground in reeds or other vegetation. The call is a slow "kwaak." Often two calls are uttered in succession. The duration of each call varies from 0.2–0.4 s, at a dominant frequency of 0.6 kHz.

BREEDING: The male holds on to the female by a sticky material secreted from glands underneath his body. The eggs are laid during rain in deep holes in the ground.

TADPOLES: Unknown.

NOTES: Known food items include termites, beetles, earwigs, spiders, cockroaches, earthworms, snails, and frogs. Bocage's tree frogs are eaten by the rhombic night adder *Causus rhombeatus* and the boomslang *Dispholidus typus*. This is one of the few African frogs known to form a thin cocoon to prevent desiccation during dry weather.

KEY REFERENCES: Loveridge 1933, Schiøtz 1999, Channing 2001.

Christy's Tree Frog

Leptopelis christyi (Boulenger, 1912)
(Plate 14.2)

This species was named for C. Christy, leader of a collecting expedition to the Belgian Congo between 1912 and 1914.

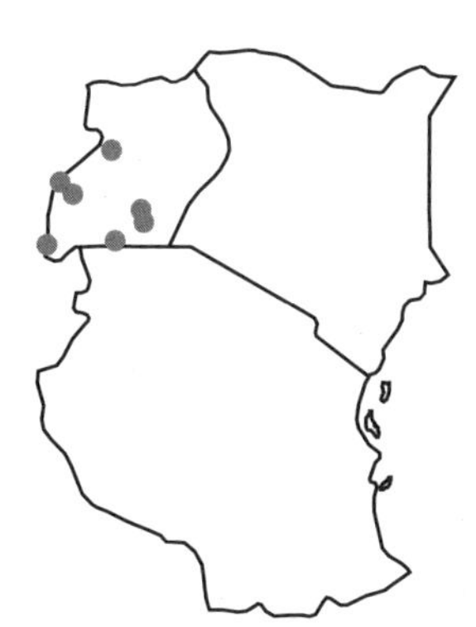

Fig. 143 *Leptopelis christyi*.

DESCRIPTION: Males reach a length of 45 mm and females, 65 mm. The tympanum is large, with a diameter about half that of the eye. Males have pectoral glands. The back usually has a dark triangle with the apex pointing forward. An irregular pale lateral line is present, and sometimes a pale yellow vertebral stripe from the snout to midbody is too. There is no white spot below the eye. The back is usually brown, although there is a green alternative phase. The ventral surface may be stippled with brown.

HABITAT AND DISTRIBUTION (FIG. 143): This species is found on grass, in low bushes, or on the ground in forests at medium altitudes in eastern central Africa. It is known from Uganda and the Democratic Republic of Congo, extending westward to Cameroon.

ADVERTISEMENT CALL: The call is a single brief clack, which resolves into a number of harmonics 0.4 kHz apart. The emphasized frequency is around 2 kHz, and the duration is 0.1 s.

BREEDING: Unknown. This species has a long breeding season—April through June or longer in the Congo Basin.

TADPOLES: Unknown.

NOTES: Recorded food items include grasshoppers.

KEY REFERENCES: Loveridge 1944, Schiøtz 1999.

Fizi Tree Frog

Leptopelis fiziensis Laurent, 1973

This species is named for the village of Fizi, in the Democratic Republic of Congo, near the north end of Lake Tanganyika, where it was first found.

DESCRIPTION: The females of this species reach 38 mm in length and the males, 35 mm. The tympanum is small, with large digital discs. The male throat is gray. The back is reddish brown with a dark pattern, and the eyes are golden. Many individuals may be a pale uniform brown. The heels and vent have fine white stripes.

Fig. 144 *Leptopelis fiziensis.*

HABITAT AND DISTRIBUTION (FIG. 144): This species is found in the forests around Lake Tanganyika.

ADVERTISEMENT CALL: Males call from low branches in dense bush. The call is a deep unmelodic clack that has not yet been recorded or analyzed.

BREEDING: Unknown.

TADPOLES: Unknown.

KEY REFERENCES: Fischer & Hinkel 1992, Schiøtz 1999.

Yellow-Spotted Tree Frog

Leptopelis flavomaculatus (Gunther, 1864)
(Plate 14.3)

The specific name *flavomaculatus* refers to the yellow spots found on the back of many young individuals.
Johnston's treefrog, brown forest treefrog, *nkewe* in Kisambara.

DESCRIPTION: The male reaches a length of 50 mm and the female, 70 mm. The width of the disc of the fourth toe is greater than the width of the outer tubercle beneath the toe. The metatarsal tubercle is about equal in length to the first toe. The male has pectoral glands. The horizontal diameter of the tympanum is more than half the diameter of the eye. The toes are webbed, with the broad web between the fourth and fifth toes reaching or passing the outer tubercle of the fifth toe. The back

Fig. 145 *Leptopelis flavomaculatus.*

is uniform green or brown, with a dark brown triangle in the center, its apex at the back of the head. Juveniles are green with yellow spots, and adults may be green with white heels (in males) or brown. The typical n-shaped markings on the back and dark bar between the eyes are present in brown individuals.

HABITAT AND DISTRIBUTION (FIG. 145): This forest form is also found in small, dry coastal forest remnants. The yellow-spotted tree frog is known from Kenya, southward through Tanzania, including the offshore islands and the Eastern Arc Mountains, to Malawi and eastern Zimbabwe.

ADVERTISEMENT CALL: The male selects dense and high vegetation from which to call, making him difficult to observe. The male sometimes calls from a burrow, or from vegetation up to 4 m above ground. The call is a soft drawn-out cry, 0.3–0.7 s long, consisting of closely spaced harmonics at 0.75, 1.00, 1.25, 1.50, 1.75, and 2.00 kHz. The call shows slight modulation in frequency, with a small rise in the beginning and a drop afterward. It is very difficult to locate.

BREEDING: Unknown.

TADPOLES: Unknown.

NOTES: Recorded food items include grasshoppers and snails.

KEY REFERENCES: Loveridge 1933, 1944 (as *L. johnstoni*); Schiøtz 1999; Channing 2001.

Karissimbi Tree Frog

Leptopelis karissimbensis Ahl, 1929

The specific name refers to Mt. Karissimbi, on the border between Rwanda and the Democratic Republic of Congo.

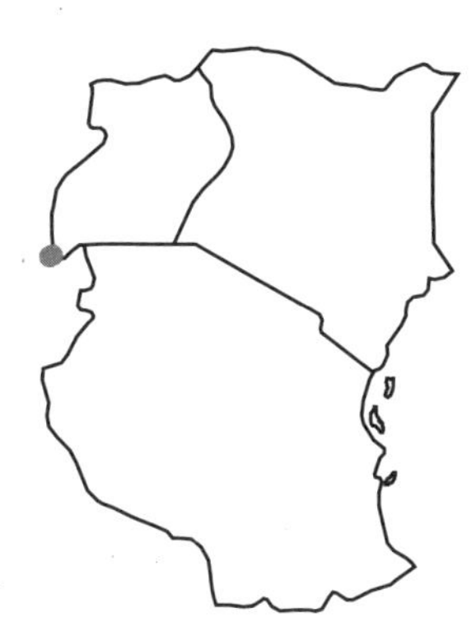

Fig. 146 *Leptopelis karissimbensis.*

DESCRIPTION: This species is small, with the females reaching 36 mm in length and the males 28 mm. The webbing is reduced and the tympanum is small. The back is usually green but may be golden brown with dark brown markings. Breeding males have a blue vocal sac.

HABITAT AND DISTRIBUTION (FIG. 146): This species is known from montane grassland and heather in eastern Democratic Republic of Congo and southwestern Uganda.

ADVERTISEMENT CALL: The call consists of two parts: a buzz followed by a slow clack. Sometimes the buzz is missing.

BREEDING: Unknown.

TADPOLES: Unknown.

NOTES: This species is very similar to the Kivu tree frog, and both are found in the same area. Field observations are required to elucidate the biology of these species.

KEY REFERENCE: Schiøtz 1999.

Kivu Tree Frog

Leptopelis kivuensis Ahl, 1929
(Plate 14.4)

The species name refers to the province of Kivu, in the Democratic Republic of Congo.

Fig. 147 *Leptopelis kivuensis.*

DESCRIPTION: This species is small, with the females reaching 36 mm in length, and the males 26 mm. The webbing is reduced and the tympanum is small. The back is pale brown with an irregular dark greenish-brown pattern. The eyes are red-orange, and there is no white spot below the eye. The males have a white gular sac.

HABITAT AND DISTRIBUTION (FIG. 147): This species is found in montane forests from eastern Democratic Republic of Congo and southwestern Uganda.

ADVERTISEMENT CALL: The males call from high vegetation. The call is a single or double clack, with emphasized frequencies around 1.3 kHz.

BREEDING: Unknown. Juveniles are bright green, and were common on the tops of sedges in swamps in the Bwindi Impenetrable Forest in Uganda.

TADPOLES: Unknown.

NOTES: This species is very similar to the Karissimbi tree frog, and field observations are required to understand their biology.

KEY REFERENCES: Drewes & Vindum 1994, Schiøtz 1999.

Plain Tree Frog

Leptopelis modestus (Werner, 1898)

The name *modestus* means "unassuming," referring to the dull coloration.

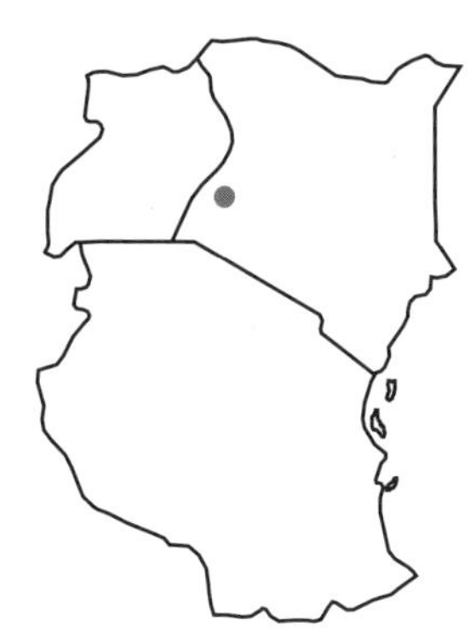

Fig. 148 *Leptopelis modestus.*

DESCRIPTION: This species is small, with males up to 35 mm in length and females 41 mm long. The digital discs are large. The back is gray-brown with a faint hourglass pattern. There is some variation in color. The East African frogs have a dark heel and vent. The throat of calling males is blue or green.

HABITAT AND DISTRIBUTION (FIG. 148): This species is known from forest at medium altitudes in eastern Democratic Republic of Congo and western Kenya, with presumed conspecific populations at high altitudes in Cameroon and Nigeria.

ADVERTISEMENT CALL: Males call from vegetation. The call is a low-pitched, slow clack, with an emphasized frequency of 2 kHz. They also produce a buzzing that may be a male spacing call.

BREEDING: Unknown.

TADPOLES: Unknown.

NOTES: It has been suggested that there are a number of species confused under this name.

KEY REFERENCE: Schiøtz 1999.

Ory's Tree Frog

Leptopelis oryi Inger, 1968

This frog was named for Albert Ory, conservator of the Garamba National Park in the Democratic Republic of Congo.

Fig. 149 *Leptopelis oryi.*

DESCRIPTION: This species is of moderate size, with the males up to 43 mm long and the females 58 mm long. The webbing is reduced. The back is pale brown with three narrow dorsal stripes and a broad dark lateral band. The males do not have pectoral glands.

HABITAT AND DISTRIBUTION (FIG. 149): This frog has been found in savanna with high grass and scattered bushes. It is only known from northeastern Democratic Republic of Congo and near Budongo forest in Uganda.

ADVERTISEMENT CALL: Males call from high in bushes and long grass. The call is a clack with a duration of 0.15 s, and an emphasized frequency of 1.5 kHz.

BREEDING: Unknown.

TADPOLES: See the chapter on tadpoles.

KEY REFERENCES: Inger 1968, Schiøtz 1999.

Cryptic Tree Frog

Leptopelis parbocagii Poynton & Broadley, 1987
(Plate 14.5)

The specific name *parbocagii* means "near *bocagii*," as this species is very similar to *L. bocagii*.
Lake Upemba forest tree frog.

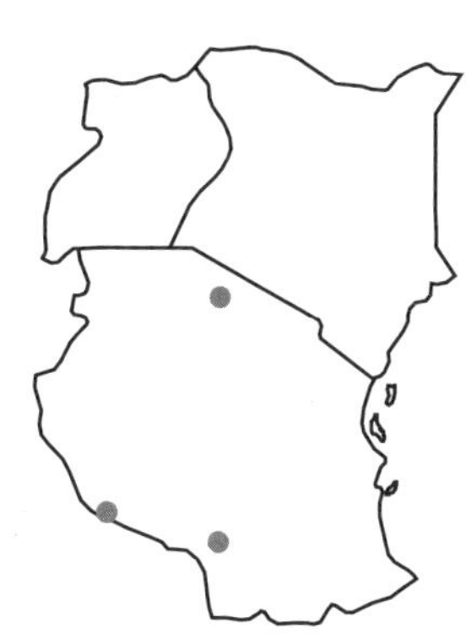

Fig. 150 *Leptopelis parbocagii.*

DESCRIPTION: These are medium to large tree frogs, with the male up to 54 mm long and the female 64 mm long. The discs are small. The width of the disc of the fourth toe is 80%–108% of the width of the distal tubercle of the toe. The metatarsal tubercle is spadelike for digging and is longer than the first toe. The male has pectoral glands. The tympanum is large. The horizontal diameter of the tympanum is half or more of the diameter of the eye. This green tree frog has a black line along the side of the white-tipped snout when young. The dorsal pattern includes a dark n- or m-shaped marking, or a darker dorsal patch extending onto the head. It can be separated from the similar *Leptopelis bocagii* by the distance between the eyes, which is 36% or more of the nostril-tympanum distance.

HABITAT AND DISTRIBUTION (FIG. 150): Cryptic tree frogs are known from savanna in Malawi and Mozambique, northern Zambia, Tanzania, and the Democratic Republic of Congo.

ADVERTISEMENT CALL: Males call from ground level or on low vegetation. The call is a loud clack, which has a duration of 0.1 s, at a dominant frequency of 1.4 kHz.

BREEDING: Unknown.

TADPOLES: Unknown.

NOTES: Bocage's tree frog and the cryptic tree frog are easily confused.

KEY REFERENCES: Stewart 1967 (as *L. bocagii*), Poynton & Broadley 1987, Schiøtz 1999, Channing 2001.

Parker's Tree Frog

Leptopelis parkeri Barbour & Loveridge, 1928
(Plates 14.6, 14.7)

This frog was named for H. W. Parker, the herpetologist of the Natural History Museum in London, 1923–1957.

Fig. 151 *Leptopelis parkeri.*

DESCRIPTION: This species is large, with males up to 46 mm and females up to 70 mm long. It is slender with extensive webbing and a small tympanum. The internarial distance is less than the nostril-eye distance, and the inner metatarsal tubercle is small. The skin is smooth above and granular beneath. The eyes are brick red to bright red. Males are gray to brown or dark olive green. Pale yellow markings form distinct transverse bands. Females are a uniform olive green. Male throats are white; females have orange throats. The sides of the body and limbs and the toes are yellow-orange. The underside is white.

HABITAT AND DISTRIBUTION (FIG. 151): Parker's tree frog is known from dense forest at high altitudes in the Eastern Arc Mountains, Tanzania.

ADVERTISEMENT CALL: The males call from high in trees. The call is a quiet buzzing at 1.5 kHz.

BREEDING: A pair was observed in amplexus in a quiet backwater of a small stream. Eggs as large as 4 mm have been recorded.

TADPOLES: Unknown.

NOTES: Reported food items include beetles, spiders, grasshoppers, cockroaches, and earwigs.

KEY REFERENCES: Barbour & Loveridge 1928a, Schiøtz 1999.

Uluguru Tree Frog

Leptopelis uluguruensis Barbour & Loveridge, 1928
(Plates 14.8, 15.1)

This species is named for the Uluguru Mountains in Tanzania.

Fig. 152 *Leptopelis uluguruensis.*

DESCRIPTION: This species is moderately sized, with males 42 mm long and females up to 59 mm long. The tympanum is small and easily overlooked. Webbing is reduced. The skin is very granular below. The back is brown or green; green-backed frogs sometimes have dorsal pale spots or rings resembling fungal growth on a log, while brown-backed individuals sometimes have a darker brown marking. Undersides are translucent white. The throat of breeding males is sky blue. Pectoral glands are absent.

HABITAT AND DISTRIBUTION (FIG. 152): This forest species is known from the Eastern Arc Mountains of Tanzania.

ADVERTISEMENT CALL: Males produce a short clack of 3 or more pulses, with an emphasized frequency of 1.3 kHz and a duration of 0.04 s.

BREEDING: Unknown. Males call from about a meter above ground, near water.

TADPOLES: Unknown.

NOTES: Recorded food items include spiders, earwigs, and grasshoppers. These tree frogs are eaten by Tornier's cat snake *Crotaphopeltis tornieri*. During the dry season individuals sit on large leaves at the edges of streams in forest, comatose, with their limbs tucked under the body and their eyes closed.

KEY REFERENCES: Barbour & Loveridge 1928a, Schiøtz 1999.

Vermiculated Tree Frog

Leptopelis vermiculatus (Boulenger, 1909)
(Plates 15.2, 15.3, 15.4)

The species name refers to the fine spots common on the backs of juveniles.

DESCRIPTION: This tree frog is one of the largest in Africa, with males up to 50 mm long and females 85 mm long. Webbing is well developed, and the tympanum is large. The males have pectoral glands. Green- and brown-backed patterns are known. Green-backed individuals have fine

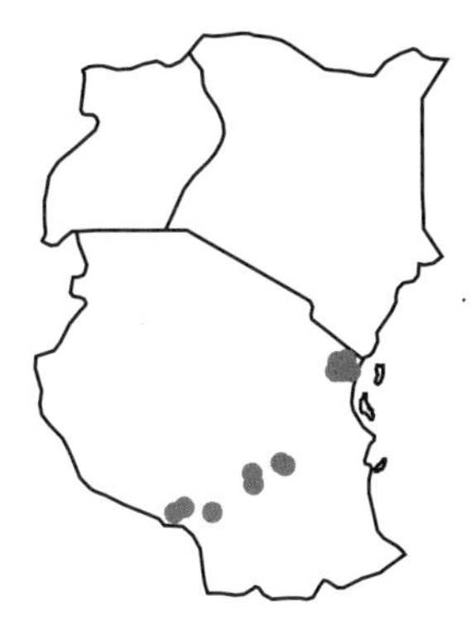

Fig. 153 *Leptopelis vermiculatus.*

dark vermiculations, with the side of the body marbled in white and black. Brown-backed animals have an irregular triangle in darker brown on the back, with the apex pointing forward. The markings are sometimes a rich olive green bordered by black. A dark blotch is found between the eyes, and the body sides are marbled. Limbs are green with darker bars. There are cream-colored spots on the upper lip. The undersurface is cream and white marbled with purple and green-brown, with the abdomen tinged with brown. The juveniles are enamel green with fine black vermiculations.

HABITAT AND DISTRIBUTION (FIG. 153): This is a forest species, known from the Usambara, Udzungwa, and Rungwe mountains in Tanzania. These tree frogs occur on low bushes and in wild bananas.

ADVERTISEMENT CALL: Males call from tree branches or low on vegetation. A male was calling from inside a hole in a tree trunk 50 cm off the ground at Rungwe. The call is a single clack with an emphasized frequency of 1.5 kHz. The duration of the call is 0.2 s, with a pulse rate of 175/s.

BREEDING: Unknown.

TADPOLES: See the chapter on tadpoles.

NOTES: They are expected to be found in the Uluguru Mountains.

KEY REFERENCES: Barbour & Loveridge 1928a, Schiøtz 1999.

Wot-wots—Genus *Phlyctimantis*

The wot-wots consist of a small group of frogs that resemble the kassinas in overall form, but there are a number of differences in the male vocal pouch. The frogs are large with long thin limbs. The fingers and toes have thickened truncated discs. The male gular flap is disclike, with the interior openings on the sides beneath the eyes. This is a forest species. The common name is onomatopoeic, from the advertisement call. Two species are found in East Africa.

KEY TO THE SPECIES

1a. Males over 45 mm long, back very warty *Phlyctimantis verrucosus*

1b. Males less than 45 mm long, back slightly warty *Phlyctimantis keithae*

Keith's Wot-wot

Phlyctimantis keithae Schiøtz, 1975
(Plates 15.5, 15.6, 15.7)

This species was named for Ronalda Keith of the American Museum of Natural History, who collected in East Africa in the 1960s.

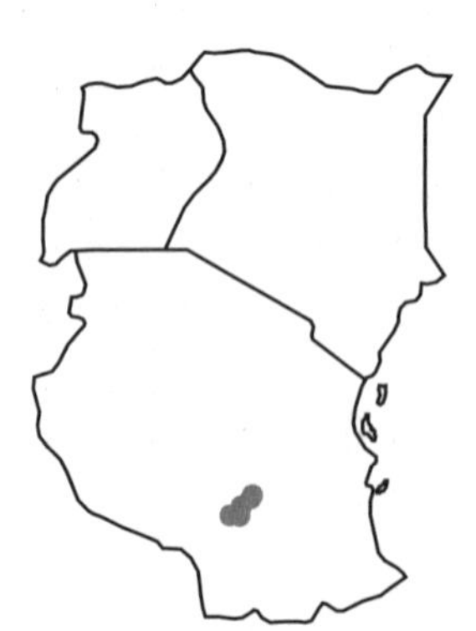

Fig. 154 *Phlyctimantis keithae.*

DESCRIPTION: The females reach 43 mm in length. The back is slightly warty to smooth and dark olive to black with small white dots. The underside is mottled in black and light blue. The inside of the limbs and posterior sides of the body are striped in black and orange. The eye is dark olive. The fingers and toes have small discs and reduced webbing.

HABITAT AND DISTRIBUTION (FIG. 154): This is a forest species, known from the Udzungwa Mountains of Tanzania.

ADVERTISEMENT CALL: Males call from concealed positions under vegetation at the edge of a pond. The calls sound like "wot-wot." The call is pulsed, rising from 0.8 to 2.2 kHz in 0.1 s, with a duration of 0.15 s. There are 8 pulses in a typical call.

BREEDING: Eggs are laid in shallow standing water with emergent vegetation. Sometimes the eggs float beneath a large bubble.

TADPOLES: Unknown.

NOTES: The defensive behavior of this species is very remarkable. When disturbed, the frog initially tucks the head forward and retracts the eyes. This is similar to the defensive behavior of many kassinas and toads. However, when further disturbed, it flicks on to the back, throwing the arms and legs out at odd angles, displaying the orange stripes and looking most unlike a frog. The return to normal posture takes place in a small number of jerks, with a pause between each movement.

KEY REFERENCES: Schiøtz 1975, 1999; Channing & Howell 2003.

Congo Wot-wot

Phlyctimantis verrucosus (Boulenger, 1912)
(Plate 15.8)

The name *verrucosus* means "warty," referring to the texture of the skin on the back in these frogs.
Warty stripe-legged frog, *Warziger Waldsteiger* in German.

DESCRIPTION: This is a large frog, with males and females up to 53 mm long. The back is dark gray or black with darker markings. Numerous white asperities cover the dorsal surface of the male. The inside of the thigh and corresponding hidden part of the flank are striped in bold yellow and black. In some animals the color is orange. The underside is dark brown with a few fine spots.

Fig. 155 *Phlyctimantis verrucosus.*

HABITAT AND DISTRIBUTION (FIG. 155): This forest species is known from the eastern part of the Democratic Republic of Congo and Uganda.

ADVERTISEMENT CALL: Males call from floating vegetation, on the ground or up to 2 m high in vegetation. The call is a very brief rising whistle, from 1.0 to 2.2 kHz, in 0.02 s. It is not pulsed like the call of Keith's wot-wot.

BREEDING: Unknown.

TADPOLES: See the chapter on tadpoles.

NOTES: When threatened, the Congo wot-wot displays the "unken reflex," where the head is kept low and the back of the body is raised to intimidate the predator. One has been observed covered with a clear latex-type liquid that had a rather noxious smell.

KEY REFERENCE: Fischer & Hinkel 1992.

Narrow-Mouthed Frogs—Family Microhylidae

Adults of the East African species in this family are small-mouthed and narrow-headed. The legs are moderately built, as most species walk, burrow, or climb. There are 14 species placed into seven genera in East Africa. Five of these genera are found only in the forests of the Eastern Arc Mountains of Tanzania. These are the warty frogs *Callulina* (two species), the three-fingered frogs *Hoplophryne* (two species), the black-banded frog *Parhoplophryne* (one species), the forest frogs *Probreviceps* (five species), and the scarlet-snouted frog *Spelaeophryne* (one species). The two widespread genera are the rain frogs *Breviceps* (two species) and the rubber frogs *Phrynomantis* (one species). Tadpoles of the rubber frogs are mid-water filter feeders; they have no jaw sheaths or labial teeth. A tadpole has been described for the Uluguru three-fingered frog, and presumably the black-banded frog has a tadpole. There is no tadpole stage in the warty frogs, the forest frogs, or the scarlet-snouted frog. However, very little is known about the biology of the microhylid frogs, and much new information is expected to be discovered and reported in the future. Molecular evidence indicates there is a need for a higher-level revision of the systematics of the family, including its relationships with the Hemisotidae.

KEY TO THE GENERA

1a. Palate with strong transverse ridges posteriorly, and a red V on the head (Fig. 156) — *Spelaeophryne*
1b. No palate ridges or red V on head — 2

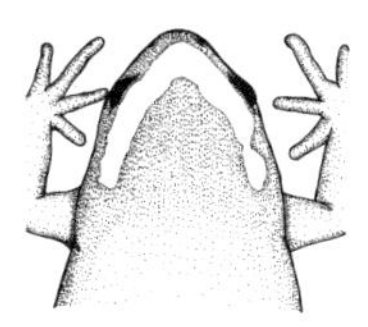

Fig. 156 Red V on head.

2a. Thumb a small swelling (Fig. 157) with a needle-like bone protruding or thumb absent — *Hoplophryne*
2b. Thumb normal — 3

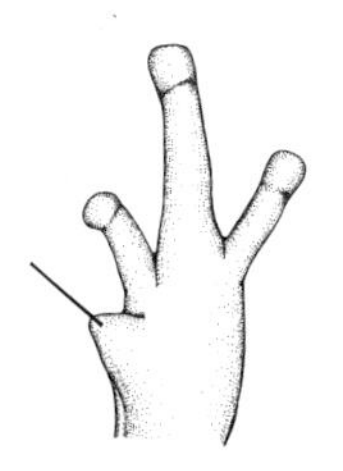

Fig. 157 Reduced thumb of male.

3a. Large white spots on belly (Fig. 158) — *Parhoplophryne*
3b. No white spots on belly, although maybe pale speckles — 4

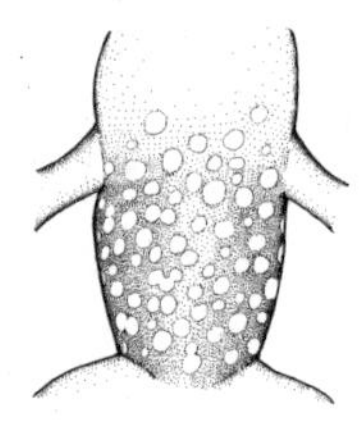

Fig. 158 Pale spots on belly.

4a. Two red or orange bands on the back (Fig. 159) *Phrynomantis*
4b. No red bands on the back 5

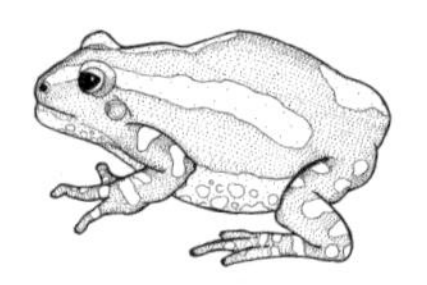

Fig. 159 Red or orange bands on back.

5a. Tips of fingers and toes tapering 6
5b. Tips of fingers and toes with squared discs (Fig. 160) *Callulina*

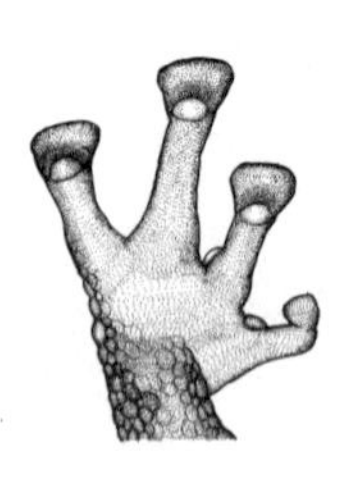

Fig. 160 Squared discs on fingers.

6a. Skin smooth *Breviceps*
6b. Skin strongly granular (Fig. 161) *Probreviceps*

Fig. 161 Granular skin.

Rain Frogs—Genus *Breviceps*

Rain frogs are specialist burrowers with small strong limbs and short rounded bodies. They burrow backward using an enlarged flattened flangelike metatarsal tubercle as a spade. Mating is very unusual in these frogs, as the male is smaller than the female, too small to be able to clasp the female in the way other frogs do. The male overcomes the size difference by gluing himself onto the back of the female. The glue is produced by skin glands on the chest. The glue persists for some days and is strong enough to permit the female to burrow backward and construct a nest underground without dislodging the male. He has to remain with the female during amplexus, so as to be in the underground nest to fertilize the eggs. Eggs develop directly into small froglets, without a free-swimming feeding tadpole stage. When conditions start to dry out, the frog is able to secrete a mucous coating that waterproofs the skin and blocks all the body orifices. With the body inflated, presumably to increase the surface area for oxygen uptake through the skin, especially the well-vascularized skin on the lower belly, the frog is able to survive long periods of dryness underground. Thirteen species are known in Africa, of which two occur in East Africa. There are probably more species of rain frogs in East Africa than the two presently recognized.

KEY TO THE SPECIES

1a. Third finger longer than the distance between the anterior corners of the eyes, metatarsal tubercles confluent *Breviceps fichus*
1b. Third finger as long as the distance between the anterior corners of the eyes, a crease present between the metatarsal tubercles *Breviceps mossambicus*

Highland Rain Frog

Breviceps fichus Channing & Minter, 2004
(Plates 16.1, 16.2)

The species name derives from the Swahili root *ficha* meaning hidden.

DESCRIPTION: Females are known to reach a length of 43 mm and the smaller males, 35 mm. The body is globular, and the limbs are short. The head is short, with a characteristic turned-down mouth. The outer finger reaches the proximal tubercle beneath the fourth finger. The basal tubercle of the first finger may be double in some individuals. There is no

Fig. 162 *Breviceps fichus.*

webbing on the toes, and the fifth toe is as long as it is wide. The third toe reaches the subarticular tubercle between the fourth and third phalanges of the fourth toe, counting from the tip. The inner and outer metatarsal tubercles fuse, without a darker area between them. The back is brown with darker mottles, the sides are speckled, and the belly is pale. The throat of females is brown, while that of breeding males is black. A black band runs from the eye, behind the angle of the jaw, to be continuous with the pigmented throat.

HABITAT AND DISTRIBUTION (FIG. 162): This species is found in grassland at altitudes above 1500 m. It is only known from the Iringa region in Tanzania. Fieldwork is required at the beginning of the short rains to determine the distribution of this species.

ADVERTISEMENT CALL: Males call during the day from well-concealed burrows at the base of dense grass. The call is a pulsed whistle with a duration of 180 ms, with the frequency rising slightly from an initial 1.9 kHz to 2.0 kHz. There are 12 pulses in a typical call.

BREEDING: Males call for only a few weeks at the start of the rains in December. Young were found in March.

NOTES: This species is very similar to the Mozambique rain frog, although the calls are quite different. As far as we know, the two species do not occur together.

KEY REFERENCE: Channing & Minter 2004.

Mozambique Rain Frog

Breviceps mossambicus Peters, 1854
(Plate 16.3)

This species is named for the island of Mozambique.
Flat faced frog, Mozambique short-headed frog, *Mopskopffrosch* or *Kurzmaulfrosch* in German, *lukumba* in Sukwa, *turye* in Nyakusa.

Fig. 163 *Breviceps mossambicus.*

DESCRIPTION: The largest female recorded had a length of 58 mm, while males are considerably smaller, reaching 30 mm. The body is globular, and the limbs are short. The head is short, with a characteristic turned-down mouth. The basal tubercle of the first finger may be double in some individuals. The outer finger reaches the proximal tubercle beneath the fourth finger. There is no webbing on the toes, and the fifth toe is distinctly longer than it is wide in most specimens. The third toe reaches the subarticular tubercle between the fourth and third phalanges of the fourth toe, counting from the tip. A darkened area of the throat may be continuous with the dark stripe from the eye to the forearm, and the dorsal color pattern is very variable from region to region.

HABITAT AND DISTRIBUTION (FIG. 163): This species is essentially a savanna form, found in leaf litter, and usually secretive for most of the year, but it is heard in large numbers for a few weeks after the start of the short rains. The Mozambique rain frog is known from Tanzania, the Mozambique coastal plain, westward to Malawi and Zimbabwe, and southward to the Eastern Cape Province of South Africa.

ADVERTISEMENT CALL: The male calls from beneath leaf litter or from a well-concealed spot at ground level. The call is a short chirp, 0.05 s long at a dominant frequency of 2.6 kHz.

BREEDING: Each egg is 0.6 mm in diameter within a 12-mm capsule. About 20 eggs are laid in a spherical chamber or nest, which is often under a stone. The female then leaves the nest and burrows nearby. The eggs hatch after 6–8 weeks, first developing into nonswimming and nonfeeding tadpoles that move around in the softened jelly and then into dark juvenile froglets, 8–9 mm long.

NOTES: Recorded food items include termites. This frog is preyed on by the vine snake *Thelotornis kirtlandii*, the white-lipped snake *Crotaphopeltis hotamboeia*, the night adder *Causus rhombeatus*, and the boomslang *Dispholidus typus*. This species is very similar to the highland rain frog, although the advertisement calls clearly separate them.

KEY REFERENCES: FitzSimons & Van Dam 1929, Loveridge 1933, Channing 2001.

Warty Frogs—Genus *Callulina*

This genus was named for its similarity to the burrowing frogs *Callula* from Asia (now *Kaloula*), with its large discs on the fingers and toes. Frogs in this genus are remarkable for their very granular, or warty, back and sides, and opposable toes. The frogs secrete a thick sticky gumlike substance when disturbed. They climb easily using the opposable toes and can be found on vegetation after dark. Males call from a meter or two above ground level. There are two species in the genus, both in East Africa.

KEY TO THE SPECIES

1a. Distance between the eye and the tympanum equal to tympanum diameter (Fig. 164) *Callulina kreffti*

1b. Distance between the eye and the tympanum equal to half the tympanum diameter *Callulina kisiwamsitu*

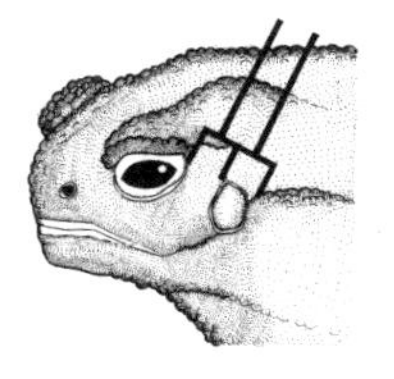

Fig. 164 Size of tympanum in relation to distance from eye corner.

Mazumbai Warty Frog

Callulina kisiwamsitu De Sá et al. 2004
(Plate 16.4)

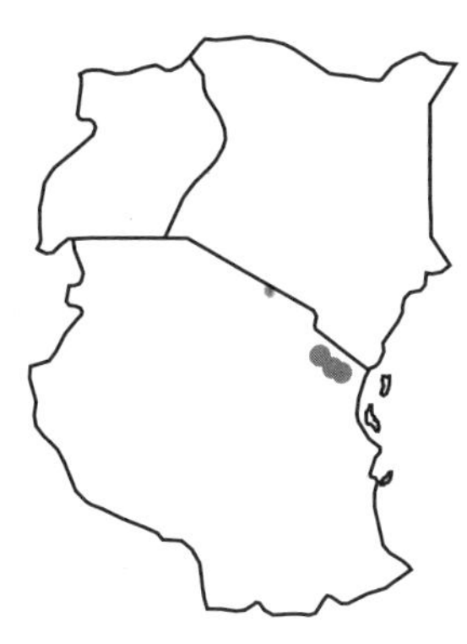

Fig. 165 *Callulina kisiwamsitu.*

DESCRIPTION: Males reach a length of 32 mm; the females are longer. The body is very glandular. The nostrils are closer to the snout tip than to the eyes. The distance between the eye and tympanum is equal to half the tympanum diameter. Distinct discs are present on the fingers and toes. There is no webbing. Subarticular tubercles are very distinct. Two metatarsal tubercles abut in the midline of the foot. The toes of the hind feet are arranged in two opposable groups, with the fourth and fifth toes together, pointing posteriorly when the animal moves on the ground. Above this frog is a gray-brown mottled with darker and lighter spots. A light vertebral stripe is often present. Below it is pale with numerous small brown spots and flecks.

HABITAT AND DISTRIBUTION (FIG. 165): This species is found in forest. During the day the frogs hide under rotting logs or stones. They are presently known from the Mazumbai Forest Reserve, and other remnants of the once widespread West Usambara and South Pare forests in Tanzania.

ADVERTISEMENT CALL: The males climb up vegetation and call while clinging to vertical branches about 15 mm in diameter. As the rains start, the males may commence calling from within their burrows but soon move above ground and climb into low bushes and other vegetation. They often call at the junction of branches, from 0.5 to 2 m off the ground. The call is a long trill or series of 12–17 notes, produced at a rate of 12 notes/s. The internote interval is equal to the duration of each note, about 0.04 s. Each note consists of 5 pulses.

BREEDING: Unknown.

KEY REFERENCE: De Sá et al. 2004.

Krefft's Warty Frog

Callulina kreffti Nieden, 1911
(Plate 16.5)

This species was named for P. Krefft, a medical doctor who carried out collecting and research in Tanzania in early 1909.

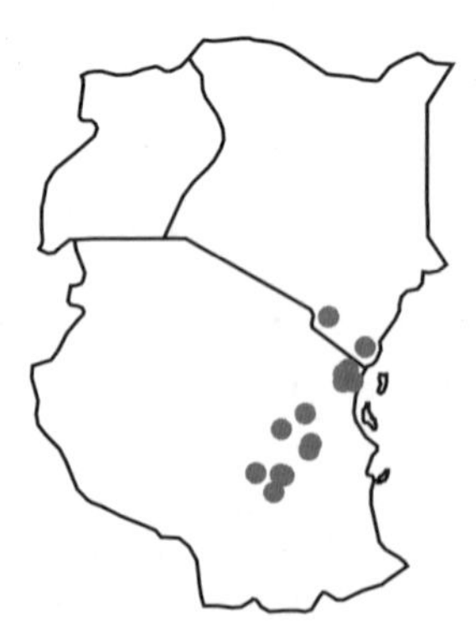

Fig. 166 *Callulina kreffti.*

DESCRIPTION: Females reach a length of 47 mm, with males about 25 mm long. The body is very glandular. The nostrils are closer to the snout tip than to the eyes. The distance between the eye and the tympanum is equal to the tympanum diameter. Distinct discs are present on the fingers and toes. There is no significant webbing, although there is some basal webbing between the fourth and fifth toes. Subarticular tubercles are very distinct. Two metatarsal tubercles are present in the midline of the foot. The toes of the hind feet are arranged in two opposable groups, with the fourth and fifth toes together, pointing posteriorly when the animal moves on the ground. Above this frog is a gray-brown mottled with darker and lighter spots. A light pinkish stripe from below the eye to the angle of the mouth is often present.

HABITAT AND DISTRIBUTION (FIG. 166): Krefft's warty frog has been found in or beneath rotting logs in forest or in cultivated or wild bananas, among rotting leaves in the upper fronds, between leaf and stem, or among debris on the ground. It is known from the Taita Hills and coastal Kenya, and the East Usambaras, Ulugurus, Ngurus, Ukagurus, and Udzungwa mountains in Tanzania, from sea level to an altitude of 2200 m.

ADVERTISEMENT CALL: During the day this frog may call after a rain from within a burrow. Usually it calls from branches of low vegetation at night. The call is a rolling trill of 3.2 s. The emphasized frequency is 2.2 kHz. Each call consists of 30 pulses, at a rate of 9/s. The calls are repeated rapidly, with a 2-s intercall interval.

BREEDING: Unknown. Eggs with a 3-mm diameter have been reported in a large female. Presumably the eggs develop directly into froglets.

NOTES: These small frogs climb well. One was seen on a leaf overhanging the pond at Amani. Known food items include millipedes, spiders, beetles, ants, and cockroaches, with ants being the main item in their diet. The frogs can exude a sticky substance from the skin glands when disturbed. They are prey of the white-lipped snake *Crotaphopeltis hotamboeia* and Tornier's cat snake *Crotaphopeltis tornieri*.

KEY REFERENCE: Barbour & Loveridge 1928a.

Three-Fingered Frogs—Genus *Hoplophryne*

These small leaf-litter frogs are characterized by the reduction of the thumb, it being represented by a group of spines, or a small bony stub. Eggs are laid in small bodies of water trapped in vegetation, like axils of banana plants, or in dead bamboo stems. The tadpoles are scavengers, with rasplike pads on either side of the mouth. Two species occur in East Africa.

KEY TO THE SPECIES

1a. Belly marbled below, a stublike thumb bone protruding in the male (Fig. 167) *Hoplophryne rogersi*
1b. Belly dusky, a rosette of spines on the thumb site in the male *Hoplophryne uluguruensis*

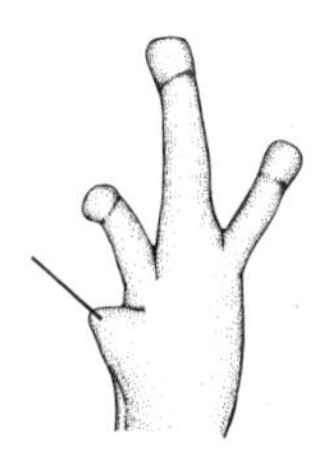

Fig. 167 Stublike thumb bone.

Rogers' Three-Fingered Frog

Hoplophryne rogersi Barbour & Loveridge, 1928
(Plates 16.6, 16.7)

Named for F. W. Rogers, custodian of the Amani Institute in the East Usambara Mountains, Tanzania.

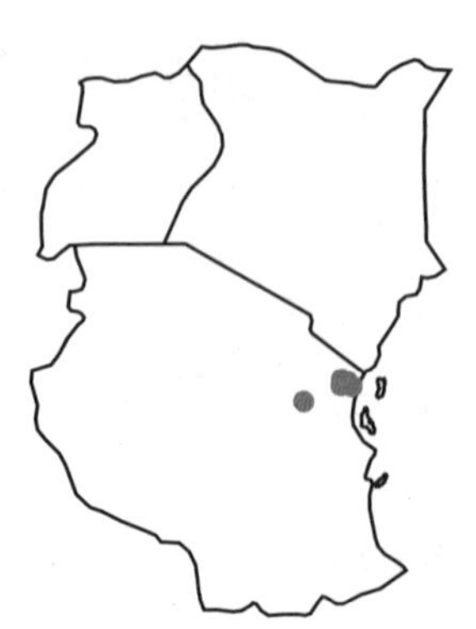

Fig. 168 *Hoplophryne rogersi.*

DESCRIPTION: Males reach a length of 26 mm and females, 32 mm. The body is stout. The first finger is a stump with a needle-like bone protruding. The fingers and toes are without discs but are broad at the tip. The skin is smooth in males, except for minute spines on the back and larger spines on the lips, throat, limbs, forearm, and chest. The back is slate blue with a black band from the snout tip to the eye that is wide at the side, then narrows toward the hind limb, and terminates on the knee. Crossbars are present on the legs. The undersurface is black, vermiculated with white to blue. The glands on the forearm and breast of males are blue.

HABITAT AND DISTRIBUTION (FIG. 168): This species is found in forest leaf litter but has also been observed in the axils of wild banana plants. It is known from the East Usambara, Magrotto, and Nguu mountains of Tanzania.

ADVERTISEMENT CALL: Unknown.

BREEDING: Small white eggs are deposited within bamboo stems or in leaf axils where water is trapped. Breeding takes place in late November to early December.

TADPOLES: See the chapter on tadpoles.

NOTES: Recorded food items include woodlice, spiders, ants, and polydesmids.

KEY REFERENCE: Barbour & Loveridge 1928a.

Uluguru Three-Fingered Frog

Hoplophryne uluguruensis Barbour & Loveridge, 1928.

This species is named for the Uluguru Mountains, where it was discovered.

DESCRIPTION: The male reaches 25 mm long, with the largest known female being 21 mm long. These are flat frogs. The first finger is only a swelling. The other three are broadened at the tips but have no discs. Toes are without webbing. The skin is smooth but in the male is covered with small tubercles, which also occur on the lips and upper chest to forearm, where they form a rosette of nine spines hidden by the forearms. Male secondary sexual characters include three spines where the first finger should be, spines on the back of the second finger, and similar

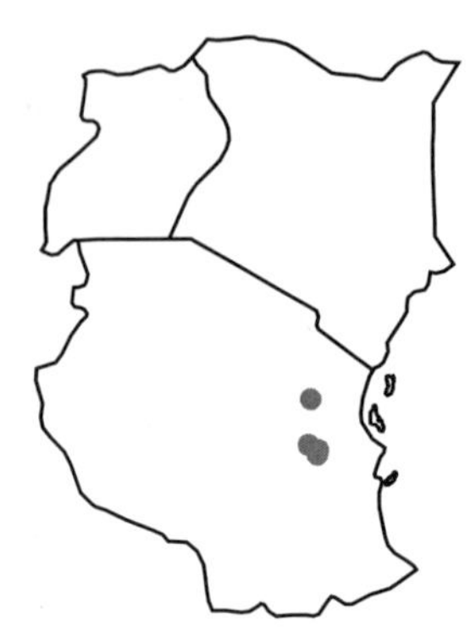

Fig. 169 *Hoplophryne uluguruensis.*

spines on the hind leg. The frogs are black above, speckled with silvery white. There are conspicuous white dots on the face. The undersurface is brownish with specklings of silver. Other color patterns include pale green to reddish backs with orange limbs.

HABITAT AND DISTRIBUTION (FIG. 169): These leaf-litter frogs are sometimes found inside fallen bamboo stems or around wild banana plants. They are known from the Uluguru and Nguu mountains in Tanzania, between altitudes of 1000 and 2500 m.

ADVERTISEMENT CALL: Unknown.

BREEDING: Breeding takes place in late September through October. Eggs are laid on the inner surface of the leaf and on the stem of banana trees, or in bamboo that is split, allowing a flattened frog to get inside. Eggs are 2.5–3.0 mm in diameter within 4.0–4.5-mm capsules, laid one deep but clumped. Although the eggs are laid out of water, they are protected in a moist environment. Clutch size is 28. The female may stay with the eggs. The tadpoles hatch and slide down into water trapped in the banana axil.

TADPOLES: See the chapter on tadpoles.

NOTES: Known food items include small beetles and ants. The only recorded predator is the Uluguru forest snake *Buhoma procterae.*

KEY REFERENCES: Barbour & Loveridge 1928a, Noble 1929.

1.1 *Arthroleptis adolfifriederici*
Adolf's Squeaker

1.2 *Arthroleptis adolfifriederici*
Adolf's Squeaker eggs

1.3 *Arthroleptis affinis*
Ahl's Squeaker (E. Harper)

1.4 *Arthroleptis reichei*
Reiche's Squeaker

1.5 *Arthroleptis stenodactylus*
Common Squeaker

1.6 *Arthroleptis tanneri*
Tanner's Squeaker

1.7 *Schoutedenella poecilonotus?*
Mottled Squeaker (M. Burger)

1.8 *Schoutedenella xenochirus*
Plain Squeaker

2.1 *Schoutedenella xenochirus*
Plain Squeaker eggs

2.2 *Schoutedenella xenodactyla*
Eastern Squeaker

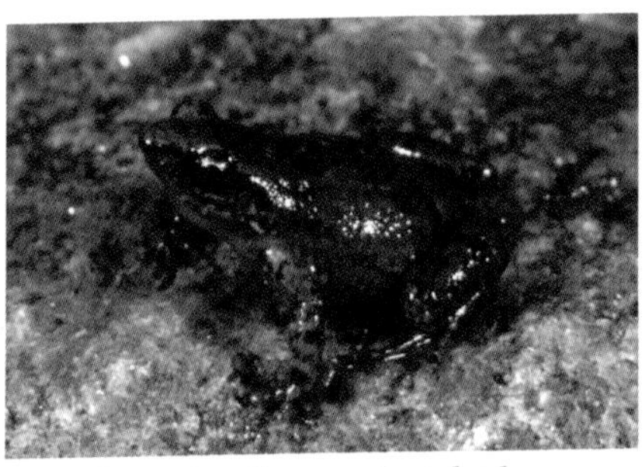

2.3 *Schoutedenella xenodactyloides*
Dwarf Squeaker

2.4 *Bufo brauni*
Braun's Toad

2.5 *Bufo garmani*
Garman's Toad

2.6 *Bufo gutturalis*
Guttural Toad

2.7 *Bufo kerinyagae*
Kerinyaga Toad (M. Tandy)

2.8 *Bufo kisoloensis*
Kisolo Toad

3.1 *Bufo lindneri*
Lindner's Toad

3.2 *Bufo lindneri*
Lindner's Toad

3.3 *Bufo lonnbergi*
Lönnberg's Toad (M. Tandy)

3.4 *Bufo lughensis*
Lugh Toad (M. Tandy)

3.5 *Bufo maculatus*
Flat-Backed Toad

3.6 *Bufo regularis*
Common Toad (M. Burger)

3.7 *Bufo steindachneri*
Steindachner's Toad

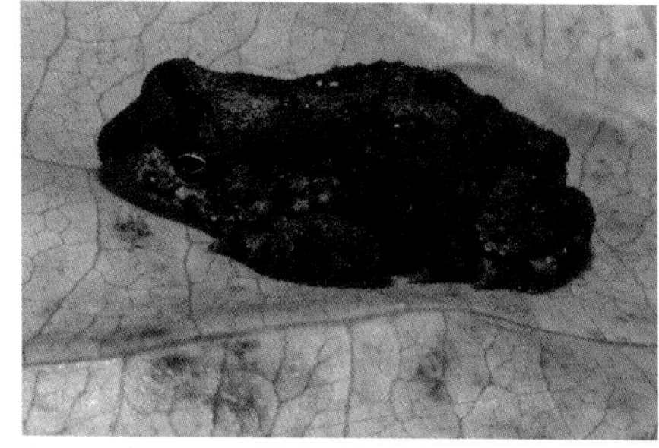

3.8 *Bufo taitanus*
Taita Toad

4.1 *Bufo uzunguensis*
Udzungwa Toad (M. Menegon)

4.2 *Bufo xeros*
Desert Toad

4.3 *Churamiti maridadi*
Beautiful Forest Toad (W. Stanley)

4.4 *Churamiti maridadi*
Beautiful Forest Toad (W. Stanley)

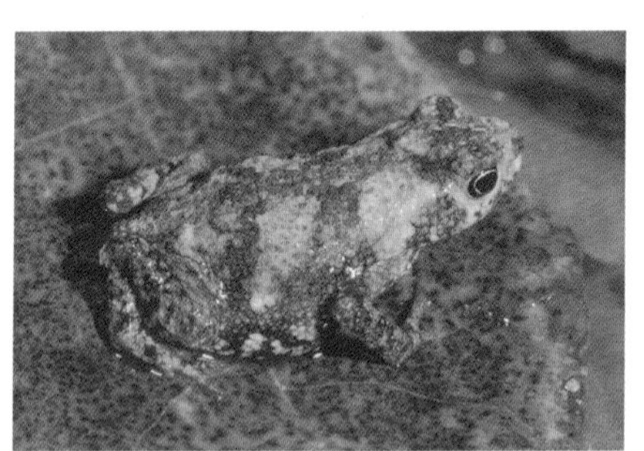

4.5 *Mertensophryne micranotis*
Woodland Toad (M. Tandy)

4.6 *Mertensophryne micranotis*
Woodland Toad in snail shell (J. Vonesh)

4.7 *Nectophrynoides asperginis*
Kihansi Spray Toad female and juvenile

4.8 *Nectophrynoides asperginis*
Kihansi Spray Toad male

5.1 *Nectophrynoides asperginis*
Kihansi Spray Toad

5.2 Sprinklers in Kihansi Gorge

5.3 *Nectophrynoides tornieri*
Tornier's Forest Toad (E. Harper)

5.4 *Nectophrynoides viviparus*
Robust Forest Toad (Tim Davenport, WCS)

5.5 *Nectophrynoides viviparus*
Robust Forest Toad (M. Menegon)

6.1 *Schismaderma carens*
Red Toad

6.2 *Stephopaedes howelli*
Mrora Forest Toad

6.3 *Stephopaedes howelli*
Mrora Forest Toad (C. & T. Stuart)

6.4 *Stephopaedes howelli*
Mrora Forest Toad

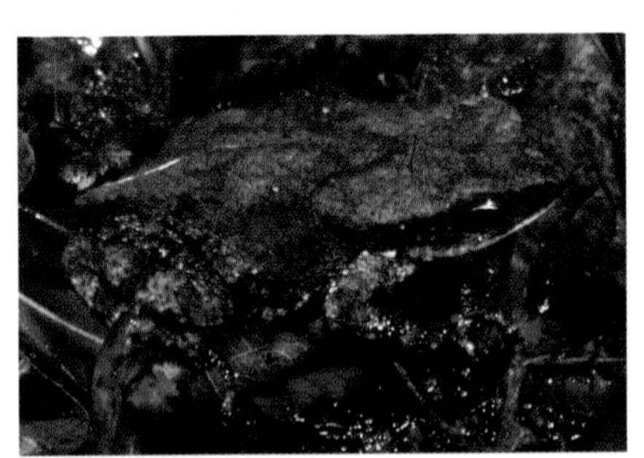

6.5 *Stephopaedes loveridgei*
Loveridge's Forest Toad

6.6 *Hemisus marmoratus*
Marbled Snout-Burrower

6.7 *Hemisus marmoratus*
Marbled Snout-Burrower burrowing

6.8 *Hemisus marmoratus*
Marbled Snout-Burrower burrowing

7.1 *Afrixalus brachycnemis*
Short-Legged Spiny Reed Frog

7.2 *Afrixalus fornasini*
Fornasini's Spiny Reed Frog

7.3 *Afrixalus fornasini*
Fornasini's Spiny Reed Frog

7.4 *Afrixalus fulvovittatus*?
Four-Lined Spiny Reed Frog

7.5 *Afrixalus laevis*
Smooth Spiny Reed Frog

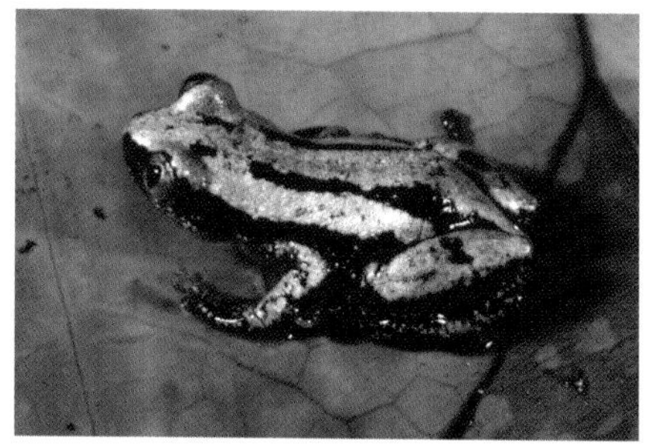

7.6 *Afrixalus morerei*
Morere's Spiny Reed Frog

7.7 *Afrixalus osorioi*
Congo Spiny Reed Frog (M. Tandy)

7.8 *Afrixalus stuhlmanni*
Stuhlmann's Spiny Reed Frog

8.1 *Afrixalus sylvaticus*
Forest Spiny Reed Frog (M. Tandy)

8.2 *Afrixalus uluguruensis*
Uluguru Spiny Reed Frog

8.3 *Afrixalus uluguruensis*
Uluguru Spiny Reed Frog eggs (J. Vonesh)

8.4 *Afrixalus wittei*
De Witte's Spiny Reed Frog

8.5 *Hyperolius acuticeps*
Sharp-Nosed Reed Frog

8.6 *Hyperolius argentovittis*
Ahl's Reed Frog

8.7 *Hyperolius argentovittis*
Ahl's Reed Frog

8.8 *Hyperolius argus*
Argus Reed Frog pair (E. Harper)

9.1 *Hyperolius bocagei*
Bocage's Reed Frog

9.2 *Hyperolius bocagei*
Bocage's Reed Frog pair

9.3 *Hyperolius castaneus*
Brown Reed Frog (M. Tandy)

9.4 *Hyperolius cinnamomeoventris*
Cinnamon-Bellied Reed Frog

9.5 *Hyperolius cinnamomeoventris*
Cinnamon-Bellied Reed Frog (M. Burger)

9.6 *Hyperolius discodactylus*
Highland Reed Frog

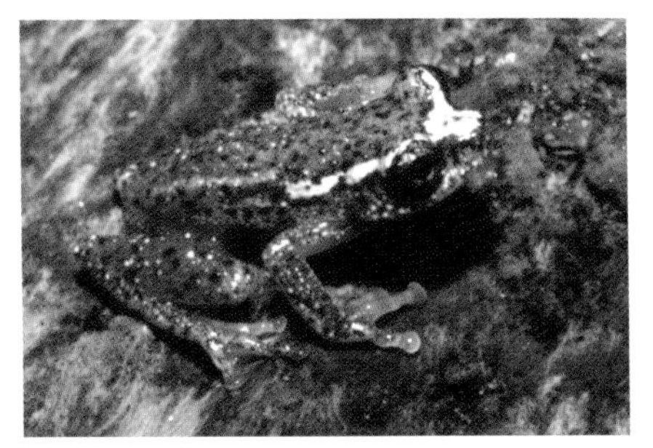

9.7 *Hyperolius frontalis*
White-Snouted Reed Frog

9.8 *Hyperolius glandicolor*
Peters' Reed Frog

10.1 *Hyperolius kihangensis*
Kihanga Reed Frog (M. Menegon)

10.2 *Hyperolius kivuensis*
Kivu Reed Frog

10.3 *Hyperolius langi*
Lang's Reed Frog

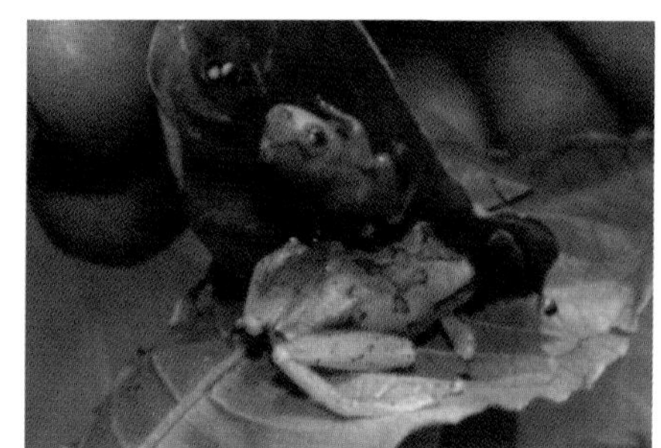

10.4 *Hyperolius langi*
Lang's Reed Frog eggs between leaves

10.5 *Hyperolius lateralis*
Side-Striped Reed Frog mottled morph

10.6 *Hyperolius lateralis*
Side-Striped Reed Frog brown morph

10.7 *Hyperolius lateralis*
Side-Striped Reed Frog green morph

10.8 *Hyperolius mariae*
Mary's Reed Frog

11.1 *Hyperolius minutissimus*
Dwarf Reed Frog

11.2 *Hyperolius mitchelli*
Mitchell's Reed Frog

11.3 *Hyperolius mitchelli*
Mitchell's Reed Frog eggs

11.4 *Hyperolius montanus*
Montane Reed Frog (W. Branch)

11.5 *Hyperolius parkeri*
Parker's Reed Frog (J. Vonesh)

11.6 *Hyperolius pictus*
Variable Reed Frog

11.7 *Hyperolius pseudargus*
Mette's Reed Frog

11.8 *Hyperolius puncticulatus*
Spotted Reed Frog

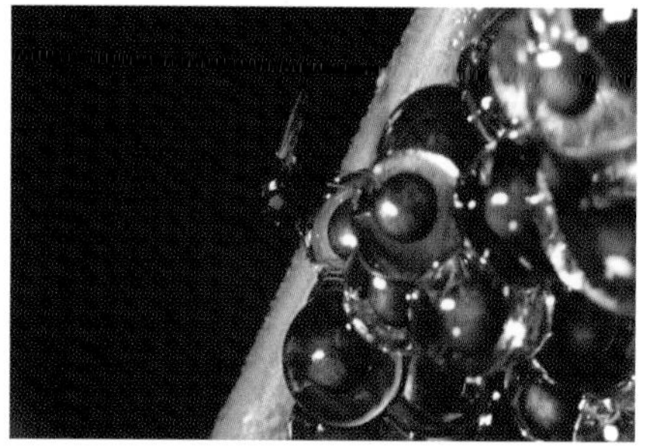

12.1 Frogfly on *Hyperolius puncticulatus* Reed Frog eggs (J. Vonesh)

12.2 *Hyperolius pusillus*
Water Lily Reed Frog

12.3 *Hyperolius quinquevittatus*
Five-Striped Reed Frog

12.4 *Hyperolius spinigularis*
Spiny-Throated Reed Frog

12.5 *Hyperolius spinigularis*
Spiny-Throated Reed Frog guarding eggs

12.6 Spider eating *Hyperolius spinigularis* (J. Vonesh)

12.7 *Hyperolius tuberilinguis*
Tinker Reed Frog

12.8 *Hyperolius viridiflavus*
Common Reed Frog

13.1 *Hyperolius viridis*
Green Reed Frog

13.2 *Kassina maculata*
Red-Legged Kassina (M.-O. Rödel)

13.3 *Kassina maculifer*
Spotted Kassina (M. Tandy)

13.4 *Kassina senegalensis*
Senegal Kassina

13.5 *Leptopelis argenteus*
Silvery Tree Frog

13.6 *Leptopelis barbouri*
Barbour's Tree Frog

13.7 *Leptopelis barbouri*
Barbour's Tree Frog

13.8 *Leptopelis barbouri*
Barbour's Tree Frog tadpole (M. Menegon)

14.1 *Leptopelis bocagii*
Bocage's Tree Frog

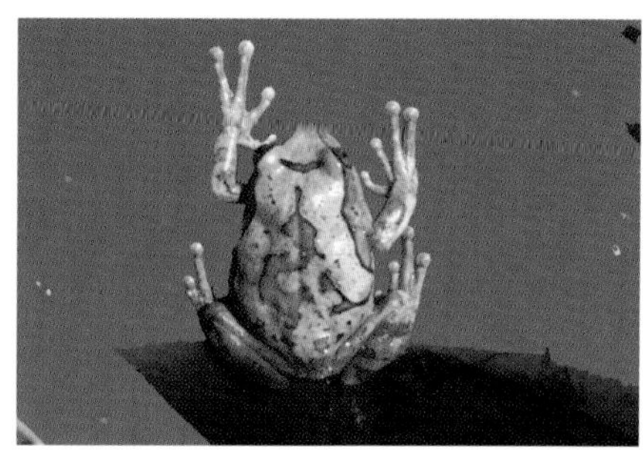

14.2 *Leptopelis christyi*
Christy's Tree Frog (J. Vonesh)

14.3 *Leptopelis flavomaculatus*
Yellow-Spotted Tree Frog

14.4 *Leptopelis kivuensis*
Kivu Tree Frog

14.5 *Leptopelis parbocagii*
Cryptic Tree Frog

14.6 *Leptopelis parkeri*
Parker's Tree Frog

14.7 *Leptopelis parkeri*
Parker's Tree Frog male

14.8 *Leptopelis uluguruensis*
Uluguru Tree Frog

15.1 *Leptopelis uluguruensis*
Uluguru Tree Frog

15.2 *Leptopelis vermiculatus*
Vermiculated Tree Frog juvenile

15.3 *Leptopelis vermiculatus*
Vermiculated Tree Frog

15.4 *Leptopelis vermiculatus*
Vermiculated Tree Frog

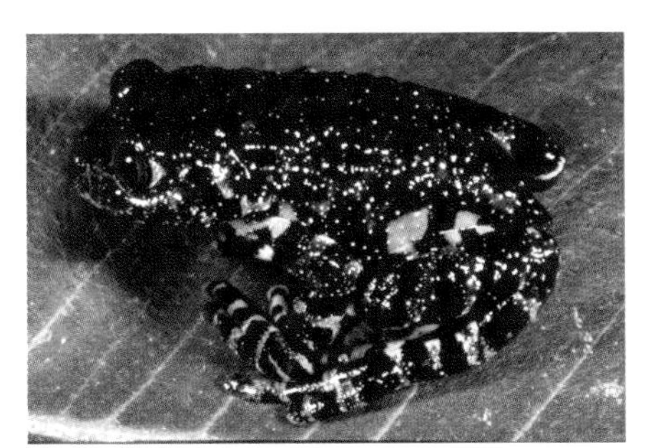

15.5 *Phlyctimantis keithae*
Keith's Wot-wot

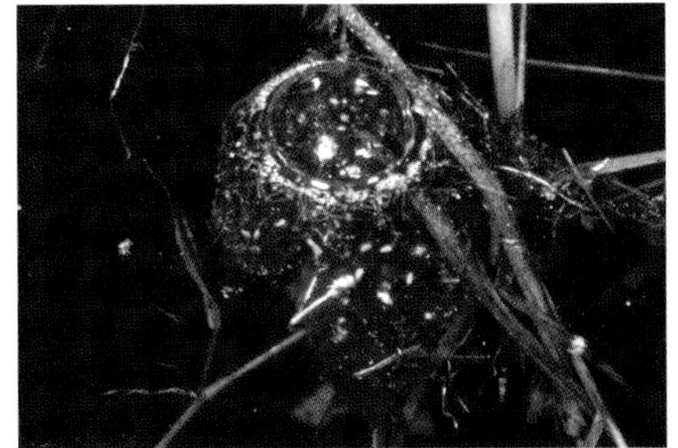

15.6 *Phlyctimantis keithae*
Keith's Wot-wot eggs

15.7 *Phlyctimantis keithae*
Keith's Wot-wot defense

15.8 *Phlyctimantis verrucosus*
Congo Wot-wot

16.1 *Breviceps fichus*
Highland Rain Frog

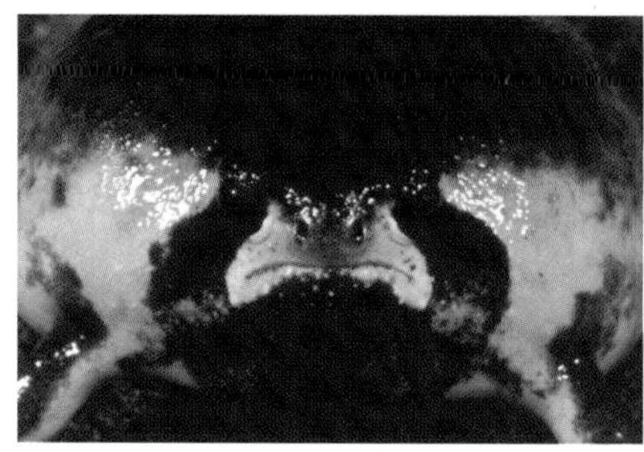

16.2 *Breviceps fichus*
Highland Rain Frog

16.3 *Breviceps mossambicus*
Mozambique Rain Frog

16.4 *Callulina kisiwamsitu*
Mazumbai Warty Frog

16.5 *Callulina kreffti*
Krefft's Warty Frog

16.6 *Hoplophryne rogersi*
Rogers' Three-Fingered Frog

16.7 *Hoplophryne rogersi*
Rogers' Three-Fingered Frog with eggs
(J. Vonesh)

16.8 *Phrynomantis bifasciatus*
Banded Rubber Frog

17.1 *Probreviceps loveridgei*
Loveridge's Forest Frog

17.2 *Probreviceps macrodactylus*
Long-Fingered Forest Frog

17.3 *Probreviceps rungwensis*
Snouted Forest Frog (Tim Davenport, WCS)

17.4 *Probreviceps uluguruensis*
Uluguru Forest Frog

17.5 *Xenopus borealis*
Northern Clawed Frog

17.6 *Xenopus muelleri*
Müller's Clawed Frog

17.7 *Afrana angolensis*
Angolan River Frog

17.8 *Amnirana albolabris*
Forest White-Lipped Frog (M. Burger)

18.1 *Amnirana galamensis*
Galam White-Lipped Frog

18.2 *Arthroleptides martiensseni*
Martienssen's Torrent Frog

18.3 *Arthroleptides yakusini*
Southern Torrent Frog

18.4 *Cacosternum* sp.
Plimpton's Dainty Frog

18.5 *Hildebrandtia macrotympanum*
Northern Ornate Frog (M. Tandy)

18.6 *Hildebrandtia ornata*
Common Ornate Frog

18.7 *Hoplobatrachus occipitalis*
Eastern Groove-Crowned Bullfrog

18.8 *Phrynobatrachus acridoides*
East African Puddle Frog

19.1 *Phrynobatrachus auritus*
Golden Puddle Frog (M.-O. Rödel)

19.2 *Phrynobatrachus bullans*
Bubbling Puddle Frog

19.3 *Phrynobatrachus dendrobates*
Climbing Puddle Frog

19.4 *Phrynobatrachus keniensis*
Upland Puddle Frog

19.5 *Phrynobatrachus krefftii*
Krefft's Puddle Frog

19.6 *Phrynobatrachus krefftii*
Krefft's Puddle Frog

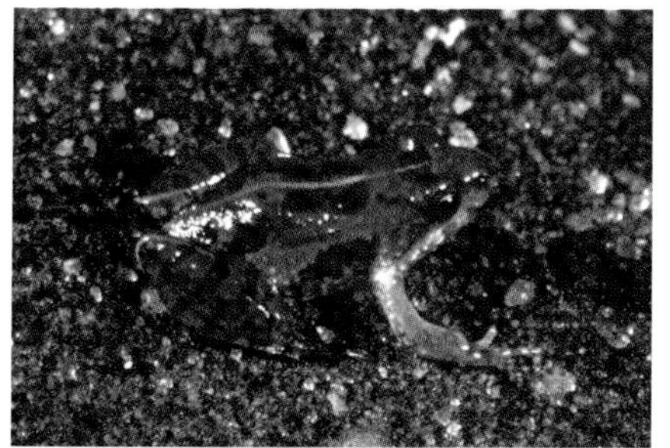

19.7 *Phrynobatrachus mababiensis*
Mababe Puddle Frog

19.8 *Phrynobatrachus natalensis*
Natal Puddle Frog

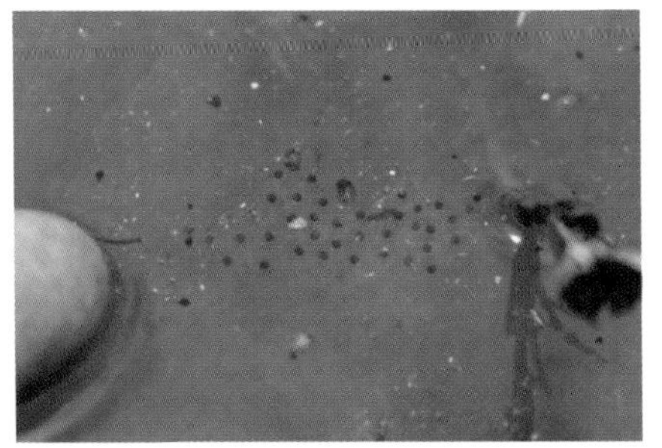

20.1 *Phrynobatrachus natalensis*
Natal Puddle Frog eggs

20.2 *Phrynobatrachus rungwensis*
Rungwe Puddle Frog

20.3 *Phrynobatrachus scheffleri*
Scheffler's Puddle Frog

20.4 *Phrynobatrachus stewartae*
Stewart's Puddle Frog

20.5 *Phrynobatrachus uzungwensis*
Udzungwa Puddle Frog (M. Menegon)

20.6 *Phrynobatrachus versicolor*
Green Puddle Frog

20.7 *Ptychadena anchietae*
Anchieta's Ridged Frog

20.8 *Ptychadena chrysogaster*
Yellow-Bellied Ridged Frog

21.1 *Ptychadena mascareniensis*
Mascarene Ridged Frog

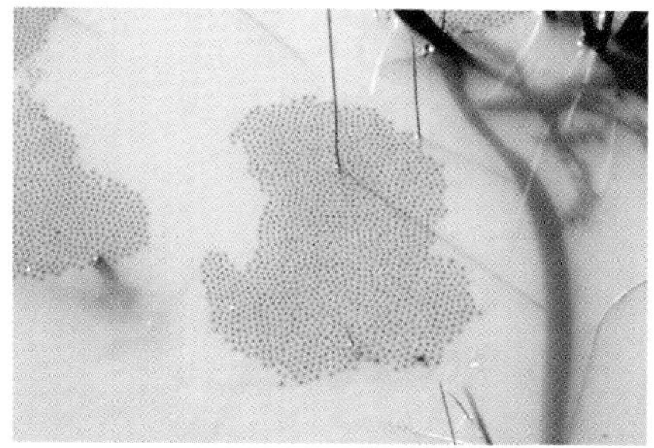

21.2 *Ptychadena mascareniensis*
Mascarene Ridged Frog eggs

21.3 *Ptychadena mossambica*
Mozambique Ridged Frog

21.4 *Ptychadena oxyrhynchus*
Sharp-Nosed Ridged Frog

21.5 *Ptychadena oxyrhynchus*
Sharp-Nosed Ridged Frog

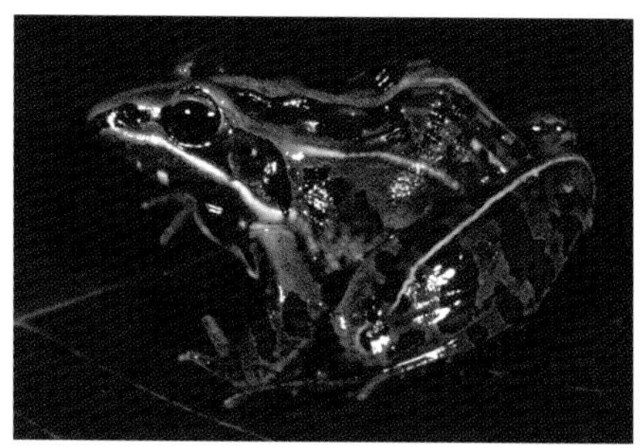

21.6 *Ptychadena porosissima*
Grassland Ridged Frog

21.7 *Ptychadena schillukorum*
Schilluk Ridged Frog

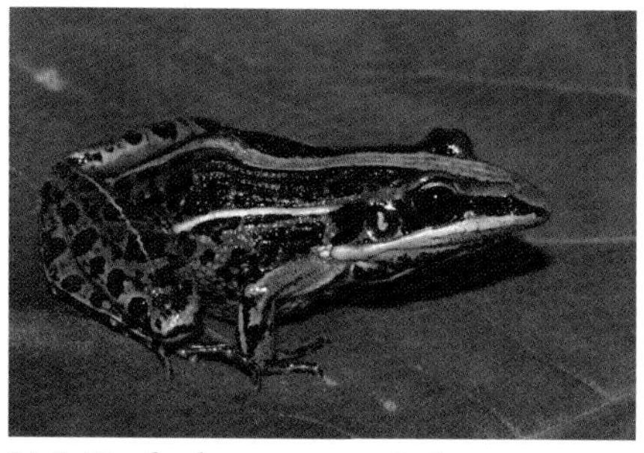

21.8 *Ptychadena stenocephala*
Narrow-Headed Ridged Frog

22.1 *Ptychadena taenioscelis*
Small Ridged Frog

22.2 *Ptychadena upembae*
Upemba Ridged Frog

22.3 *Ptychadena uzungwensis*
Udzungwa Ridged Frog

22.4 *Pyxicephalus adspersus*
African Bullfrog

22.5 *Pyxicephalus adspersus*
African Bullfrog tadpoles

22.6 *Pyxicephalus edulis*
Edible Bullfrog

22.7 *Strongylopus kitumbeine*
Kitumbeine Stream Frog

22.8 *Strongylopus merumontanus*
Mt. Meru Stream Frog

23.1 *Tomopterna cryptotis*
Cryptic Sand Frog

23.2 *Tomopterna luganga*
Red Sand Frog

23.3 *Tomopterna tandyi*
Tandy's Sand Frog

23.4 *Tomopterna tuberculosa*
Rough Sand Frog

23.5 *Chiromantis petersii*
Peters' Foam-Nest Frog

23.6 *Chiromantis rufescens*
Western Foam-Nest Frog (M. Largen)

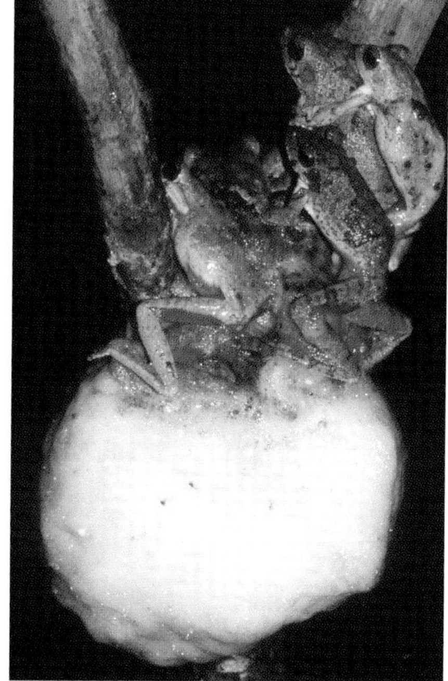

23.7 *Chiromantis xerampelina*
Southern Foam-Nest Frog (M. Burger)

24.1 *Boulengerula boulengeri*
Boulenger's Caecilian

24.2 *Boulengerula taitana*
Taita Hills Caecilian (J. Measey)

24.3 *Schistometopum gregorii*
Flood Plain Caecilian (J. Measey)

24.4 *Scolecomorphus kirkii*
Kirk's Caecilian (J. Visser)

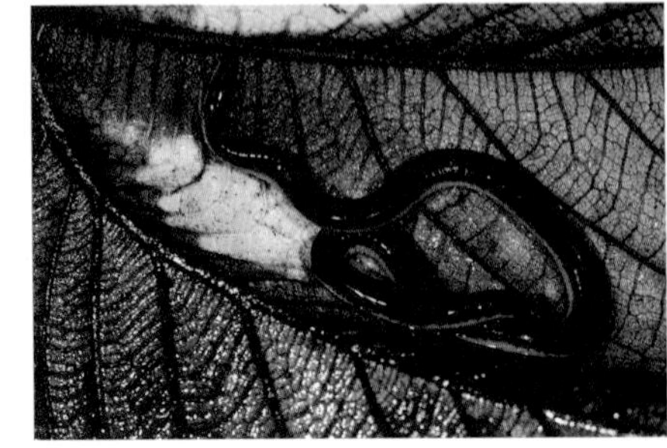

24.5 *Scolecomorphus vittatus*
Ribbon Caecilian (M. Menegon)

Black-Banded Frog—Genus *Parhoplophryne*

The frogs are small, flattened, and only known from one locality on the East Usambara Mountains of Tanzania. They are similar in overall morphology to the blue-bellied frogs, except that they do not have a reduced first finger. There is only one species known in this genus.

Usambara Black-Banded Frog

Parhoplophryne usambarica Barbour & Loveridge, 1928.

This species is named for the Usambara Mountains.

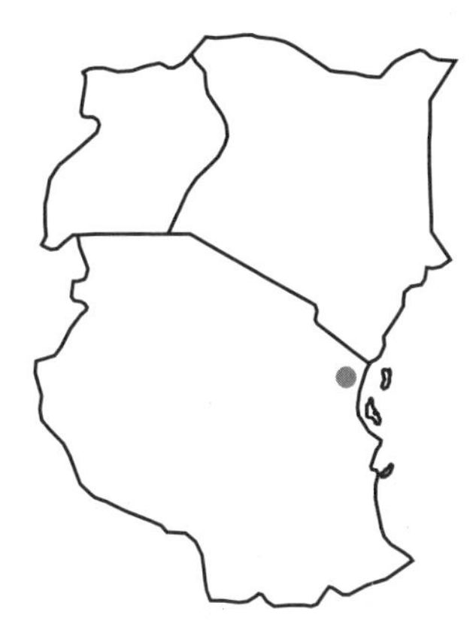

Fig. 170 *Parhoplophryne usambarica.*

DESCRIPTION: These frogs are stout. Females reach a length of 23 mm. There is no webbing between the toes and no dilated toe tips. The skin is smooth with rugose folds. The back is slate blue with a black band from the snout tip to the eye, which is wide at the side, then narrows to the hind limb, and terminates on the knee. There are dark crossbars on the legs. The undersurface is black, vermiculated with white tinged with blue. The glands on the forearm and breast of males are blue.

HABITAT AND DISTRIBUTION (FIG. 170): This forest species inhabits leaf litter and is presently only known from the East Usambara Mountains of Tanzania.

ADVERTISEMENT CALL: Unknown.

BREEDING: Unknown.

TADPOLES: Unknown. Tadpoles originally described as this species were subsequently identified as *Hoplophryne rogersi.*

NOTES: This monotypic genus differs from the blue-bellied frogs by the presence of a clavicle.

KEY REFERENCES: Barbour & Loveridge 1928a, Noble 1929.

Rubber Frogs—Genus *Phrynomantis*

Rubber frogs get their name from the texture of their skin. They do not jump or hop, but rather are excellent walkers and climbers. They are usually black with red markings. Tadpoles are large, midwater filter feeders, without jaw sheaths or labial teeth. The genus is known from West Africa to Ethiopia, southward to South Africa. The species are distributed in moist and dry tropical regions, with one species endemic to the Namib Desert. There are five species, one of which occurs in East Africa.

Banded Rubber Frog

Phrynomantis bifasciatus (Smith, 1847)
(Plate 16.8)

The species name *bifasciatus* refers to the two bright orange or red bands down the back.
Red-banded frog, red-banded rubber frog, *Bindenfrosch* or *Zweistreifen-Wendelhalsfrosch* in German, *Turye* in Kinyakusa.

Fig. 171 *Phrynomantis bifasciatus.*

DESCRIPTION: The male reaches a length of 53 mm and the female, 65 mm. The skin is smooth. The head is flat and narrow, with a tympanum just smaller than the eye. The toes and fingers have small discs, which are truncated, and very little webbing. As in many climbing frogs, the last segment of the fingers and toes is not in line with the rest of the finger or toe. The back has an orange or red patch above the vent, and two broad bands of the same color running from the snout over the eyes to the leg. The lower surface is smooth, with white spots and blotches on a gray background.

HABITAT AND DISTRIBUTION (FIG. 171): This frog is known from the moist savannas of Kenya west to Angola and southward to Namibia and northern South Africa. It has been found in small-mammal burrows and beneath the sheath leaves of banana plants, and will climb into any small crevice during the day.

ADVERTISEMENT CALL: The male calls from the water's edge or while concealed under vegetation, down rodent holes, or often inside logs. The call is a long melodious trill, lasting up to 3 s, with emphasized frequencies at 1.2 and 2.4 kHz. Each note lasts only 0.02 s, with a similar interval between notes. The call is loud and can be heard for some distance.

BREEDING: This species breeds during the rainy season in summer, when huge choruses form after heavy rains. Each egg is 1.3 mm in diameter within a 4-mm capsule. The eggs are deposited either in a 75-mm mass on vegetation or sunk. Clutch size is 300–1500. It is not uncommon to find tadpoles in water trapped in animal prints in mud, especially those of an elephant. The tadpole hatches after 4 days.

TADPOLES: See the chapter on tadpoles.

NOTES: When disturbed, the frog inflates and arches its body, with the head tucked in. The bright contrasting color of the back is believed to serve as a warning to predators. It also gives off a skin toxin that causes irritation. Although these frogs can be handled with no ill effects, if the toxin enters through scratches on your hands, painful swellings, difficulty breathing, headache, faster heart rate, and nausea may result. These symptoms only last about 4 hours. Water from a container that had held these frogs later produced further mild symptoms when it was splashed onto scratches. Other species of frogs placed in a container with the banded rubber frog will die.

Known food items include ants. The banded rubber frog is eaten by the rhombic night adder *Causus rhombeatus* and the hammerkop *Scopus umbretta*. Maximum recorded longevity in captivity is 6 years 7 months.

KEY REFERENCES: Loveridge 1933, Jaeger 1971, Colley 1987, Channing 2001.

Forest Frogs—Genus *Probreviceps*

The forest frogs are robust, short-limbed burrowing species that are found in old mountain forests. The males have larger tympani than the females. Five species are known, of which four are found in the high mountain forests of Tanzania. We expect that more species remain to be discovered, so the following key and species descriptions should be treated as a starting point and not as a definitive treatment. All the species are remarkably similar, and we have found it necessary to use geographic information to separate many of the species in the key. In the field the advertisement calls are distinctive. A newly recognized species from the Ukaguru Mountains is presently being described, which is included here as *Probreviceps* sp.

KEY TO THE SPECIES

1a. Tympanum hidden — *Probreviceps uluguruensis*
1b. Tympanum distinct — 2

2a. Snout protrudes strongly (Fig. 172) — 4
2b. Snout not protruding — 3

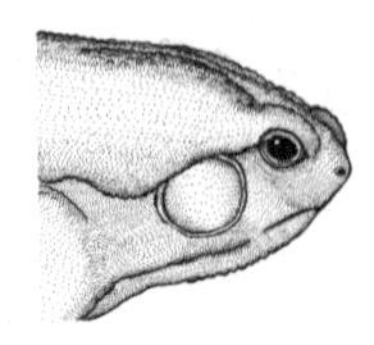

Fig. 172 Protruding snout.

3a. Known from the Pares and Usambara mountains of Tanzania — *Probreviceps macrodactylus*
3b. Known from mid altitudes on the Uluguru mountains of Tanzania — *Probreviceps loveridgei*

4a. Found on the Ukaguru mountains *Probreviceps* sp.
4b. Found on Mt. Rungwe and the Udzungwa Mountains of Tanzania *Probreviceps rungwensis*

Loveridge's Forest Frog

Probreviceps loveridgei Parker, 1931
(Plate 17.1)

This species was named for Arthur Loveridge, who made important contributions to East African amphibian studies.

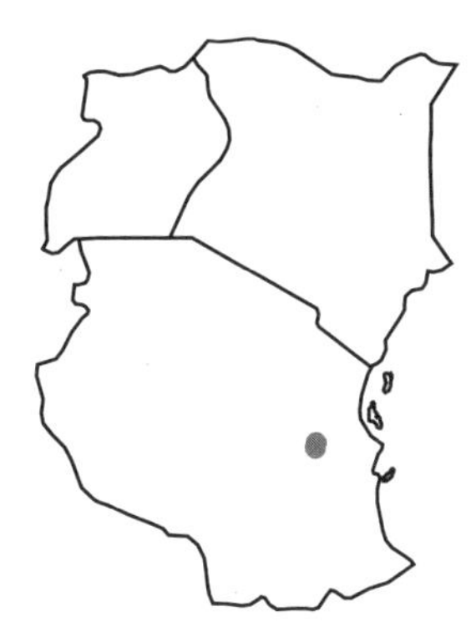

Fig. 173 *Probreviceps loveridgei.*

DESCRIPTION: The males reach a length of 33 mm; large females reach a length up to 45 mm. The body is short and stout, with a width almost two-thirds of the body length. The mouth is down-turned and small, with the angle of the mouth below the eye. The diameter of the tympanum is 28%–41% of the eye diameter in females and 39%–66% of that in males. The snout projects beyond the lower lip, which is nearly vertical. The head is wide, slightly more than one-third of the body length. The eye is small, with a diameter around 10% of the body length. Tibia length is 38%–45% of snout-urostyle length. The palms of the hands have large folds, and the feet have a large pebble-like inner metatarsal tubercle. The skin is covered with fine granular warts. The back is a uniform brown. The vent is posteroventral.

HABITAT AND DISTRIBUTION (FIG. 173): This species is found in forest and forest remnants at mid altitudes on the Uluguru Mountains, Tanzania.

ADVERTISEMENT CALL: Males call from burrows or leaf litter after rain. The call consists of 4 or 5 notes in 0.6 s. Each note consists of 2–5 pulses. The emphasized frequency is 1.68 kHz.

BREEDING: Unknown. Presumably the eggs are laid in a burrow and develop directly into small froglets.

NOTES: Field studies are required to determine the range of this species, as it may occur on mountains near the Ulugurus.

KEY REFERENCES: Parker 1931, Mkonyi et al. 2004.

Long-Fingered Forest Frog

Probreviceps macrodactylus (Nieden, 1926)
(Plate 17.2)

The species name *macrodactylus* refers to the long fingers.
Turye in Nyakusa, *kikolwe* in Shambala.

Fig. 174 *Probreviceps macrodactylus.*

DESCRIPTION: The males reach a length of 40 mm; large females reach a length up to 65 mm. The body is short and stout, with a width almost two-thirds of the body length. The mouth is down-turned and small, with the angle of the mouth below the eye. The diameter of the tympanum is 43%–51% of the eye diameter in males and females. The head is wide, slightly more than one-third of the body length. The snout projects slightly beyond the lower lip, which is nearly vertical. The eye is small, with a diameter around 10% of the body length. Tibia length is 37%–44% of snout-urostyle length. The palms of the hands have large folds, and the feet have a large pebble-like inner metatarsal tubercle. The back is a uniform brown; dorsolateral markings may be present in some animals. The underside is dark with white spots. The vent is posteroventral.

HABITAT AND DISTRIBUTION (FIG. 174): This forest species is found under logs and leaf litter. It is known from the North Pares and Usambara mountains.

ADVERTISEMENT CALL: Males call from burrows or while concealed beneath vegetation after rain. The call, a low-pitched chirp at 1.9 kHz, consists of 12–14 notes in 0.3 s, at a repetition rate of 38 notes/s. The call is slightly frequency modulated, with a peak in the middle.

BREEDING: The long-fingered forest frog breeds during the small rains in November. Eggs are deposited in a burrow, where they develop directly into juveniles without a free-swimming tadpole stage.

NOTES: Recorded food items include ants and beetles. This species is eaten by Tornier's cat snake *Crotaphopeltis tornieri*.

KEY REFERENCES: Barbour & Loveridge 1928a, Mkonyi et al. 2004.

Snouted Forest Frog

Probreviceps rungwensis Loveridge, 1932
(Plate 17.3)

This species is named for Mt. Rungwe in southern Tanzania.

Fig. 175 *Probreviceps rungwensis.*

DESCRIPTION: The males reach a length of 48 mm; larger females can be up to 60 mm long. The body is short and stout, with a width almost two-thirds of the body length. The mouth is down-turned and small, with the angle of the mouth below the eye. The tympanum is large, 63%–145% of the eye diameter in males and 43%–83% in females. The head is wide, about one-third of the body length. The snout is pointed and projects strongly beyond the lower lip, with a flat area stretching from the tip to behind the nostrils. The eye is small, with a diameter around 10% of the body length. Tibia length is 39%–48% of snout-urostyle length. The palms of the hands have large folds, and the feet have a large pebble-like inner metatarsal tubercle. The uniform brownish dorsum often has a darker dorsolateral stripe. The underside is dark with white spots. The vent is posteroventral.

HABITAT AND DISTRIBUTION (FIG. 175): This frog is found in forest on Mt. Rungwe and the Udzungwa Mountains in Tanzania.

ADVERTISEMENT CALL: Unknown.

BREEDING: This species breeds during the small rains in November. The eggs are deposited in a burrow, where they develop directly into juveniles without a free-swimming tadpole stage.

NOTES: Known food items include ants and beetles. The hard pointed snout suggests that this species actively burrows.

KEY REFERENCE: Loveridge 1932a.

Ukaguru Forest Frog

Probreviceps sp.

DESCRIPTION: Females are known to reach 35 mm in length, while the males are much smaller. The flat head has a strongly protruding snout with a blue-gray, hardened tip. The eyelid extends back as a skin fold to above the arm. The tympanum is distinct. Males have a very large tympanum, ranging from 13–16% of the snout-urostyle body length, while females have a smaller tympanum that is 4–7% of body length. The foot is about as long as the tibia. The back is brown with five gray transverse bands, that angle backwards from the midline.

HABITAT AND DISTRIBUTION: This species is found in leaf litter, in forest on the Ukaguru Mountains, Tanzania.

ADVERTISEMENT CALL: Males call during the day, with a little activity at night. Males are always concealed under forest debris while calling. The call is a slow series of clicks, produced at about 30/minute.

BREEDING: Unknown.

Uluguru Forest Frog

Probreviceps uluguruensis (Loveridge, 1925)
(Plate 17.4)

This species is named for the Uluguru Mountains.
Kimbofu in Kami.

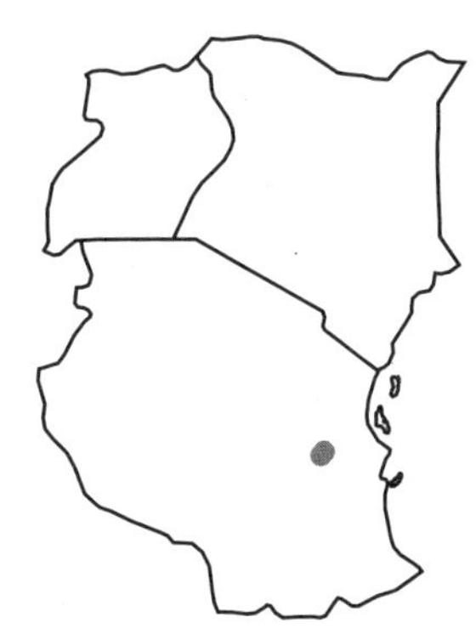

Fig. 176 *Probreviceps uluguruensis.*

DESCRIPTION: The males reach a length of 30 mm; the larger females reach up to 43 mm. The tympanum is not visible. The body is short and stout, with a width almost two-thirds of the body length. The mouth is down-turned and small, with the angle of the mouth below the eye. The head is wide, about one-third of the body length. The eye is small, with a diameter around 10% of the body length. The legs are short; tibia length is 28%–39% of snout-urostyle length. The palms of the hands have large folds, and the feet have a large pebble-like inner metatarsal tubercle. Some specimens have white-tipped spinules on the tubercles below the fourth and other fingers. The color is variable, from cream to terra-cotta, with purplish sides and belly. Large females may have a deep red band on each side of the body, which may be pale in some individuals. The underside is dark with white spots. The vent is ventral, unlike the other species.

HABITAT AND DISTRIBUTION (FIG. 176): This frog is found on the grasslands between altitudes of 2000 and 2500 m on the Uluguru Mountains. It hides at the base of grass tussocks between the stems and in forest patches.

ADVERTISEMENT CALL: Males call from grass tussocks or under leaf litter. The call is a low-pitched chirp, consisting of a series of around 8 notes in 0.4 s. The note repetition rate is 21/s. The emphasized frequency is 1.4 kHz.

BREEDING: The female deposits about 20 eggs in a burrow at the start of the long rains. The eggs are creamy white, 4 mm in diameter. The eggs develop directly into small adults without going through a free-swimming tadpole stage.

NOTES: Recorded food items include termites, ants, woodlice, millipedes, beetles, ticks, caterpillars, and beetle larvae. When disturbed, these small frogs jump and run surprisingly quickly.

KEY REFERENCES: Barbour & Loveridge 1928a, Mkonyi et al. 2004.

Scarlet-Snouted Frog—Genus *Spelaeophryne*

This robust frog has short thick limbs and a characteristic broad red V on the head. It remains mysterious, as no breeding data are available despite the species being known to science for 80 years. There is only one species in this genus.

Scarlet-Snouted Frog

Spelaeophryne methneri Ahl, 1924

This species was collected in a cave by Methner, a German civil servant in East Africa from 1902 to 1916. The frog is named after him.
Scarlet-snouted black frog.

Fig. 177 *Spaeleophryne methneri.*

DESCRIPTION: The largest male known is 52 mm long and the female, 53 mm long. The pupil is horizontal. The tongue is large, the palate has no vomerine teeth, but there are transverse dermal ridges posteriorly. There are no small teeth along the upper jaw. The fingers and toes are short and thick with undilated tips. The nostril is situated at the tip of the snout. Snout length exceeds eye diameter. The tympanum is round, equal in size to the eye. The skin is variable in texture, being pitted, rugose, granular, or smooth. An indistinct fold runs from the eye to the shoulder. The dorsal surface is black, with a bright scarlet V-shaped band following the outline of the snout to behind and above the tympanum, close above the eyes. Sometimes an additional red bar runs between the eyes; rarely, no red is present on the snout. The undersurface is brownish.

HABITAT AND DISTRIBUTION (FIG. 177): This forest species is found in leaf litter. It is known from the Taita Hills in Kenya, the Uluguru and Udzungwa mountains, and southeastern Tanzania.

ADVERTISEMENT CALL: Unknown.

BREEDING: Unknown. The eggs are large, suggesting that they are laid in a terrestrial nest and develop directly into small frogs.

NOTES: The frog secretes a gummy substance that is more tenacious than the upper layers of skin. If held firmly, the frog may secrete the skin gum and then twist out of its skin, leaving it attached to the predator while the naked frog scuttles away. Known food items are termites.

KEY REFERENCE: Barbour & Loveridge 1928a.

Clawed Frogs—Family Pipidae

These frogs spend nearly their whole life in water and are adapted to an aquatic existence. They have smooth skin, a streamlined shape, large webbed feet, and sensory lateral line organs. Calling, feeding, and breeding take place in water. They feed on aquatic prey. They are also known to feed on their own tadpoles, in so doing making indirect use of the algal food supply in the water on which the tadpoles feed. They do not possess a tongue, unlike most other frogs. This family has a very good fossil record, with the earliest fossils known from the Lower Cretaceous of the Middle East 120 million years ago. Three genera are found in Africa: *Hymenochirus*, *Pseudhymenochirus*, and *Xenopus*. Only the genus *Xenopus* is known in East Africa.

Clawed Frogs—Genus *Xenopus*

The name *Xenopus* means "strange foot," referring to the claws on three of the five toes. The frogs in this genus are found in all kinds of water bodies, natural and man-made, including fairly salty water. Many are known to have a courtship behavior that involves the clasped pair swimming in loops. The eggs are deposited at the top of the loop while the pair is inverted. The tadpoles are transparent with large fins, and they form large schools in mid-water, where they filter out small food items.

The frogs vary from 35 to 135 mm in length and are smooth-skinned, with extremely streamlined shapes. The head is flattened, and the eyes are small, positioned on top of the head. The arms are delicate, with four long fingers. The strong hind legs have toes that are fully webbed. Three toes on each foot are supplied with black claws (sometimes with an additional claw on the prehallux) that are used for defense and for tearing at large food items. Sensory organs that look like "stitches" are arranged around the eyes and along the sides. The legs are attached to the body with a sliding mechanism involving the pelvic girdle. Lengthening increases the efficiency of movement by increasing the swimming stroke, enabling the frog to quickly swim away from a predator. The body can also be shortened as an escape mechanism.

This endemic African genus is known worldwide as an experimental animal, and much has been discovered about its genetics and development. Thousands of these frogs are used in the world's laboratories every year and might be regarded as one of Africa's most important scientific exports. Many are bred in Europe and North America to supply eggs for genetic studies. This all started when it was discovered in the early 1930s that they could be used as reliable tests for human pregnancy; a few drops of urine from a pregnant woman, injected under the skin of the female frog, caused eggs to be laid. Other reasons why this genus is widely used in laboratories are the same reasons why it is so widespread and successful in nature: they are hardy, able to live under a variety of conditions, and breed easily with high rates of success. They have unique compounds in the skin secretions that protect them from disease; these antimicrobials are also useful in human disease control. Finally, these animals are able to develop a generation in a year or less, which enables them to rapidly pioneer any areas that are newly flooded.

The various species of clawed frogs are major prey items for many large birds and other animals.

There are about 20 species in Africa, with 6 in East Africa.

KEY TO THE SPECIES

1a. Claws present on the first three toes — 2

1b. Claws present on the first three toes plus the prehallux (Fig. 178) — *Xenopus ruwenzoriensis*

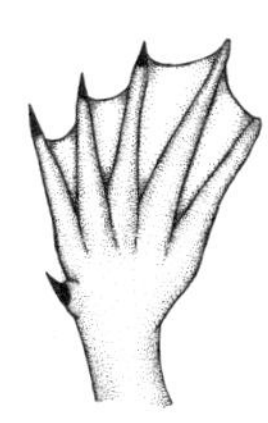

Fig. 178 Four claws on foot.

2a. Prehallux cone-shaped, prominent, pointed — 3

2b. Prehallux small or absent — 4

3a. Subocular tentacle longer than half the eye diameter (Fig. 179) — *Xenopus muelleri*

3b. Subocular tentacle half or less than half the eye diameter — *Xenopus borealis*

Fig. 179 Subocular tentacle.

4a. Fifth toe longer than the tibia *Xenopus victorianus*
4b. Fifth toe equal to tibia length 5

5a. Back color uniform with no spots, head rounded *Xenopus wittei*
5b. Back marbled, silvery spots present, head wedge-shaped *Xenopus vestitus*

Northern Clawed Frog

Xenopus borealis Parker, 1936
(Plate 17.5)

The name *borealis* is Greek for "northern," referring to the species being found in the northern parts of East Africa.

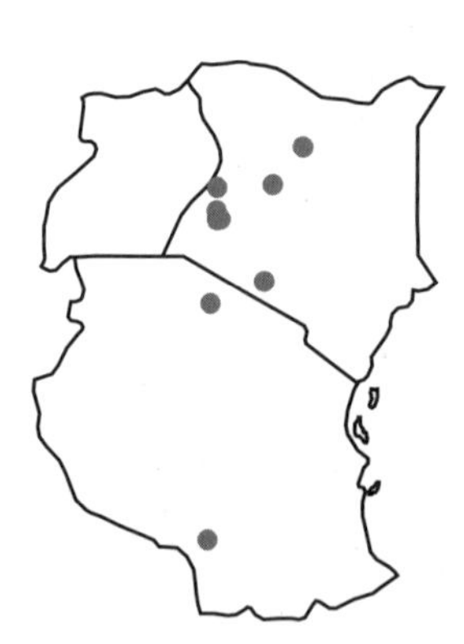

Fig. 180 *Xenopus borealis.*

DESCRIPTION: The females are 75 mm long but are known to reach 95 mm in extreme cases; most males are around 60 mm long. A distinct cone-shaped prehallux is present. The small tentacle below the eye is no more than half as long as the eye diameter. The back is brown to gray with a large number of irregular dark markings, which become more dense on the hind limbs and adjacent back area. The underside is pale with some spotting. There are 13–17 lateral line organs around the eye, and 23–30 along the back. The lower eyelid covers three-fourths of the eye. The female has two lobes around the cloaca.

HABITAT AND DISTRIBUTION (FIG. 180): This species is found mostly in pools and slow streams at altitudes above 1500 m. It is known from the highlands around Njombe, through northern Tanzania and Kenya.

ADVERTISEMENT CALL: The male produces a slow series of clacks, at a rate of 2/s, with occasional bursts of rapid notes up to 12/s. The call duration varies from 5 to 35 s. The emphasized frequency is 1.2 kHz.

BREEDING: Unknown. The eggs are small, 1.1 mm in diameter.

TADPOLES: The tadpoles are not distinguishable from those of the other species. See the chapter on tadpoles.

NOTES: This species has 36 chromosomes.

KEY REFERENCES: Vigny 1979a, 1979b; Tinsley & Kobel 1996.

Müller's Clawed Frog

Xenopus muelleri (Peters, 1844)
(Plate 17.6)

This species is named for the anatomist Johannes Müller of Friedrich-Wilhelms University in Berlin.
Müller's smooth clawed frog, *hamuru* in Nyaturu, *kisisulu* in Nilamba, *Müllers Krallenfrosch* in German.

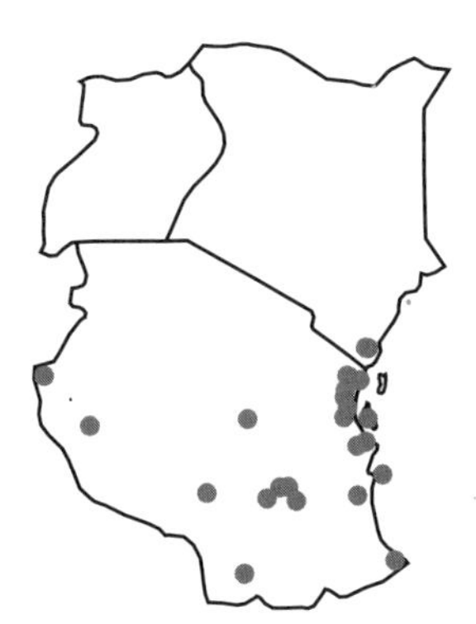

Fig. 181 *Xenopus muelleri.*

DESCRIPTION: This is a large frog, with males up to 50 mm long and average females up to 65 mm; very large females can be 90 mm long. The tentacle below the eye is at least half as long as the eye diameter. A protective membrane covers more than half of the eye. There are 22–27 lateral line bars between the eye and the vent and 11–18 lateral line organs around the eye. The inner metatarsal tubercle is small and pointed. The three inner toes are clawed, with the inner metatarsal tubercle distinctly projecting. The back is olive to brown with five to eight large dark spots, while the underside is pale yellowish with varying amount of dark pigmentation. The underside of the legs is yellow or orange with pale silvery gray spots.

HABITAT AND DISTRIBUTION (FIG. 181): This frog may be found in very small water bodies and occurs widely in wooded savanna and grassland. It is known from low-lying areas in southeastern Kenya, Tanzania, and the offshore islands, south through Zambia and Malawi into northern South Africa.

ADVERTISEMENT CALL: The male calls underwater. The call consists of only one component, with a duration of 0.2 s and 5–7 pulses at a pulse rate of 26–32/s. The emphasized frequency ranges from 774 Hz to 1182 Hz. These frogs have a complex vocal system, producing more than just the advertisement call.

BREEDING: Eggs, 1.0 mm in diameter, are laid in small groups attached underwater.

TADPOLES: See the chapter on tadpoles.

NOTES: The adults eat toad tadpoles, fish, conspecific tadpoles, frog eggs, termites, beetles, and ants. In turn they are preyed on by the hammerkop, the rufous beaked snake *Rhamphiophis rostratus* in Malawi, and the southeastern green snake *Philothamnus hoplogaster*. The latter snake also eats tadpoles. These frogs may be restricted to permanent ponds during the dry season. When the rains come, the frogs move into temporary water bodies, where conditions for breeding are better. The skin contains a number of substances, one of which is a peptide called *xenopsin*, which is known to cause contraction in the rat stomach. Another peptide found in the skin is *caerulein*. This causes spasms in the gallbladder and stimulates secretion of the pancreatic juice. Together these skin defenses cause the predator to vomit and to feel ill for some time. Predators soon learn to exclude *Xenopus* from their diet. In captivity the frogs have been known to live for 9 years. This species is a 36-chromosome diploid, consisting of two sets of 18.

KEY REFERENCES: Inger 1968; Vigny 1979a, 1979b; Tinsley & Kobel 1996.

Ruwenzori Clawed Frog

Xenopus ruwenzoriensis Tymowska & Fischberg, 1973

Thc specific name refers to the Ruwenzori Mountains between Uganda and the Democratic Republic of Congo.

Fig. 182 *Xenopus ruwenzoriensis.*

DESCRIPTION: This is a medium-sized species, with the females reaching a length of 57 mm and the males 44 mm. The head is moderately small, and the arm-snout distance is three and a half to four times the snout-vent length. The lower eyelid covers two-thirds of the eye. Three inner toes and the inner metatarsal tubercle have black, horny claws. The back is brownish with a few large spots. The throat, chest, and abdomen are brown without spots, or yellow with black spots. Ventral surfaces of the limbs are yellow with bold, black spots.

HABITAT AND DISTRIBUTION (FIG. 182): This species is found in rain forest. It is known from the foothills of the Ruwenzori Mountains, on the border between Uganda and the Democratic Republic of Congo.

ADVERTISEMENT CALL: The call is described as short high-pitched metallic trills. Each trill is 0.05–0.1 s long, with 7–10 pulses. The emphasized frequency is 1.5 kHz, with many higher harmonics.

BREEDING: Unknown.

TADPOLES: Unknown.

NOTES: This species has 108 chromosomes, made up of six sets of 18.

KEY REFERENCES: Inger 1968, Tinsley & Kobel 1996.

Jacketed Clawed Frog

Xenopus vestitus Laurent, 1972

The name *vestitus* means "garment," referring to the color pattern of the back, which makes the animal look like it is wearing a coat.

Fig. 183 *Xenopus vestitus.*

DESCRIPTION: This is a medium-sized species, with females up to 55 mm long and males 43 mm long. The head is wedge-shaped with small eyes. The head is pale, with a dark transverse band separating it from the back. The back is brown with a characteristic marbling of silvery bronze pigment. The belly is heavily spotted.

HABITAT AND DISTRIBUTION (FIG. 183): This species is known from rain forest in the highlands forming the border of Uganda, Rwanda, and the Democratic Republic of Congo.

ADVERTISEMENT CALL: The male produces a series of trills 0.2–1.2 s long, consisting of 26–114 pulses. The dominant frequency is 2.0 kHz.

BREEDING: Unknown.

TADPOLES: Unknown.

NOTES: This species has 72 chromosomes, made up of four sets of 18.

KEY REFERENCE: Tinsley & Kobel 1996.

Lake Victoria Clawed Frog

Xenopus victorianus Ahl, 1924

The specific name refers to the type locality, Lake Victoria.

Fig. 184 *Xenopus victorianus.*

DESCRIPTION: These frogs are large, with the female reaching up to 78 mm long and the male 60 mm. The head is large, and the lower eyelid covers half of the eye. The orbital tentacle is about one-third the diameter of the eye. The fifth toe is slightly longer than the tibia. The inner metatarsal tubercle is without a black horny claw and does not protrude. The back is yellowish to green, with many small spots and a few irregular large spots. Ventral surfaces are yellowish brown with a few black spots.

HABITAT AND DISTRIBUTION (FIG. 184): This frog is found in arid, savanna, and forested habitats. It is the most common clawed frog in northern Tanzania, Burundi, Rwanda, eastern Democratic Republic of Congo, Uganda, and adjacent Sudan, to Kenya.

ADVERTISEMENT CALL: The call is a rapid series of very short trills, only 0.02–0.05 s long. Each trill consists of 2–6 pulses, with a dominant frequency of 2.2–2.5 kHz. The trills are produced at a rate of 9/s.

BREEDING: Unknown.

TADPOLES: Unknown.

NOTES: This species is a 36 chromosome diploid, consisting of two sets of 18.

KEY REFERENCE: Tinsley & Kobel 1996.

De Witte's Clawed Frog

Xenopus wittei Tinsley et al. 1979

This species was named for G.-F. De Witte, herpetologist at the Institut Royal des Sciences Naturelles in Brussels until 1951.

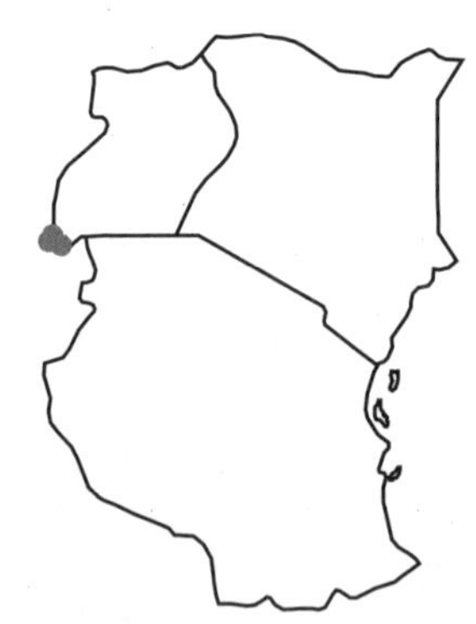

Fig. 185 *Xenopus wittei.*

DESCRIPTION: This is a medium-sized species, with females reaching up to 61 mm long and males 50 mm. The head is rounded with large eyes, with a short tentacle under each eye. The back is chocolate brown with no spots. Underneath is yellowish and finely spotted. The throat may be heavily spotted.

HABITAT AND DISTRIBUTION (FIG. 185): This species is known from shallow standing water, at altitudes up to 2300 m. It has been recorded from the volcanic highlands forming the border of Uganda, Rwanda, and the Democratic Republic of the Congo.

ADVERTISEMENT CALL: The call consists of short trills produced without interruption, resulting in a continuous long call. Each individual trill is 0.3–0.6 s, with 9–24 pulses. The dominant frequency is 1.5 kHz.

BREEDING: Many small eggs are attached individually to the substrate under water.

TADPOLES: Unknown.

KEY REFERENCES: Fischer & Hinkel 1992, Tinsley & Kobel 1996.

Common Frogs—Family Ranidae

This family of common frogs has representatives in practically all parts of the world. This is the largest family in East Africa, with 53 species, placed in 11 genera. The smallest frog in this family in East Africa is the Mababe puddle frog, which may reach a length of 23 mm. The largest frog in this family, and in East Africa, is the African bullfrog, which may reach a length of 230 mm and a weight of over a kilogram! Although most have a typical life history with free-swimming tadpoles that develop from eggs laid in water, many species here show a number of interesting adaptations to life in forests where standing pools of water are uncommon.

KEY TO THE GENERA

1a.	Vomerine teeth present (Fig. 186)	4
1b.	Vomerine teeth absent	2

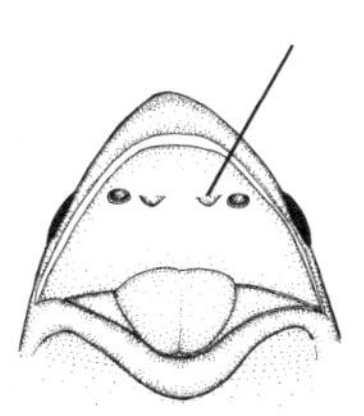

Fig. 186 Vomerine teeth.

2a.	Bifid discs on fingers and toes (Fig. 187)	*Arthroleptides*
2b.	Discs, if present, not bifid	3

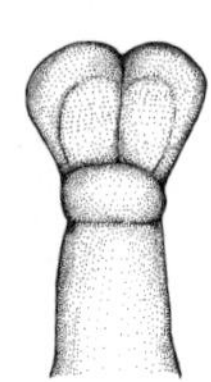

Fig. 187 Bifid discs.

3a. Belly with distinct dark spots (Fig. 188) *Cacosternum*
3b. Belly not spotted *Phrynobatrachus*

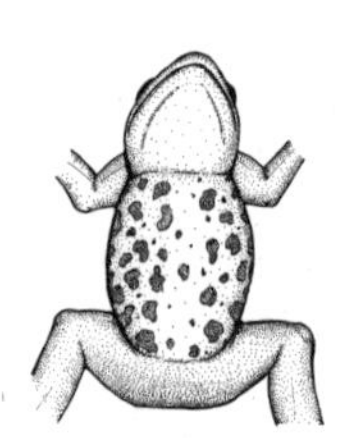

Fig. 188 Spotted belly.

4a. Outer metatarsals separated from rest of sole by a web (Fig. 189) 5
4b. Outer metatarsals bound into a fleshy sole, inner metatarsal tubercle flanged 9

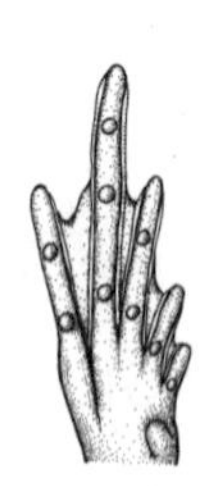

Fig. 189 Outer metatarsal separated from sole.

5a. Vomerine projections between the internal nostrils and not reaching their margins 6
5b. Vomerine projections abutting onto anterior margins of internal nostrils *Ptychadena*

6a. Foot at least as long as distance from tip of urostyle to tympanum (Fig. 190) *Strongylopus*
6b. Foot shorter than distance from tip of urostyle to tympanum 7

Fig. 190 Length of foot in proportion to urostyle-tympanum distance.

7a. Transverse skin groove behind eyes (Fig. 191) *Hoplobatrachus*
7b. No transverse skin groove behind eyes 8

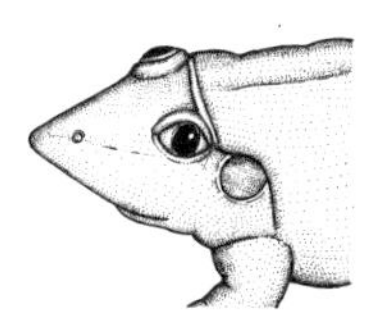

Fig. 191 Transverse groove behind eyes.

8a. A golden, green, or brown band, as wide as the back, with a light margin running from snout to vent *Amnirana*
8b. No band as wide as the back, running from snout to vent *Afrana*

9a. Subarticular tubercles not prominent on fleshy digits, more than eight dorsal skin ridges distinct *Pyxicephalus*
9b. Subarticular tubercles prominent; digits not fleshy; less than six dorsal skin ridges, if present (Fig. 192) 10

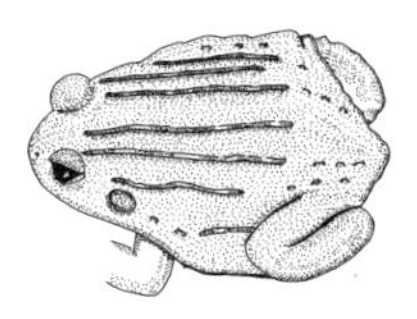

Fig. 192 Longitudinal skin ridges.

10a. Paired vocal pouches open behind angle of jaw; tympanum distinct, with a diameter two-thirds or more of eye diameter (Fig. 193) *Hildebrandtia*
10b. No paired vocal pouch openings; tympanum indistinct, with a diameter smaller than half the eye diameter *Tomopterna*

Fig. 193 Large tympanum.

River Frogs—Genus *Afrana*

This large genus of typical frogs with long legs is associated with water and active throughout the year. They occur from forest to arid savanna and occupy all kinds of man-made impoundments. The tadpoles are large with long muscular tails. This genus is known as *akalenga* or *olulenga* in Luganda, and as *orutoori* in Runyankole and related languages in Uganda. It is known as *eretretre* in Lugbara. There are 10 species in Africa, of which 3 are found in East Africa.

KEY TO THE SPECIES

1a. Two phalanges free of web on one side of the fourth toe — 2
1b. Three phalanges free of web on the fourth toe — *Afrana ruwenzorica*

2a. Nostril closer to the eye than the snout tip — *Afrana wittei*
2b. Nostril midway between the eye and the snout tip — *Afrana angolensis*

Angolan River Frog

Afrana angolensis (Bocage, 1866)
(Plate 17.7)

The species was named for Angola.
Dusky-throated river frog, Nutt's frog, Chapin's frog, *senyamiganda* in Kiga, *akidot* in Karamojong, *lungula* in Logooli and Tiriki, *chula* in Taita and Kinga, *miula* in Hehe, *isodo* in Gisu, *lulenga* in Ganda.

DESCRIPTION: These are large robust frogs, with the males reaching a length of 74 mm and the females 110 mm. The head is about as broad as long, with the nostril midway between the eye and the snout tip. The tympanum diameter is half to three-fourths the eye diameter. The first finger is longer than the second. The tibia length is more than half the body length in adults, although it may be slightly less in some individ-

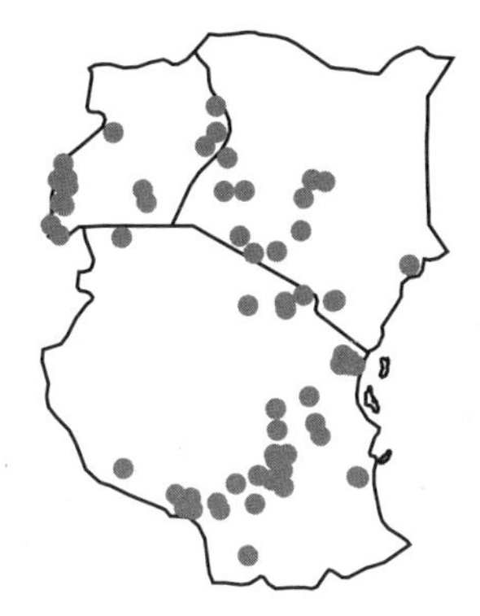

Fig. 194 *Afrana angolensis.*

uals. The web is dark. On the fourth toe two phalanges are free of web on the outside, and one is free on the inside, although there is variation, and some individuals may have two and a half phalanges free of web on one side. The fifth toe may be webbed to the very tip, or have up to one phalanx free. Vomerine teeth are in two oblique rows from the anterior borders of the internal nostrils to the level with the posterior borders. The distance between the vomerine teeth varies from one-third to equal to the length of one tooth. A narrow but distinct dorsolateral fold is present, sometimes broken into a series of ridges.

Coloration is variable. These frogs are commonly brown with green on the head. The legs and back of thighs are spotted, the undersides are yellow, and the throat, chest, and sides of the belly are marbled. Variations include a uniform back between the dorsolateral folds, or heavily spotted. Some individuals have bright green or yellow vertebral bands. Overall color may be green, tan, brown, or reddish. A dark band extends from the eye through the tympanum. Breeding males have minute spinosities on the back and sides that do not extend across the belly.

HABITAT AND DISTRIBUTION (FIG. 194): This species is a generalist, present in ponds, streams, and rivers in grassland and forest, from sea level to an altitude of 2800 m. It is known from Ethiopia south to southern Democratic Republic of Congo, Rwanda, Burundi, and Angola, through East Africa to South Africa.

ADVERTISEMENT CALL: Males call from the water, often floating near emergent vegetation. Calls may be heard throughout the year, and at all times of the day, with a peak after dark. The call is biphasic, consisting of a series of clicks and a number of croaks. There is some variation (see Notes below), but a typical call is 10–15 single or double clicks in 1.5 s, followed by a pause of about the same length, and then a croak or "wauk." The croak has a duration of about 0.5 s and consists of a large number of pulses, initially at a rate of 40/s, then quickly speeding up to 200/s. The dominant energy of the call is at 1.5 kHz. Some individuals may produce fewer clicks, while other click sequences may continue for a few seconds.

BREEDING: This species breeds throughout the year in East Africa, with a peak in December. The eggs are deposited in shallow water along the edges of ponds and flowing water.

TADPOLES: See the chapter on tadpoles.

NOTES: Although the populations of Angolan river frogs in East Africa cannot be separated based on morphology, the variation in advertisement calls suggests that there are cryptic species present. A molecular study and detailed call observations are required. Food records include beetles, bugs, caterpillars, millipedes, snails, spiders, crabs, moths, ants, flies, wasps, crickets, and lacewings. This large common frog is an important item of diet for many animals, and has been preyed on by Battersby's green-snake *Philothamnus battersby*, the black-lined green snake *Hapsidophrys lineatus*, green-snakes *Philothamnus* sp., the olive marsh snake *Natriciteres olivacea*, and the southern striped skaapsteker *Psammophylax tritaeniatus*. Apart from many snakes they are also eaten by banded mongoose *Mungos mungo*, the genet *Genetta* sp., the intermediate egret *Mesophoyx intermedius*, and the hammerkop *Scopus umbretta*. Driver ants *Dorylus* sp. will attack frogs they come across in vegetation.

KEY REFERENCES: Loveridge 1933, 1936.

Ruwenzori River Frog

Afrana ruwenzorica (Laurent, 1972)

This species is named for the Ruwenzori Mountains.

DESCRIPTION: This is the smallest of the East African river frogs. Maximum female length is 62 mm, with males known to be up to 55 mm long. The body is streamlined, with the head slightly longer than wide. The snout is longer than the eye width, with the nostril closer to

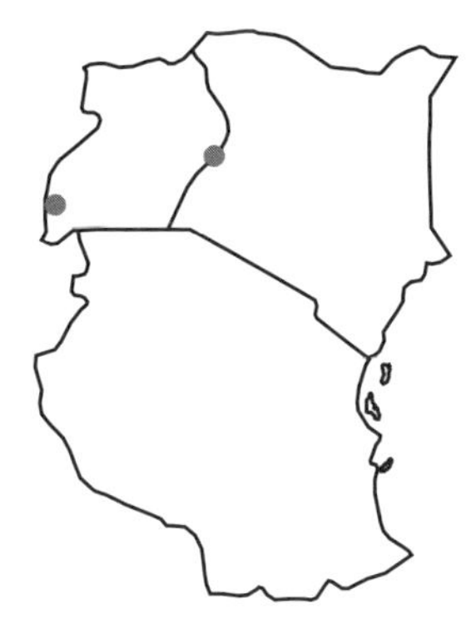

Fig. 195 *Afrana ruwenzorica.*

the eye than the snout tip. Internostril distance is equal to half the distance between the anterior corners of the eyes. The tympanum is distinct, its diameter being about two-thirds the diameter of the eye. Three phalanges on the fourth toe are free of webbing. The inner metatarsal tubercle is about 20% of the length of the first toe. The tibia length is more than half the body length. The skin is smooth or sometimes spiny in sexually active males. A dorsolateral fold runs from the eye to the inguinal region. The back is dark green to brown, with the head very spotted and a V mark between the eyes. A yellow vertebral stripe is found in many individuals. A dark band encloses the tympanum, and a dark line runs from the jaw to form a subocular mark reaching the anterior of the eye.

HABITAT AND DISTRIBUTION (FIG. 195): This species is associated with fast-flowing streams. It is known from the mountains between Uganda and the Democratic Republic of Congo, and Mt. Elgon between Uganda and Kenya.

ADVERTISEMENT CALL: Unknown.

BREEDING: Unknown.

TADPOLES: Tadpoles with high numbers of labial tooth rows have been collected but have not been positively associated with the adults. See the chapter on tadpoles.

KEY REFERENCE: Laurent 1972b.

De Witte's River Frog

Afrana wittei (Angel, 1924)

Named for G.-F. De Witte, herpetologist at the Institut Royal des Sciences Naturelles in Brussels until 1951.

Fig. 196 *Afrana wittei.*

DESCRIPTION: Males reach a length of 55 mm and females, 62 mm. The head is longer than broad, and vomerine teeth are present in oval groups between the internal nostrils. Internostril distance is equal to the width of the upper eyelid, with the nostril closer to the eye than the snout tip. The tympanum is distinct, its diameter being two-thirds to four-fifths the diameter of the eye. The first finger is slightly shorter than the second. Subarticular tubercles are small. Webbing is well developed, indented, but does not reach the tip of the third and fifth toes. Two phalanges of the fourth toe are free of web. The inner metatarsal tubercle is about a third of the inner toe length. There is no outer metatarsal tubercle. Tibia length is 40%–57% of body length. In males the gular sacs are dark and the throat is yellow.

HABITAT AND DISTRIBUTION (FIG. 196): This frog occurs at a high altitude in grasslands. It is known from the Kenyan highlands.

ADVERTISEMENT CALL: Unknown.

BREEDING: Unknown.

TADPOLES: Unknown.

KEY REFERENCE: Loveridge 1936.

White-Lipped Frogs—Genus Amnirana

This group of large water frogs has glandular skin. Many are characterized by a broad green or gold vertebral band, and all have a white glandular upper lip. Some species call while floating, although the species covered here may call from terrestrial sites. The spotted tadpoles have remarkable glandular patches on the body. Only two species are known from East Africa, although others may be expected in western Uganda.

KEY TO THE SPECIES

1a. Expanded discs on toes — *Amnirana albolabris*
1b. Toe tips not expanded — *Amnirana galamensis*

Forest White-Lipped Frog

Amnirana albolabris (Hallowell, 1856)
(Plate 17.8)

The species name refers to the white lip.
White-lipped hylarana, *mote* in Lega.

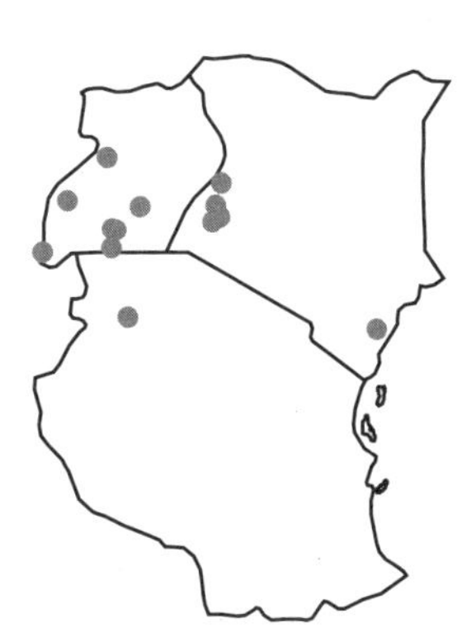

Fig. 197 *Amnirana albolabris.*

DESCRIPTION: Males reach a length of 57 mm and females, 74 mm. These frogs have large discs on the fingers and toes. Dorsal skin glands are distinct, with slight glands at the angles of the jaw. Webbing is moderate, with two to three phalanges on the fourth toe free of web. The back has distinct longitudinal glandular skin folds, and is brown with black spots. The back of the thigh is variable. A pale interrupted longitudinal band on each side is characteristic.

HABITAT AND DISTRIBUTION (FIG. 197): This forest species is known from West Africa, through the Congo Basin to Uganda, Kenya, and around Lake Victoria in Tanzania.

ADVERTISEMENT CALL: The males produce a soft call in chorus with others. The call increases in tempo and volume, then reaches a crescendo before ending.

BREEDING: This species breeds in temporary pools that form during the rainy season. Large masses of eggs are deposited that float at the surface. Each egg is 2 mm in diameter and pale with a dark pole.

TADPOLES: See the chapters on tadpoles.

NOTES: Recorded food items include crickets, stone flies, slugs, and snails. Tadpoles of this species are taken by the fishing spider *Thalassius spinosissimus*. Adults have been found perched on small branches just off the surface of the water in swampy situations. In the Ivory Coast this species is associated with the dwarf crocodile *Osteolaemis tetraspis*, from which it is protected by its skin secretions. Even catfish refuse to prey on the tadpoles. There is variation in the advertisement calls from different parts of its range, suggesting the presence of undescribed cryptic species.

KEY REFERENCES: Loveridge 1944, Perret 1977, McIntyre 1999, Rödel & Ernst 2001, Rödel 2003.

Galam White-Lipped Frog

Amnirana galamensis (Duméril & Bibron, 1841)
(Plate 18.1)

This species was named for the Galam lakes in Senegal, where it was discovered.
Golden-backed frog, *malondi* in Pokomo.

DESCRIPTION: The male grows to a length of 78 mm and the female, up to 86 mm. The distance between the eyes is equal to the distance from the nostril to the eye. The tympanum is slightly smaller than the eye. The tips of the fingers and toes are not expanded. Webbing is moderate, with two and a half to three segments of the fourth toe free of web. The breeding male has a pair of baggy throat pouches opening through a pair

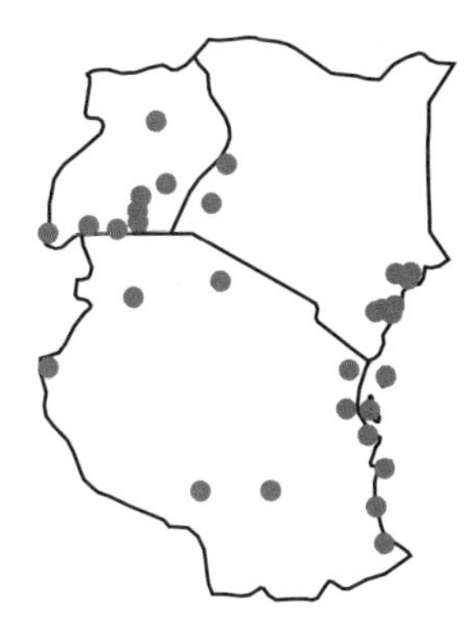

Fig. 198 *Amnirana galamensis.*

of slits parallel to the lower jaw. The color pattern is attractive; dark sides border a broad golden vertebral band, edged with a pair of flattened ridges running from eye to leg. The upper lip is lighter than the lower one. Males are yellowish, overlaid with brown. The flanks and thighs are marbled with brown on cream. Nuptial pads on the upper arms are cream, and the undersurfaces of the hands and feet are gray.

HABITAT AND DISTRIBUTION (FIG. 198): This frog is found in permanent ponds in savanna and coastal forest. It occurs from Mozambique to northern Zambia and northward to Somalia, then westward to Senegal. It has been found on the islands off Tanzania.

ADVERTISEMENT CALL: Males call while floating in water or resting on submerged vegetation. The males gather in permanent pools, calling within 75 cm of one another. The call resembles a nasal bleat, 0.6 s long, with emphasized frequencies between 2.0 and 3.0 kHz. The males chorus, with the calls becoming faster and faster until they reach a peak before pausing. A shorter call is also produced.

BREEDING: Females deposit around 5000 eggs, laid in clutches of 1500 or more. The eggs are 1.5–2.0 mm in diameter within a jelly capsule 6.0–7.0 mm. The eggs float on the surface in a single layer, with the dark pole on top. They are usually attached to vegetation, where they are shaded.

TADPOLES: The tadpoles hatch and soon exhibit the characteristic spotted color pattern. See the chapter on tadpoles.

NOTES: Known food items include grasshoppers, caterpillars, dragon flies, spiders, crickets, wingless wasps, beetles, and scorpions. This frog uses termitaria as refuges during the dry season and has been found emerging at the start of the rains. The skin contains bradykinins, substances that mimic mammalian hormones by slowing the heart rate and making any mammal predator feel very ill. Local villagers in West Africa eat this species. Cooking destroys the activity of the skin secretions. In West Africa the males call from terrestrial positions.

KEY REFERENCES: Loveridge 1936 (as *Rana albolabris*), Roseghini et al. 1988 (as *Hylarana galamensis*), Rödel 2000 (as *Hylarana galamensis*).

Torrent Frogs—Genus *Arthroleptides*

This genus consists of frogs associated with fast-flowing streams. They have large discs on the fingers and toes that are divided into two. The tadpoles are able to develop on rocks that have a film of water running over them. The pupil is horizontal. All three species in the genus occur in East Africa.

KEY TO THE SPECIES

1a. Base of glandular supratympanic ridge thickened, animals never over 35 mm in snout-vent length — *Arthroleptides dutoiti*
1b. Base of glandular ridge never thickened, mature animals exceed 50 mm in snout-vent length — 2

2a. Web reaching or passing the proximal subarticular tubercle of the fourth toe (Fig. 199) — *Arthroleptides yakusini*
2b. Web never reaching the proximal subarticular tubercle of the fourth toe (Fig. 200) — *Arthroleptides martiensseni*

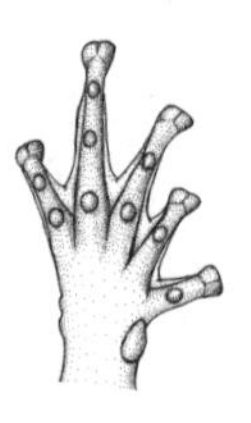

Fig. 199 Web passing proximal tubercle.

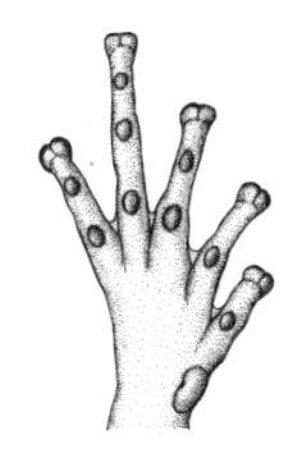

Fig. 200 Web not passing proximal tubercle.

Mt. Elgon Torrent Frog

Arthroleptides dutoiti Loveridge, 1935

This species was named for C. A. du Toit of Stellenbosch University, in South Africa, who first collected it.
Elgon montane torrent frog.

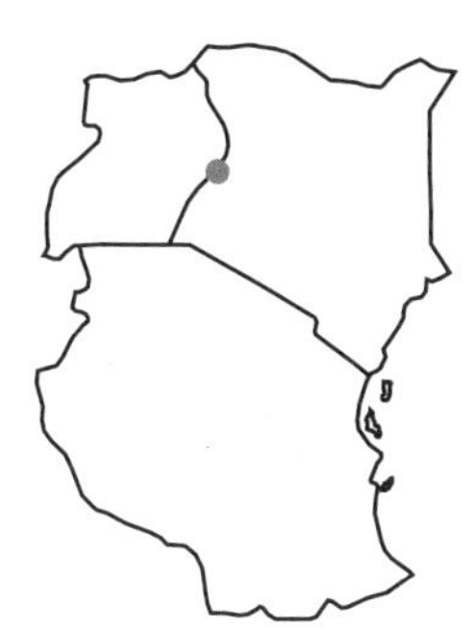

Fig. 201 *Arthroleptides dutoiti.*

DESCRIPTION: This is the smallest of the three species, with females reaching a length of 31 mm and males 25 mm. The head is slightly broader than long, and vomerine teeth are absent. The nostrils are situated midway between the eye and the snout tip. The tympanum is nearly two-thirds of eye width. The interorbital space is about equal to the width of upper eyelid. The base of the supratympanic ridge is thickened. The fingers and toes have distinct discs, each with an upper median groove. The toes have slight webbing. The skin is rough, especially in sexually active males. The back is brown with darker markings. The tips of the digits are edged with white.

HABITAT AND DISTRIBUTION (FIG. 201): This species has been found in streams on Mt. Elgon on the Kenya side, but probably occurs wherever suitable streams are present on the mountain. It is only known from Mt. Elgon, on the border between Kenya and Uganda.

ADVERTISEMENT CALL: Unknown.

BREEDING: Unknown.

TADPOLES: Unknown.

NOTES: This species has not been seen in the field for a number of years.

KEY REFERENCES: Loveridge 1935, Channing et al. 2002b.

Martienssen's Torrent Frog

Arthroleptides martiensseni Nieden, 1910
(Plate 18.2)

This species is named for Martienssen, the station commander at Tanga in Tanzania, who presented a collection of amphibians and reptiles to the Berlin Zoological Museum.
Usambara montane torrent frog.

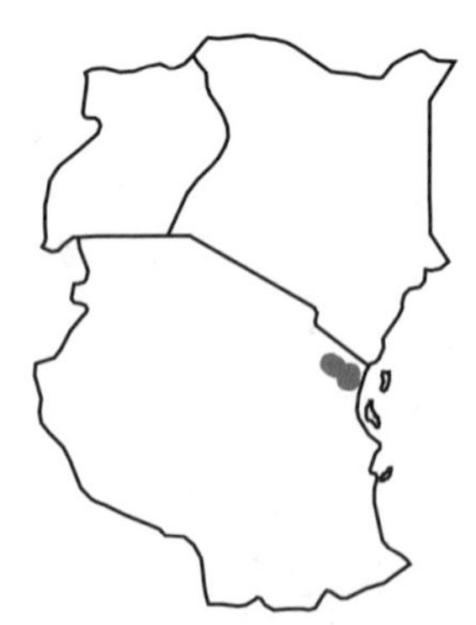

Fig. 202 *Arthroleptides martiensseni.*

DESCRIPTION: The males reach a length of 74 mm; the largest female known is 67 mm. The interorbital space equals the width of the upper eyelid in adults, and the nostril is one and a half to twice as far from the eye as the tip of the snout. The canthus rostralis is angular. The tympanum is round and about two-thirds the size of the eye. The skin is smooth in adults, but may be slightly rough in juveniles. The back is dark olive with a brown interorbital bar and a dark band from the nostril through the eye. The lips are dark flecked with white, and the throat is off-white flecked with brown. The belly is white. The palms and soles are dark brown. Webbing reduced, not reaching subarticular tubercle of the fourth toe. Breeding males develop thickened forearms and a dark papilla on the tympanum.

HABITAT AND DISTRIBUTION (FIG. 202): These frogs are associated with forest; some have been found at the sides of streams, others beneath vegetation debris on the forest floor. This species is known from the northern Eastern Arc, the Usambara, and the Magrotto mountains.

ADVERTISEMENT CALL: Unknown.

BREEDING: Unknown. A female full of black eggs was collected 15 December.

TADPOLES: The tadpoles are able to live and develop in a thin film of water. See the chapter on tadpoles.

NOTES: Adults can be found around waterfalls on ledges or sitting in very shallow, fast-flowing water. They escape by jumping from boulder to boulder across the stream. Food items include grasshoppers, wasps, beetles, flies, and spiders.

KEY REFERENCES: Barbour & Loveridge 1928a, Loveridge 1944, Channing et al. 2002.

Southern Torrent Frog

Arthroleptides yakusini Channing et al., 2002
(Plate 18.3)

The name *yakusini* is derived from Swahili for "of the south."

DESCRIPTION: These are large frogs, with the females reaching a length of 68 mm and exceptional males reaching 73 mm. The head is broad. The horizontal diameter of the tympanum is slightly less than half the distance between the anterior corners of the eyes. In breeding males a black tympanic papilla protrudes from the upper half of the tympanum, which is dark, round, and ringed with small tubercles. The

Fig. 203 *Arthroleptides yakusini.*

supratympanic ridge obscures the upper posterior margin of the tympanum. The terminal discs of the fingers and toes are large, each divided with a pair of dorsal scutes. The main webbing passes the proximal subarticular tubercle of the fourth toe on the inside and reaches it on the outside. The main webbing reaches the middle subarticular tubercle of the fifth toe. The dorsal skin is granular, with small white-tipped warts on the side of the head. The upper and lower jaws are edged with minute dark-tipped spines. The back and upper limbs are gray-brown with darker mottling. The back of the thigh is speckled with white on a dark background. The throat is dark with pale speckling, while the chest and belly have a paler background with white speckles. The underside of the limbs is pale.

HABITAT AND DISTRIBUTION (FIG. 203): These frogs are found along rocky streams, where adults may sit on wet rocks or retire into cracks beneath huge boulders. The species occurs on the Udzungwa and Uluguru mountains in Tanzania.

ADVERTISEMENT CALL: Males call from September, peaking after the start of the short rains in November. Calling takes place after dark, from the forest floor, although males may climb onto rocks or fallen tree trunks to call. The call is a series of short "wauks" repeated at long intervals of up to 25 s. Calls consist of 4–8 pulses produced in 60 ms. The emphasized frequency varies from 1.2 to 1.4 kHz.

BREEDING: Breeding takes place from September, if there are rains, with a peak in December and January. Eggs are laid in a clear jelly on wet rock faces, camouflaged against the dark reflective surface. The rock faces may be horizontal to vertical, either with a small amount of moisture trickling over them or with a thin film of flowing water. Egg masses vary from about 200 to small clutches of 10 or less. The eggs hatch into tadpoles within 3 or 4 days. The tadpoles remain on the same rock, with eggs and tadpoles frequently being found together. Each egg is 2 mm in diameter within a 4-mm capsule.

TADPOLES: The tadpoles are specialized for life on wet rock faces. See the chapter on tadpoles.

NOTES: The adults forage on the forest floor away from rivers. At night the juveniles may be found on vegetation at the edge of streams or concealed in crevices in stream banks or tree roots.

KEY REFERENCES: Klemens 1998 (as *Arthroleptides martiensseni*), Channing et al. 2002.

Dainty Frogs—Genus *Cacosternum*

These small frogs are typically slender with thin limbs. Five species have been described, ranging from the southern tip of Africa to western Somalia. A number of other cryptic species are also present. At the time of writing, a study to determine the status of these populations is in progress. Only one species is recorded in East Africa.

Plimpton's Dainty Frog

Cacosternum sp.
(Plate 18.4)

This frog does not yet have a scientific name.

Fig. 204 *Cacosternum* sp.

DESCRIPTION: This small frog is not known to exceed 25 mm in length. The body shape is characteristic, with a narrow head widening to the belly region. The arms and legs are thin and long. The feet are unwebbed. The back pattern is very variable, from green through brown, with spots or stripes. The undersurface has discrete, small gray or black spots.

HABITAT AND DISTRIBUTION (FIG. 204): This species is found in grassland and the arid acacia savanna in northern Tanzania to central Kenya. Records of dainty frogs from Ethiopia and Somalia attributed to *Cacosternum boettgeri* may be this species.

ADVERTISEMENT CALL: The male calls from water level. Usually it is concealed in cracks or under vegetation. The call is a rapid series of high-pitched clicks. Each click consists of pairs or triplets of pulses. Four to seven clicks are uttered at a rate of 10/s, at an emphasized frequency of 4.2–4.8 kHz.

BREEDING: This species breeds in flooded depressions, especially where grass is growing. Breeding takes place during the short rains from October to April in the East African highlands, although a second bout of calling and breeding may occur during the long rains. Amplexus is axillary. The eggs are laid underwater attached to grass stems, with some eggs sinking to the bottom. Each egg is 0.6 mm in diameter. Clutch size is 250.

TADPOLES: See the chapter on tadpoles.

NOTES: At the time of writing, this species is being described.

Ornate Frogs—Genus *Hildebrandtia*

These medium-sized frogs are stocky and brightly patterned, with greens, golds, pinks, blacks, and browns being the dominant colors. The males have paired vocal pouches, with openings below and behind the angle of the jaw. Ornate frogs are found throughout moist savannas. There are three species, of which two are found in East Africa.

KEY TO THE SPECIES

1a. Throat lightly speckled, tympanum about the same size as the eye, simple subarticular tubercles under fingers and toes (Fig. 205) *Hildebrandtia macrotympanum*

1b. Light and dark longitudinal markings on the throat, tympanum smaller than the eye *Hildebrandtia ornata*

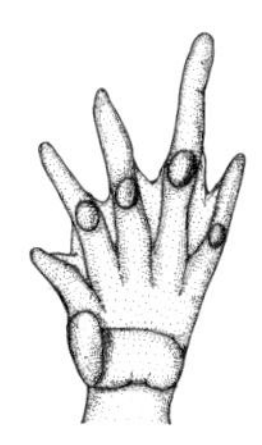

Fig. 205 Enlarged inner metatarsal tubercle.

Northern Ornate Frog

Hildebrandtia macrotympanum (Boulenger, 1912)
(Plate 18.5)

The specific name refers to the large tympanum.

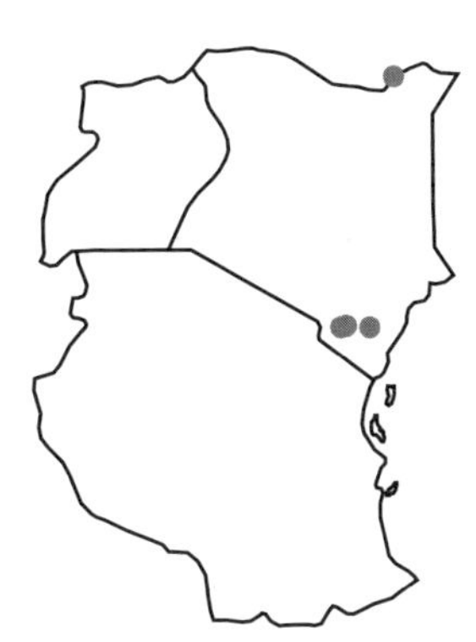

Fig. 206 *Hildebrandtia macrotympanum.*

DESCRIPTION: The females can exceed a length of 45 mm. The body is stout, and the head is longer than wide. The snout is nearly twice as long as the eye diameter, about equal to the distance between the anterior corners of the eyes. The nostril is nearer to the eye than to the snout tip. The tympanum is elliptical, its larger vertical diameter just less than the diameter of the eye. An inconspicuous white glandular ridge runs below the tympanum. Each finger has a simple basal subarticular tubercle. There is no outer metatarsal tubercle, and the inner metatarsal tubercle is strongly compressed and flangelike. Three and a half to three and two-thirds of the phalanges on the fourth toe are free of web. The back is brown to olive green, with a pattern that includes dark mottling on the sides and limbs, a dark dot between the eyes, and a darker gray band from the side of the snout passing below the eye, including the tympanum, and ending at the hind leg. It is edged with black above. The upper lip is white. The underside is white with gray mottling on the jaw and slight speckling on the throat. The eye is golden above, and copper below with darker mottlings.

HABITAT AND DISTRIBUTION (FIG. 206): This species is found in sandy wooded savanna, from southern Ethiopia and Somalia to southern Kenya.

ADVERTISEMENT CALL: Males call from flooded seasonal pools. The call is a brief hoot. It has a duration of 0.1 s, at an emphasized frequency of 0.62 kHz. The calls are produced in rapid succession.

BREEDING: Unknown.

TADPOLES: See the chapter on tadpoles.

KEY REFERENCE: Balletto et al. 1978 (as *Tomopterna scorteccii*).

Common Ornate Frog

Hildebrandtia ornata (Peters, 1878)
(Plate 18.6)

The name refers to the ornate back with green, brown, gold, and white markings.
Hildebrandt's burrowing frog.

DESCRIPTION: This stocky frog may reach a length of 65 mm in the male and 70 mm in the female. Webbing is not extensive, just reaching the middle tubercle beneath the fourth toe. The inner metatarsal tubercle is longer than the first toe, and flattened for digging. The diameter of the tympanum is about two-thirds of the eye diameter, and equal to or greater than the eye-nostril distance. A glandular dorsolateral ridge

Fig. 207 *Hildebrandtia ornata.*

runs from above the tympanum to the sacral region. The fourth toe has three to three and a half phalanges free of web. The outer metatarsal tubercle is feeble or absent. Breeding males have nuptial pads that are light brown, covering the upper three inner fingers and extending onto the forearm. The frog is very beautifully marked, with a broad golden brown or green band down the back. Other green, golden, or brown patterns are common. A pair of characteristic Y-shaped light bands is found on the throat, against a dark background.

HABITAT AND DISTRIBUTION (FIG. 207): This species is found in savanna, from West Africa through to eastern and southern Africa.

ADVERTISEMENT CALL: The male starts calling later than most species during the evening, often not before 22h30min. It calls from shallow pans or some meters from the edge of water. The call is a harsh bellow, with a duration of 0.3–0.4 s and a dominant harmonic at 1 kHz rising to 1.8 kHz, with a second harmonic at 2.5 kHz rising to 2.8 kHz. Each call consists of very rapidly pulsed notes.

BREEDING: The eggs are 1.4 mm in diameter within 3-mm capsules and are scattered singly in shallow water.

TADPOLES: The remarkable tadpoles of this species are carnivorous, attacking the tadpoles of other species. See the chapter on tadpoles.

NOTES: Small toads (*Bufo gutturalis*) are a large part of the diet of this frog.

Other food items include ants, crickets, beetles, bugs, grasshoppers, snails, caterpillars, myriapods, earthworms, spiders, and termites. Individuals produce a cocoon during the dry season, which they shed after the first rains.

KEY REFERENCES: Schmidt & Inger 1959, Inger 1968, Rödel 2000.

Groove-Crowned Bullfrogs—Genus *Hoplobatrachus*

These large frogs are similar in overall shape to the bullfrogs, but have a transverse groove behind the eyes. The tadpoles are carnivorous. At least two species are known, of which only one reaches East Africa.

Eastern Groove-Crowned Bullfrog

Hoplobatrachus occipitalis (Günther, 1859)
(Plate 18.7)

The name *occipitalis* refers to the transverse pale groove behind the eyes. Giant swamp frog, African tigrine frog, tiger frog, *chula* in Fipa, *lunda* in Manyema, *Riesen-Wasserfrosch* or *Afrikanischer Tigerfrosch* in German.

Fig. 208 *Hoplobatrachus occipitalis.*

DESCRIPTION: This large robust frog may grow to 160 mm long in females and 104 mm in males. A characteristic skin groove passes across the head just behind the eyes; it may appear only as a pale line in preserved specimens. The webbed feet and delicate hands show the aquatic nature of this frog, and are similar to those of the clawed frogs *Xenopus*.

Webbing is extensive, even on the outside of the toes. There is one subarticular tubercle at the base of the fingers, with two or more on each toe. The eyes are positioned on top of the head. The male has a small round opening to each vocal pouch, situated near the angle of the jaw. This vocal pouch opening is approximately the same size as the eye. The tympanum is equal in size to the eye diameter. Three toothlike protuberances are found on the lower jaw. The back is generally a shade of gray with darker markings, with a pale occipital patch behind and between the eyes. The lips may be yellow to red. A black stripe runs from the eye to the angle of the jaw. The underside is white with large dark irregular spots. There is sometimes a greenish tinge in the layer of mucus on the skin, caused by chains of spherical algae. This is especially noticeable in older specimens, which might be bright green.

HABITAT AND DISTRIBUTION (FIG. 208): This species is found in lakes, deep permanent ponds, swamps, and rivers. At night it can be found floating, even in very shallow water. This widespread frog is known from northern Angola and Zambia, through northwestern Tanzania, Uganda, and Kenya, to Ethiopia and North Africa, and westward to Senegal.

ADVERTISEMENT CALL: The call consists of a series of deep notes, each 0.15 s long with emphasized harmonics at 0.8 and 2.5 kHz. The internote interval is 0.2 s.

BREEDING: This frog breeds throughout the year. Each female produces 469–3752 eggs that are white with a black pole and measuring from 2.9 to 3.7 mm in diameter. The eggs are attached singly to the pond bottom.

TADPOLES: The tadpoles are carnivorous. See the chapter on tadpoles.

NOTES: These frogs are able to escape disturbance by bounding rapidly across the surface of the water into vegetation on the opposite side of the pool. They feed on many smaller frogs, with evidence that they are cannibalistic. Feeding records include the sharp-nosed ridged frog *Ptychadena oxyrhynchus*, the guttural toad *Bufo gutturalis*, the desert toad (and their tadpoles) *Bufo xeros*, the water lily reed frog *Hyperolius pusillus*, fish, snakes, lizards, birds, tadpoles, river crabs *Potamonautes* sp., beetles, ants, spiders, dragonflies, dragonfly nymphs, termites, crickets, bees, grasshoppers, and a few other insects. They are able to live in warm springs with water between 36 °C and 40 °C, where they maintain a core temperature between 28 °C and 32.7 °C, while the sympatric common toad *Bufo regularis* maintains temperatures between 21.0 °C and 25.6 °C. Recent molecular work indicates that the African species of *Hoplobatrachus* are closely related to a number of Asian species, and suggests that the ancestors of the African species migrated out of Asia in the Miocene.

KEY REFERENCES: Lamotte & Züber-Vogeli 1954; Guibé & Lamotte 1958; Schiøtz 1963; Inger 1968; Micha 1975; Salvador 1996 (all as *Rana occipitalis* or *Dicroglossus occipitalis*); Spieler & Linsenmair 1997, 1998; Rödel 2000; Kosuch et al. 2001.

Puddle Frogs—Genus *Phrynobatrachus*

This is a large genus of small brown frogs. They are common in practically all damp habitats in the warmer areas and are an important component of the tropical ecosystems, as they breed continuously and reach maturity at 5 months. Many different color patterns are known. Savanna species share one set of color patterns, while rain forest species share a different set. The shared patterns in different habitats probably serve as camouflage against common predators. There are 20 species presently known from East Africa. Be aware that there are undescribed species, many of which will key out here as *Phrynobatrachus natalensis*.

KEY TO THE SPECIES

1a. Tympanum visible — 9
1b. Tympanum not visible — 2

2a. Male throat white — *Phrynobatrachus uzungwensis*
2b. Male throat speckled, gray to yellow — 3

3a. Male throat deep yellow anteriorly — *Phrynobatrachus scheffleri*
3b. Male throat not yellow — 4

4a. Tips of fingers having very small swellings — 5
4b. Tips of fingers not expanded — 6

5a. Male throat black — *Phrynobatrachus ukingensis*
5b. Male throat gray — *Phrynobatrachus rungwensis*

6a. Back of thigh blotched *Phrynobatrachus keniensis*
6b. Back of thigh with pale band 7

7a. Male throat black *Phrynobatrachus parvulus*
7b. Male throat gray 8

8a. Broad web between the third and fourth toes extending beyond proximal subarticular tubercle of the fourth toe *Phrynobatrachus stewartae*
8b. Broad web between the third and fourth toes not extending beyond proximal subarticular tubercle of the fourth toe *Phrynobatrachus mababiensis*

9a. Discs on fingers absent 18
9b. Discs present, or fingers slightly swollen 10

10a. Discs distinct (Fig. 209) 11
10b. Discs very small, only swellings 12

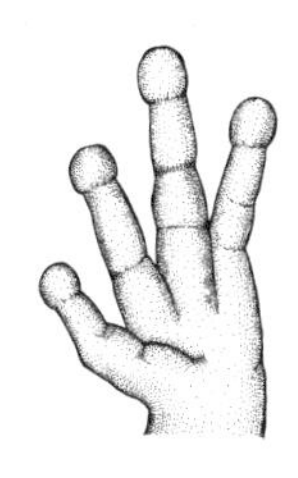

Fig. 209 Distinct discs.

11a. Tympanum smaller than eye 19
11b. Tympanum larger than eye *Phrynobatrachus dendrobates*

12a. Tympanum larger than eye *Phrynobatrachus versicolor*
12b. Tympanum smaller than eye 13

13a. Less than two phalanges of the fourth toe free of web *Phrynobatrachus perpalmatus*
13b. Two or more phalanges of fourth toe free of web 14

14a. Two to three phalanges of fourth toe free of web 15
14b. More than three phalanges of fourth toe free of web 16

15a. Male throat speckled *Phrynobatrachus kinangopensis*
15b. Male throat black (solid color) *Phrynobatrachus acridoides*

16a. Spines on tips of the first two toes (Fig. 210), two ridges from eye to shoulder, body over 50 mm long *Phrynobatrachus irangi*
16b. No spines on first two toes, a single ridge from eye to shoulder, less than 30 mm long 17

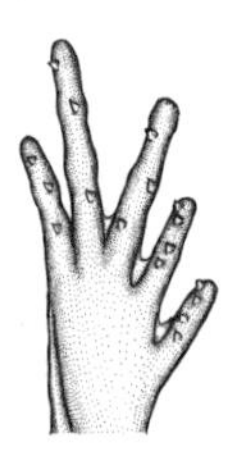

Fig. 210 Spines on first two toe tips.

17a. Base of hind limb yellow, tympanum half the size of the eye, male throat black *Phrynobatrachus graueri*
17b. No yellow on hind limb, male throat brown *Phrynobatrachus rouxi*

18a. Males up to 34 mm long *Phrynobatrachus natalensis*
18b. Males up to 24 mm long *Phrynobatrachus bullans*

19a. Male throat yellow *Phrynobatrachus kreffti*
19b. Male throat black *Phrynobatrachus auritus*

East African Puddle Frog

Phrynobatrachus acridoides (Cope, 1867)
(Plate 18.8)

The specific name *acridoides* means "cricket-like," referring to the advertisement call.
Small puddle frog, Zanzibar puddle frog, Zanzibar river frog, *koko* in Nyakusa.

DESCRIPTION: The male reaches 28 mm long and the female, 30 mm. The male has small folds in the vocal pouch running parallel to the jaw. The tympanum is visible. Small discs are present as swellings on the tips of the fingers and toes. Webbing is variable, but the main web reaches the outer tubercle beneath the third toe, without passing the middle

Fig. 211 *Phrynobatrachus acridoides.*

tubercle of the fourth toe. A glandular ridge runs from each eye back to the shoulder region, sometimes continuing backward as a border to the dorsal band. Other paired warts may occur on the back. Males have black throats that become roughened in the breeding season. The back is gray-brown with darker markings, sometimes with a pale or green vertebral band. The upper jaw has light speckling. The male has an evenly shaded throat region, which is speckled in the female. A pale stripe occurs on the back of the thigh. The belly is white with a yellowish tinge in the groin.

HABITAT AND DISTRIBUTION (FIG. 211): This species is found in savanna and forest wherever there is water, and on the coastal plain or below an altitude of 700 m with a coastal fauna. It is known from the islands of Pemba and Zanzibar, where it was first reported. This species is distributed from Somalia southward to northeastern South Africa.

ADVERTISEMENT CALL: The call is a harsh creak repeated continuously day and night. Males call while resting on vegetation at the surface in shallow water. Each call has a duration of about 0.9 s, with a pulse rate of 50/s. The emphasized frequency is 2 kHz. They call from the start of the short rains through to mid-April or longer.

BREEDING: The breeding season is March to May. The eggs are small, dark, and 0.8 mm in diameter and may be attached to grass at the surface or floating in single-layered mats 200 mm across. Eggs were found in the Udzungwa Mountains in groups of 10–200.

TADPOLES: Unknown.

NOTES: Recorded food items include beetles, moth larvae, and caterpillars. They are preyed on by the floodplain viper *Atheris superciliaris*, the olive sand snake *Psammophis sibilans*, and the savanna vine snake *Thelotornis capensis*. The small discs may improve this frog's ability to move over soft mud. We are unable to separate *Phrynobatrachus pakenhami* found on Pemba Island from *P. acridoides* and therefore regard the former as synonymous with the latter.

KEY REFERENCES: Loveridge 1936, Dudley 1978, Dudley et al. 1979, Channing 2001.

Golden Puddle Frog

Phrynobatrachus auritus Boulenger, 1900
(Plate 19.1)

The species name means "golden."

Fig. 212 *Phrynobatrachus auritus.*

DESCRIPTION: The females reach a length of 38 mm, with the males being slightly shorter. This species possesses a conical papilla on the tongue. The snout is shorter than the eye diameter. Nostrils are slightly closer to the tip of the snout than the eye. The tympanum is distinct, half the size of the eye. The fingers are long with distinct discs; the first finger is shorter than the second. The subarticular tubercles are large. The tibia is 53%–57% of the body length. The toes have distinct discs, and webbing reaches the disc of the fifth toe. The inner metatarsal tubercle is small and oval. The tubercle on the side of the tarsus is conical. The skin is smooth or finely granular, with a glandular ridge above the tympanum but no glandular ridge on the head. The ventrum is smooth, and breeding males have dark gular pouches with distinct folds. The upper surface is brown, with small black spots in some individuals.

HABITAT AND DISTRIBUTION (FIG. 212): This forest species is known from Nigeria to the Congo Basin and western Uganda.

ADVERTISEMENT CALL: The males call from the ground or at the edge of standing water. The call is a harsh trill of about 20 pulses produced at a rate of 60/s, with an emphasized frequency of 1.6 kHz.

BREEDING: Unknown.

TADPOLES: Unknown.

KEY REFERENCES: Boulenger 1919 (as *Phrynobatrachus discodactylus*), Perret 1966 (as *Phrynobatrachus plicatus auritus*), Márquez et al. 2000.

Bubbling Puddle Frog

Phrynobatrachus bullans Crutsinger et al., 2004
(Plate 19.2)

The species name *bullans* means "bubbling," referring to the air that this species expels when diving.

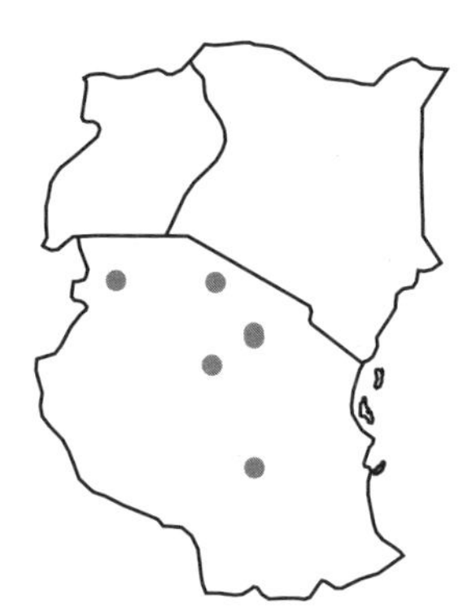

Fig. 213 *Phrynobatrachus bullans.*

DESCRIPTION: Males reach a length of 24 mm, with females up to 27 mm long. The snout is short and rounded. A median papilla is present on the tongue. The tympanum is round, darker than surrounding pigmentation, and 64% the size of the eye. The tympanum diameter is 30% of the tympanum-snout length and 33% the jaw width. The distance between the anterior corners of the eye is 81% of the eye width and equal to the tympanum diameter. The gular region of the male is dark gray with minute asperities located anteriorly on the throat. The gular sac is not baggy and has folds running parallel to the jaw line. The nostrils are located much closer to the snout than the eyes and have a slightly raised ridge bordering them.

The tibia is 46% and the femur, 43% of the body length. No femoral glands are present. The dorsum is mostly smooth with small, round asperities on the lower back and legs. The belly is white and is not speckled. Finger and toe tips are rounded but not swollen or expanded into discs. The foot length is 47% of the body length. Three phalanges of the fourth toe are free of web. The back is gray to brown. The upper jaw is mottled and the lower jaw is banded. A light eye-to-arm band is present. The anterior face of the thigh and posterior part of the tibia are banded. The posterior face of the thigh is speckled but contains no lateral banding.

HABITAT AND DISTRIBUTION (FIG. 213): This frog is found in savanna on the upland plateau of central Tanzania from the Rukwa Valley to the Serengeti Plain.

ADVERTISEMENT CALL: The call is a long series of notes produced over 20 s or more. The call starts softly, gradually building in volume over the first 10 s. The call proper appears to stop at 16 s, with a few longer, irregular notes thereafter. Some individuals only produce the regular series of notes. The notes are produced at a rate of 3.5/second. Each note is pulsed, with a typical note consisting of 20 pulses at a rate of 125/s. The emphasized frequency is 2.9 kHz.

BREEDING: Unknown, although males call during both short and long rains.

TADPOLES: Unknown.

NOTES: This species is very similar to the Natal puddle frog. The most reliable character for distinguishing the two species is the very different advertisement call.

KEY REFERENCE: Crutsinger et al. 2004.

Climbing Puddle Frog

Phrynobatrachus dendrobates (Boulenger, 1919)
(Plate 19.3)

The specific name refers to the tree frog–like size of the large discs. Disk-toed puddle frog.

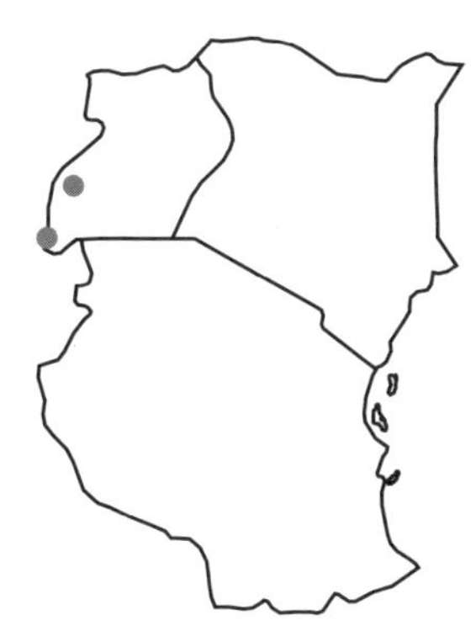

Fig. 214 *Phrynobatrachus dendrobates.*

DESCRIPTION: The males may reach a length of 31 mm and the females, up to 33 mm. The head is as long as wide. The snout overhangs the lower jaw and is longer than the eye width. The nostrils are closer to the snout than the eyes. The tympanum is twice as large as the eye. The fingers have characteristic, well-developed discs, with the largest twice the width of the base of the finger. The tibia is about 60% of the body length. Three phalanges of the fourth toe are free of web. The outer edge of the fourth and fifth toes, and to a degree the third toe, has spiny tubercles. The back is brown with a yellow band in the eye. The limbs are barred, with small pale spots behind the thigh. The undersurfaces are yellow with brown vermiculations, with a thin pale midventral stripe under the jaw, sometimes continuing backward, and often with a transverse pale stripe from arm to arm. The male throat is yellow.

HABITAT AND DISTRIBUTION (FIG. 214): This species is found on damp leaves on the forest floor or perched on low leaves along the banks of small rivers. It is known from eastern Democratic Republic of the Congo and western Uganda.

ADVERTISEMENT CALL: Males call from elevated positions in vegetation. The vocalization consists of a twittering resulting from a series of calls. Each call consists of 3 pulses in 150 ms. The dominant energy is at 3.8 kHz.

BREEDING: Clutch size is about 38. The eggs have been found in a rolled dead leaf, 2 m above water. The male remains with the eggs and calls to attract other females to his position. One nest contained 17 eggs with well-developed tadpoles, while another 9 had been fertilized and were just starting to develop. This suggests that the male attracts different females to his nest site. Other clutches are laid on leaves above water, with the male nearby.

TADPOLES: Unknown.

NOTES: This frog is known to be eaten by the black-lined green snake *Hapsidophrys lineatus*.

KEY REFERENCES: Boulenger 1919, Loveridge 1944.

Grauer's Puddle Frog

Phrynobatrachus graueri (Nieden, 1911)

This frog is named for the collector, Rudolf Grauer, who worked for the Zoological Museum of Berlin on an expedition led by H. Schubotz.

Senyamiganda in Kiga, but not specific; *ntoli* in Tooro; *etoli* in Konjo; *Kashakara* in Lega, although this term is used for other small frogs as well.

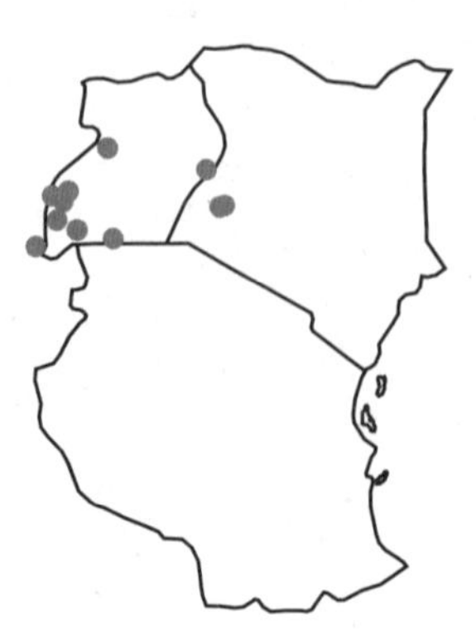

Fig. 215 *Phrynobatrachus graueri.*

DESCRIPTION: These are small frogs, with the male reaching 23 mm long and the females 25 mm or a little more. The snout is short, slightly longer than the horizontal eye diameter. The nostrils are positioned midway between the snout tip and the eyes. The tympanum is distinct, equal to half the eye width. The fingers and toes have small discs. There is slight webbing at the base of the toes, with three and a half phalanges of the fourth toe free of web. A glandular skin fold runs from behind the eye toward midline to the level behind the shoulder. Females are dark olive above with a darker band from the nostril to the forearm. A narrow or wide vertebral stripe is present, and a pair of light lines flank the vent. The throat is black in males, and the base of the hind limb is often yellow.

HABITAT AND DISTRIBUTION (FIG. 215): This forest species is found in damp leaf litter near streams. It is known from Rwanda, eastern Democratic Republic of the Congo, and Uganda, to western Kenya.

ADVERTISEMENT CALL: The advertisement call is described as "tink-tink" or clicking. It has not been recorded and analyzed.

BREEDING: Unknown.

TADPOLES: Unknown.

NOTES: This species is recorded as food of Günther's green tree snake *Dipsadaboa unicolor.*

KEY REFERENCES: Loveridge 1936, 1944.

Irangi Puddle Frog

Phrynobatrachus irangi Drewes & Perret, 2000

This species is named for the Irangi Forest in Kenya.

DESCRIPTION: This is the largest East African puddle frog, with females exceeding 50 mm in length. It is distinguished from the other large species (Krefft's puddle frog *Phrynobatrachus kreffti*) by less webbing (three phalanges of the fourth toe are free), a snout that is rounded in profile, and no bright yellow under the throat of breeding

Fig. 216 *Phrynobatrachus irangi.*

males. The frog is robust, with an oval tympanum three-fourths the diameter of the eye. The toe tips are slightly expanded. Males have one or two spines on the underside of the tips of the first two toes, as well as many spines on the walking surface of the foot. The tarsal tubercle is distinct with a pale spine. A pair of glandular ridges runs from behind the eye slightly toward the midline. A second pair of ridges continues backward running away from the midline. The frogs are pale tan to reddish brown with darker brown markings. A pale area covers the top of the snout from between the eyes, with the same color in a patch behind the tympanum. The underside is pale, sometimes with darker mottling.

HABITAT AND DISTRIBUTION (FIG. 216): This high-altitude grassland species is known from the Aberdare Mountains and the slopes of Mt. Kenya.

ADVERTISEMENT CALL: Males call during the day. The call is described as a series of rasping notes but has not been analyzed.

BREEDING: Unknown.

TADPOLES: Unknown.

NOTES: This species is active during the day.

KEY REFERENCE: Drewes & Perret 2000.

Upland Puddle Frog

Phrynobatrachus keniensis Barbour & Loveridge, 1928
(Plate 19.4)

The species name *keniensis* means "of Kenya."

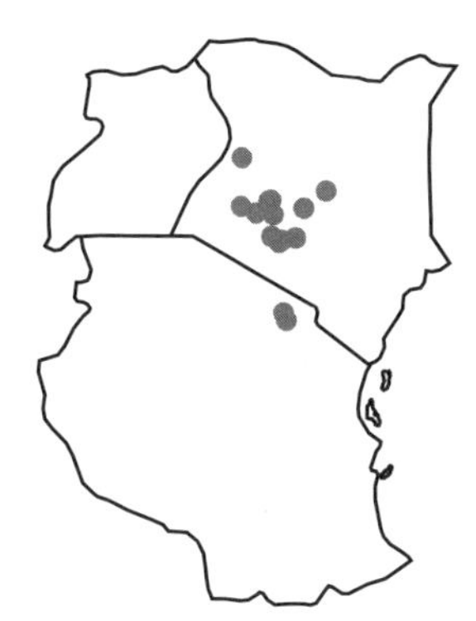

Fig. 217 *Phrynobatrachus keniensis.*

DESCRIPTION: The males reach a length of 21 mm, with the females up to 26 mm long. The nostril is midway between the snout tip and the eye. There is an indentation or fold at the anterior end of the upper eyelid. The tympanum is hidden. The toes are about one-third webbed, with three phalanges of the fourth toe free of web. The tips of the fingers and toes are not dilated. The dorsal skin is smooth with flattened warts. The back can be a uniform gold to brown, through to heavily blotched. A yellow vertebral line is sometimes present. Often there is a dark lateral band below a line of glandular warts. The undersurface is white with brown spots, especially in a broad band across the chest. The chin and throat are speckled with reddish brown. The back of the thigh is blotched.

HABITAT AND DISTRIBUTION (FIG. 217): This frog is found on the edge of water bodies like swamps and slow streams, at altitudes up to 3000 m. It is known from the highlands of Mts. Kikuyu, Molo, Kinangop, and Kenya in Kenya and Mt. Meru in Tanzania.

ADVERTISEMENT CALL: Males call during the day, from vegetation in slow-flowing water. The call is a quick series of clicks that resembles a coin dropping.

BREEDING: Unknown.

TADPOLES: Unknown.

NOTES: Known food items include spiders, myriapods, flies, beetles, ants, bugs, springtails, and mites.

KEY REFERENCES: Barbour & Loveridge 1928b, Loveridge 1936, Razzetti & Msuya 2002.

Kinangop Puddle Frog

Phrynobatrachus kinangopensis Angel, 1924

This species was named for the Kinangop Plateau in Kenya.
Kenya highlands puddle frog.

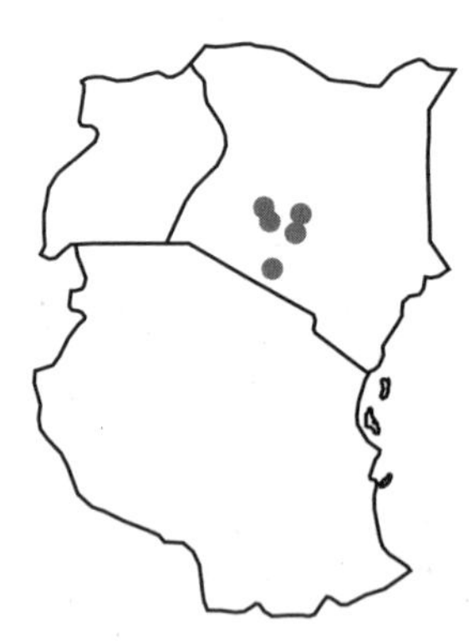

Fig. 218 *Phrynobatrachus kinangopensis.*

DESCRIPTION: The males reach a length of 19 mm and females, 24 mm. The frog is stocky with dorsal asperities in both sexes. Very small discs are present on the fingers and toes, and two phalanges of the fourth toe, two phalanges on the fifth toe, and almost two on the outside of the second toe are free of web. The foot is longer than the tibia. The nostrils are equidistant from the snout tip to the eye. The posterior half of the body of sexually active males has small white pustules and strong lateral folds. A dark band from the nostrils to the tympanum, bordered by a silvery streak, is characteristic. The underside has fine, dark, even stippling over the throat and abdomen. The male throat is speckled.

HABITAT AND DISTRIBUTION (FIG. 218): This species is known at an altitude of 3100 m in rain-filled pools on montane grassland, from the Kenyan highlands.

ADVERTISEMENT CALL: Unknown.

BREEDING: Unknown.

TADPOLES: Unknown.

KEY REFERENCES: Loveridge 1944, Grandison & Howell 1983.

Krefft's Puddle Frog

Phrynobatrachus kreffti Boulenger, 1909
(Plates 19.5, 19.6)

Named for P. Krefft, a medical doctor from Braunschweig, Germany, who carried out collecting and research in Tanzania in early 1909.

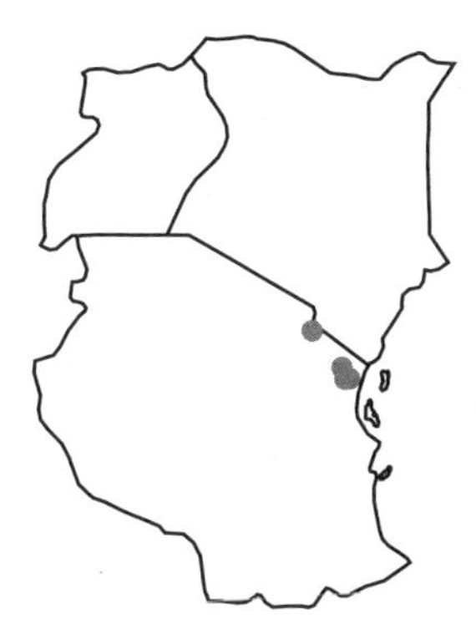

Fig. 219 *Phrynobatrachus kreffti.*

DESCRIPTION: Males reach a length of 36 mm and females, 41 mm. Digital discs are strongly developed. Two phalanges of the fourth toe are free of web. Males have a thickened thumb-pad, and the toes have plantar tubercles developed as sharp spines. The first and second toes have small spines on the inside. Males are olive with brown markings, including a pale band between the eyelids, and are white below with a black line around the jaw. The throat is yellow with a brighter yellow margin below the lower jaw. There is a dusky band across the chest. Females have white throats.

HABITAT AND DISTRIBUTION (FIG. 219): This large puddle frog is found in small streams, marshes, and pools, sometimes in the middle of forest. It is known from the North Pares, Usambaras, and Magrotto mountains in Tanzania.

ADVERTISEMENT CALL: Males call from the edge of shallow water or while concealed under vegetation in water. The call consists of a series of low-pitched notes that are produced quite rapidly at first. A typical call consists of about 13 notes, with a duration of 2.7 s at an initial rate of about 8 notes/s. The final notes are slower. The emphasized frequency is 2.3 kHz.

BREEDING: These frogs start breeding when the long rains commence in November.

TADPOLES: See the chapter on tadpoles.

NOTES: Recorded food items include beetles, spiders, ants, bugs, weevils, millipedes, crabs, and cockroaches.

KEY REFERENCES: Barbour & Loveridge 1928a, Loveridge 1944, Drewes & Perret 2000.

Mababe Puddle Frog

Phrynobatrachus mababiensis (FitzSimons, 1932)
(Plate 19.7)

This frog was named for the Mababe Depression in Botswana.
Common puddle frog, dwarf puddle frog, common cricket frog, Mababi puddle frog.

Fig. 220 *Phrynobatrachus mababiensis.*

DESCRIPTION: These frogs are very small. The males grow up to 19 mm long and females are slightly larger at 21 mm. The gular sac of males has a transverse fold. Femoral glands are present. The toe tips and fingertips are swollen but discs are not present. Broad web reaches the proximal subarticular tubercle of the third toe and usually reaches the proximal tubercle of the fourth toe. The tarsal tubercle is prominent. The dorsum is mainly brown with darker and lighter markings and has small warts. The light and dark barring on the upper and lower jaws correspond. Many different color patterns are known, with vertebral stripes or bands being common.

HABITAT AND DISTRIBUTION (FIG. 220): This species occurs in savanna at high and low altitudes. It is found in wet areas with emergent vegetation, like roadside ditches and grassy fields. It is known from Angola and Namibia, to Tanzania and eastern South Africa.

ADVERTISEMENT CALL: Males call from concealed positions at the base of grass near pools and small bodies of water. The call consists of a number of long buzzes, typically interspersed with a few separate clicks. The buzzing may continue for seconds, at a pulse rate of 80/s and at an emphasized frequency of 4.1 kHz. Males were calling in mid-April on the University of Dar es Salaam campus.

BREEDING: The small black eggs are 0.9 mm in diameter within 1.1-mm capsules. They are laid in a single layer, about 50 mm across, which floats on the surface. Some calling may be heard for most months of the year, and egg clutches have been recorded from September to February in Tanzania.

TADPOLES: See the chapter on tadpoles.

NOTES: In captivity these frogs eat mosquitoes. They are eaten by the white-lipped snake *Crotaphopeltis hotamboeia*, the Mt. Rungwe bush viper *Atheris rungwensis*, and the cattle egret *Bubulcus ibis*.

KEY REFERENCES: Stewart 1967, Channing 2001.

Natal Puddle Frog

Phrynobatrachus natalensis (Smith, 1849)
(Plates 19.8, 20.1)

This species was named for the province of Natal (now KwaZulu-Natal) in South Africa.
Snoring puddle frog, *lulenga* in Ganda.

Fig. 221 *Phrynobatrachus natalensis.*

DESCRIPTION: The male reaches a length of 34 mm and the female, 40 mm. The nostrils are nearer the end of the snout than the eyes. The male throat sacs have longitudinal folds. Femoral glands are absent. There are no discs on the fingers and toes, although a slight swelling may be present. The first and second fingers are of equal length. The tympanum diameter is two-thirds the diameter of the eye. Webbing in this species is quite variable, and the skin may be rough or smooth (partly reflecting the confused taxonomy). Broad web reaches the outer tubercle below the third toe, passes midway between the inner and middle tubercles of the fourth toe, and reaches the outer tubercle of the fifth toe. Three phalanges of the fourth toe are free of web. Coloration is variable. A rich brown background is typical, although a thin vertebral stripe, darker speckling, and a pale dorsal band are some of the more common patterns.

HABITAT AND DISTRIBUTION (FIG. 221): This frog is found in grassland and thorn savanna, associated with streams and temporary pools. It is widely distributed in sub-Saharan Africa.

ADVERTISEMENT CALL: The male calls during day and night in wet weather. The call is a slow quiet snore, with a dominant harmonic of 2 kHz, a pulse rate of about 60–80/s, and a duration of 0.5 s. Two or more males may call in chorus, but more commonly the calls are heard without a pattern from around small bodies of water.

BREEDING: Breeding takes place from spring through summer. The mating pair swim while depositing eggs. Eggs may be laid in deep water, or floating in a single layer on the surface in very shallow water, frequently associated with emergent vegetation. They are small, 0.6–1.0 mm in diameter, and dark brown on top. About 200 eggs are laid in each clutch. The time to metamorphosis is 27–40 days.

TADPOLES: See the chapter on tadpoles.

NOTES: The maximum recorded longevity in captivity is 5 years. This frog is known to eat termites, ants, caterpillars, earthworms, snails, arthropods, and other frogs. In turn, it is preyed on by the black-necked spitting cobra *Naja nigricollis* and the lizard buzzard *Kaupifalco monogrammicus*. The name *Phrynobatrachus natalensis* is applied to a complex of similar species that are presently being investigated.

KEY REFERENCES: Loveridge 1944, Stewart 1974, Channing 2001.

Dwarf Puddle Frog

Phrynobatrachus parvulus (Boulenger, 1905)

The specific name *parvulus* means "very small."
Luanda puddle frog, *bungulula* in Kinga.

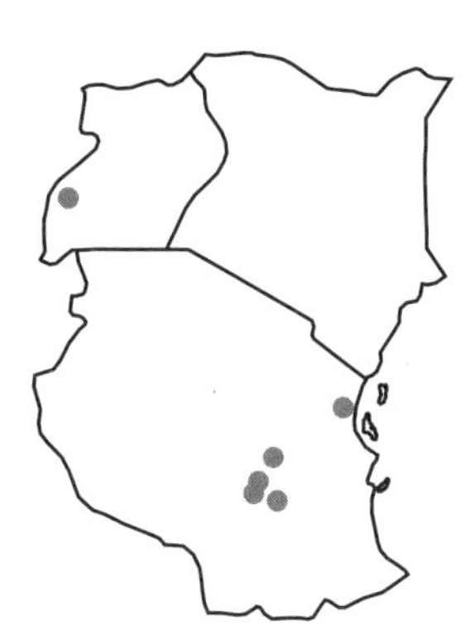

Fig. 222 *Phrynobatrachus parvulus.*

DESCRIPTION: The male in this species grows to a length of 16 mm and the female, up to 25 mm. The male has a black throat sac with a transverse fold, and a flattened gland on the upper leg. Digital discs are absent. Webbing is not extensive, with the broad web only reaching the inner tubercle below the third toe and not or only just reaching the inner tubercle of the fourth toe. The metatarsal tubercle is rounded with a conical profile. The back is mostly brown, with a lighter band below the tympanum that runs from the eyelid to the arm. The back of the thigh has a lighter band running from knee to knee.

HABITAT AND DISTRIBUTION (FIG. 222): This species is found in the higher nonforested areas, recorded from Angola eastward through the Democratic Republic of Congo and Uganda to Tanzania, and southward to Botswana.

ADVERTISEMENT CALL: Males call from the surface of shallow water, sometimes floating or otherwise supported by vegetation. The call is a series of buzzing notes. Each note is about 175 ms long, with the notes produced at a rate of 8/s. Each note consists of a clearly audible series of pulses. Analysis shows about 9 pulses/note. Frequencies between 3.5 and 4.5 kHz are broadly emphasized.

BREEDING: These small frogs breed during the wet seasons, and probably throughout the year.

TADPOLES: Unknown.

NOTES: Food items that have been reported include earthworms, snails, and arthropods.

KEY REFERENCES: Stewart 1967, Channing 2001.

Webbed Puddle Frog

Phrynobatrachus perpalmatus Boulenger, 1898

The specific name *perpalmatus* means "much webbing."

DESCRIPTION: This frog is small, with the male being up to 25 mm long and the female up to 29 mm. The tympanum is two-thirds the size of the eye. The first finger is shorter than the second. No femoral glands are present in the male. The throat sac of the male is unpigmented, without distinct folds. Small discs are present on the fingers and toes.

Fig. 223 *Phrynobatrachus perpalmatus.*

The webbing is well developed, with the broad web extending past the outer subarticular tubercle of the third toe and middle tubercle of the fourth toe. One to one and a half phalanges of the fourth toe are free of web. The heel tubercle is flaplike, and the metatarsals are nearly completely separated by web. The skin is quite smooth, with a few scattered small warts. A dark, light-edged streak runs from the eye to the groin through the tympanum. There is a dark crossbar on the thigh and one on the tibia. The back of the thighs have a wavy dark band on white. A pale band below the vent runs from knee to knee, with a second band interrupted at the vent. The throat has some brown dots.

HABITAT AND DISTRIBUTION (FIG. 223): This species is found in permanently wet sites from both forest and savanna. It is reported from Sudan to western Tanzania, Zambia, Malawi, and adjacent Mozambique.

ADVERTISEMENT CALL: The males call at night from flooded grasslands. The call is a metallic ticking.

BREEDING: Unknown.

TADPOLES: Unknown.

NOTES: This species is known to eat earthworms, snails, ants, beetles, spiders, waterbugs, bugs, dragonfly larvae, and fly larvae. The olive sand snake *Psammophis sibilans* and the savanna vine snake *Thelotornis capensis* prey on it.

KEY REFERENCES: Loveridge 1933, 1944; Channing 2001.

Roux's Puddle Frog

Phrynobatrachus rouxi (Nieden, 1912)

This frog was named for the herpetologist J. Roux of the Natural History Museum in Basel, Switzerland.

Fig. 224 *Phrynobatrachus rouxi.*

DESCRIPTION: Males and females reach a length of 23 mm. The nostrils are situated midway between the eyes and the tip of snout. The tympanum is just visible. The fingers and toes have small discs, and the first finger is shorter than the second. The toes are half-webbed, with three phalanges of the fourth toe free of webbing. The inner metatarsal tubercle is oval, and the outer metatarsal tubercle is small and round. The tarsal tubercle is joined to the inner metatarsal tubercle by a thin skin fold. The tarsal tubercle is slightly closer to the metatarsal tubercles than the metatarsal tubercles are to each other. A small glandular fold and a row of small warts are present behind each eye. Fine spinules cover the tibia, tarsus, and underside of the foot, and small granules are seen on the flanks. The throat and chest is brown, and the posterior region of the belly is yellow. A thin vertebral stripe is often present.

HABITAT AND DISTRIBUTION (FIG. 224): This forest species is known from the western shores of Lake Victoria to Mt. Kenya.

ADVERTISEMENT CALL: Unknown.

BREEDING: Unknown.

TADPOLES: Unknown.

KEY REFERENCE: Nieden 1912.

Rungwe Puddle Frog

Phrynobatrachus rungwensis (Loveridge, 1932)
(Plate 20.2)

The specific name refers to Mt. Rungwe in southern Tanzania.
Rungwe river frog.

Fig. 225 *Phrynobatrachus rungwensis.*

DESCRIPTION: The male of this small species reaches a length of 23 mm and the female, 24 mm. The nostrils are nearer the snout tip than the eyes. The tympanum is not visible. Glands are present on the upper leg. The gray throat sac in the male is folded at the back. Small discs are present on the tips of the middle toes, but the fingers are only slightly swollen. Webbing is not extensive. The broad web reaches midway between the tubercles below the third toe and reaches the inner subarticular tubercle of the fourth toe. Three phalanges of the fourth toe are free of web, but only two on the other toes are. The tarsal tubercle is conical, with no transverse fold. No light band is present below the tympanum. The back is brown with darker markings. The skin glands on the back are not arranged in longitudinal rows. A light band on the back of the thigh is edged with a dark border. The hind limbs have crossbars.

HABITAT AND DISTRIBUTION (FIG. 225): This species has been found in open grassland and is known from southern Democratic Republic of the Congo, southwestern Tanzania, and northern Malawi.

ADVERTISEMENT CALL: Unknown.

BREEDING: Unknown.

TADPOLES: Unknown.

KEY REFERENCE: Loveridge 1932a.

Scheffler's Puddle Frog

Phrynobatrachus scheffleri (Nieden, 1911)
(Plate 20.3)

This species was named for Mr. Scheffler of Kibwesi in Kenya who collected the specimen around the turn of the last century.
Kasambara in Nyakusa, *lungalla* in Logooli and Tiriki.

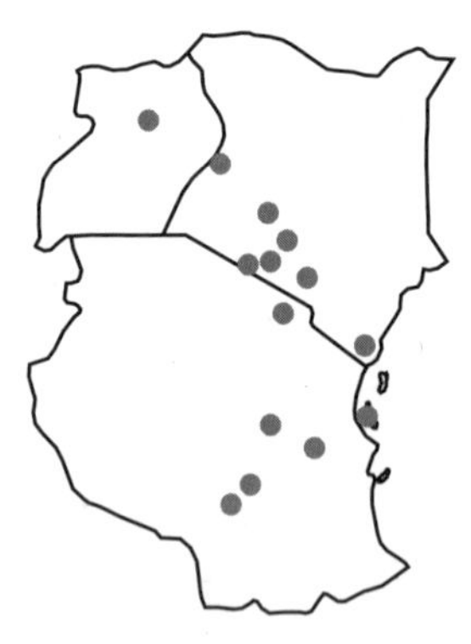

Fig. 226 *Phrynobatrachus scheffleri.*

DESCRIPTION: The males grow to 19 mm long and the females, 22 mm. The nostrils are situated midway between the eyes and the snout tip. The tympanum is not visible. The fingers and toes have slightly expanded tips, and the first finger is shorter than the second. The toes have slight webbing. Most large skin glands are in the shoulder region. The back is brown with a dark diagonal mark between the eyes. Many have a fine pale vertebral line. The belly is pale with pigment flecks on the throat and breast. Breeding males have a yellow throat, a white or pale yellow belly, and a few lateral specklings. Females are white below.

HABITAT AND DISTRIBUTION (FIG. 226): This species is found in leaf litter or on mud near water holes, but not in the water. It is widespread in East Africa, but its distribution is not well known.

ADVERTISEMENT CALL: The males call from vegetation in shallow water. The call is a long trill, lasting 2.2 s. The pulse rate is 41/s. The dominant frequency is 4.0 kHz, with harmonics present at 4-kHz intervals.

BREEDING: Unknown.

TADPOLES: Unknown.

NOTES: This species has been confused with *Phrynobatrachus minutus* from Ethiopia. These frogs are known to be preyed on by the olive marsh snake *Natriciteres olivacea* and the olive sand snake *Psammophis sibilans*.

KEY REFERENCES: Loveridge 1933, 1936.

Stewart's Puddle Frog

Phrynobatrachus stewartae Poynton & Broadley, 1985
(Plate 20.4)

This species was named for the collector, Margaret Stewart, a professor at the State University of New York, Albany.
Stewart's river frog.

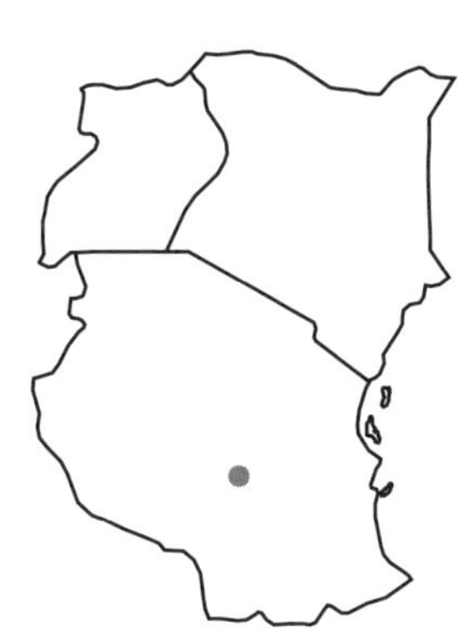

Fig. 227 *Phrynobatrachus stewartae.*

DESCRIPTION: The males grow up to 21 mm long and the females, 23 mm. The male throat sac has a flap at the back. The glands on the upper leg are yellow, elongated, and flattened. Discs are absent. Webbing is moderate, with the broad web extending midway between the tubercles beneath the third toe, and passing the inner tubercle of the fourth toe on the inside. The heel tubercle is spurlike on a fold. The back is brown, with a darker triangle between the eyes. The warts on the back are edged with black. There is no light band below the tympanum.

HABITAT AND DISTRIBUTION (FIG. 227): This frog is restricted to grassy areas with vegetation in water. It is known from northern Malawi and southern Tanzania.

ADVERTISEMENT CALL: Calling takes place under vegetation in marshy areas. The call is similar to that of the Mababe puddle frog *Phrynobatrachus mababiensis*, but is louder, faster, and longer. It consists of a long, buzzy, low cricket-like trill followed by a crisp "chip," with the "chip" repeated one or more times. The call has yet to be recorded and analyzed.

BREEDING: Unknown.

TADPOLES: Unknown.

NOTES: A female contained eggs that were 0.6 mm in diameter and darkly pigmented. More observations are required of this interesting frog.

KEY REFERENCES: Stewart 1967, Poynton & Broadley 1985.

Ukinga Puddle Frog

Phrynobatrachus ukingensis (Loveridge, 1932)

This species was named for the Ukinga area in southern Tanzania.

DESCRIPTION: This frog is very small. The male grows to 19 mm long and the female is slightly longer, at 21 mm. The nostril is nearer the snout tip than the eye. The tympanum is hidden, and the black throat sac of the male has a transverse fold. Glands are present on the upper leg. The toe tips and fingertips are clearly expanded into small discs. The broad web reaches the inner tubercle beneath the third toe and usually reaches the inner tubercle of the fourth toe. The tarsal tubercle is promi-

Fig. 228 *Phrynobatrachus ukingensis.*

nent. The skin is smooth with flattened warts. The back is mainly brown with darker and lighter markings, often with a light vertebral line and lines on the tibia and thigh. Many small warts are present on the back. The light and dark barring on the upper and lower jaws correspond. Many different color patterns are known, with vertebral stripes or bands being common.

HABITAT AND DISTRIBUTION (FIG. 228): This species is found in wooded savanna and is known from the forests of southern and eastern Tanzania and northern Malawi.

ADVERTISEMENT CALL: The male calls from a concealed position at the base of grass near pools and small bodies of water. The call consists of a number of long buzzes, typically interspersed with a few separate clicks. The buzzing may continue for seconds, at a pulse rate of 80/s and at an emphasized frequency of 4.1 kHz.

BREEDING: This frog breeds in summer, although some calling may be heard for most months of the year. The small black eggs are 0.9 mm in diameter within 1.1-mm capsules. They are laid in a single layer, about 50 mm across, which floats on the surface.

TADPOLES: Unknown.

NOTES: This species has been confused with *Phrynobatrachus mababiensis,* which is very similar but does not have expanded toe discs.

KEY REFERENCE: Loveridge 1932a.

Udzungwa Puddle Frog

Phrynobatrachus uzungwensis Grandison & Howell, 1983
(Plate 20.5)

This species was named for the Udzungwa Mountains of Tanzania.

Fig. 229 *Phrynobatrachus uzungwensis.*

DESCRIPTION: Females reach a length of 25 mm and males, 21 mm. The tympanum is not easily visible externally. The nostrils are nearer the snout tip than the eyes. Small discs are present on the fingers and toes. Webbing is well developed, with two to two and a half phalanges of the fourth toe free. The tibia length is more than half of the body length. The upper surfaces of sexually mature males are covered with small white spines, and a nuptial pad is present on the thumb. The vocal sac is unpigmented, although the gular region may have a scattering of darker marks on a pale background. The lower lips are black. The dorsal color is similar to that in many other puddle frogs, with various cross-barring and darker markings on the back, sometimes with an orange or red band behind the eyes. The tips of the fingers are rose red, and the upper arms are pale orange. The throat is white, usually with a pair of pectoral blotches and a pale stripe on each side of the vent. Breeding males have gray-brown nuptial pads but no femoral glands.

HABITAT AND DISTRIBUTION (FIG. 229): This species occurs near water in rain forest and may be found just above ground on leaves. It is known from the Udzungwa and Uluguru mountains in Tanzania.

ADVERTISEMENT CALL: Unknown.

BREEDING: Unknown.

TADPOLES: Unknown.

KEY REFERENCE: Grandison & Howell 1983.

Green Puddle Frog

Phrynobatrachus versicolor Ahl, 1924
(Plate 20.6)

The specific name refers to the greenish color of the back.
Western rift puddle frog, *miusi* in Lega, *Bunter Krötenfrosch* in German, *grenouille-crapaud coloré* in French.

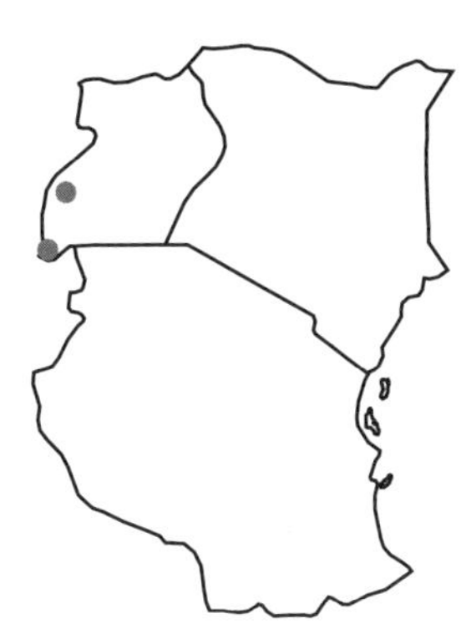

Fig. 230 *Phrynobatrachus versicolor.*

DESCRIPTION: Males reach a length of 28 mm and females, 34 mm. The tympanum is larger than the eye.

These frogs are delicately built, and the toes and fingers have small discs. Three to three and three-fourths phalanges on the fourth toe are free of web. Minute spines are present on the soles of the feet in both sexes but are bigger in males. The skin is smooth, with warts in longitudinal rows, although the dorsal skin texture is variable, from smooth to warty. Coloring is variable, often a brownish back with broad lateral golden bands and a dark lower edge. A light vertebral band is present in some animals.

HABITAT AND DISTRIBUTION (FIG. 230): This species is found associated with water, at altitudes above 2000 m, in seeps in montane forests, but also in many deforested areas. It is known from the Democratic Republic of Congo to Rwanda, Burundi, and southwestern Uganda.

ADVERTISEMENT CALL: The call consists of a rapid series of trills. A typical call consists of four trills, each with 7–12 pulses. The duration of each trill is 70 ms, and the total length is 0.48 s. The dominant frequency rises within each trill, from an initial 2.1 kHz to a peak of 2.5 kHz, before descending for the last 2 pulses of each trill.

BREEDING: Unknown.

TADPOLES: Unknown.

NOTES: This species is known to eat bugs and other arthropods.

KEY REFERENCES: Loveridge 1955, Fischer & Hinkel 1992.

Ridged Frogs—Genus *Ptychadena*

This is a large group of specialist jumping frogs. They have characteristic longitudinal skin folds on the back. Eggs are laid in a floating mat in quiet pools. They occur throughout sub-Saharan Africa. In East Africa 13 species are present. This genus is known as *akalenga* or *olulenga* in Luganda, and as *orutoori* in Runyankole and related languages in Uganda. They are known as *eretretre* in Lugbara.

KEY TO THE SPECIES

1a.	A pale triangle on snout (Fig. 231)	2
1b.	No pale triangle, although a pale vertebral stripe may be present	3

Fig. 231 Pale triangle on snout.

2a. Distance from nostril to tip of snout greater than internarial distance *Ptychadena oxyrhynchus*
2b. Distance from tip of snout to nostril not more than internarial distance *Ptychadena anchietae*

3a. Two to two and one-third phalanges of fourth toe free of web 10
3b. Two and one-half phalanges or more on fourth toe free of web 4

4a. Length of foot more than half body length (Fig. 232) 5
4b. Length of foot not more than half body length 8

Fig. 232 Proportions of foot length and body length.

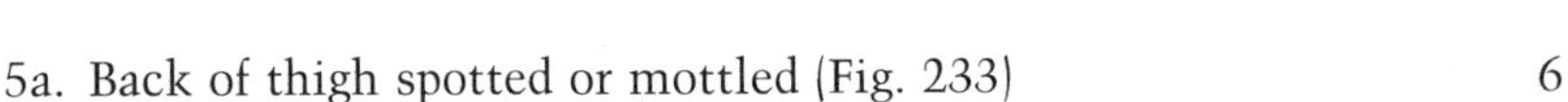

5a. Back of thigh spotted or mottled (Fig. 233) 6
5b. Back of thigh with light and dark longitudinal bands (Fig. 234) 9

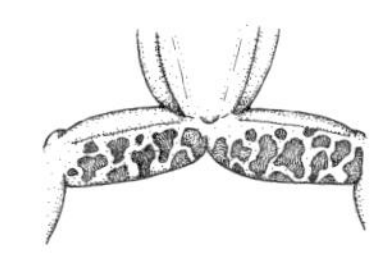

Fig. 233 Back of thigh mottled.

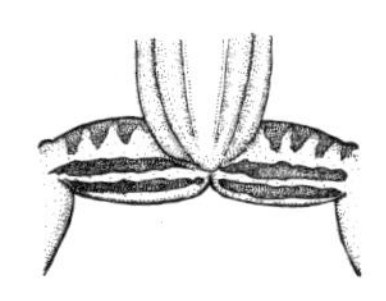

Fig. 234 Back of thigh banded.

6a. Outside of thigh marbled, inside uniform *Ptychadena mahnerti*
6b. Spotting or marbling on inside and outside of thigh 7

7a. Upper surface of tibia with a light longitudinal line (Fig. 235) 11
7b. No light line on upper surface of tibia; a pair of skin ridges on snout (Fig. 236) *Ptychadena uzungwensis*

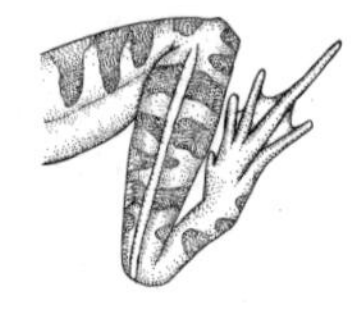

Fig. 235 Light line on tibia.

Fig. 236 Pair of skin ridges on snout.

8a. Paravertebral skin folds continuous from head to sacrum — 12
8b. No continuous skin folds from head to sacrum — *Ptychadena schillukorum*

9a. Vocal pouch slits in line with the arm — *Ptychadena taenioscelis*
9b. Vocal pouch slits aligned below arm — *Ptychadena stenocephala*

10a. Back of thigh banded — *Ptychadena mascareniensis*
10b. Back of thigh marbled — *Ptychadena christyi*

11a. Back of thigh with rows of light spots — *Ptychadena porosissima*
11b. Back of thigh marbled — *Ptychadena chrysogaster*

12a. Light line on tibia — *Ptychadena mossambica*
12b. No light line on tibia — *Ptychadena upembae*

Anchieta's Ridged Frog

Ptychadena anchietae (Bocage, 1867)
(Plate 20.7)

This frog was named for the collector, José Alberto de Oliveira Anchieta, who worked in Angola from 1864 to 1897.
Savanna ridged frog, plain grass frog.

DESCRIPTION: These are medium-sized frogs, with males up to 51 mm long and females up to 62 mm. The distance from the nostril to the snout tip is equal to or less than the distance between the nostrils. Throat pouch slits end level with the lower edge of the arm in males. The tym-

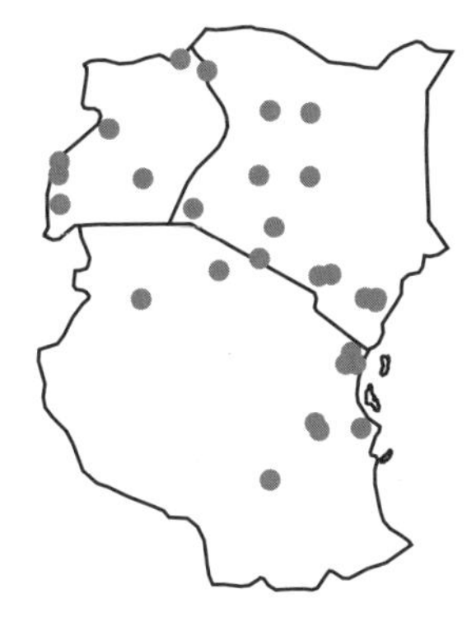

Fig. 237 *Ptychadena anchietae.*

panum is slightly smaller than the eye and is edged with white. No row of tubercles is present under the fourth metatarsal. One and a half to two segments of the fourth toe are free of web. The top of the snout is light, forming a pale triangle. There is no light vertebral band or light line on the upper surface of the tibia. The back of the thigh shows light parallel longitudinal bands. The back is often uniform brown without markings. A dark stripe runs from the nostril through the eye and the tympanum to the arm.

HABITAT AND DISTRIBUTION (FIG. 237): This species is a savanna form, usually found in grassland near water. It is known from Ethiopia through East Africa to eastern South Africa and Angola.

ADVERTISEMENT CALL: The males call from open ground near water. The call is a high-pitched trill, repeated rapidly. Each trill consists of 10–12 pulses, with a duration of about 0.2 s and harmonics at 1.8 and 3.5 kHz, rising respectively to 1.9 and 4 kHz by the end of the call. The duration of the call is 0.25 s. This species also possesses a faster chorus call and a harsher territorial call.

BREEDING: These frogs breed during the summer rainy season. Mating and egg laying take place in shallow temporary ponds. The eggs float in a single layer on the water surface, with up to 300 eggs in a raft. Each egg is white with a gray top half and 1 mm in diameter within a 3-mm capsule.

TADPOLES: See the chapter on tadpoles.

NOTES: This species is preyed on by the northern stripe-bellied sand snake *Psammophis sudanensis*, the cattle egret *Bubulcus ibis*, and the yellow-billed kite *Milvus migrans*. Perhaps the most remarkable predator is the large slit-faced bat *Nycteris grandis*, which finds the male frog by listening for its advertisement call. The bat is tuned into the call of this species only, as it does not take other calling frogs. Unlike many other *Ptychadena* species, this one jumps away from water when disturbed.

KEY REFERENCES: Lamotte & Perret 1961b, Channing 2001.

Christy's Ridged Frog

Ptychadena christyi (Boulenger, 1919)

This species was named for C. Christy, leader of a collecting expedition to the Belgian Congo between 1912 and 1914.

Fig. 238 *Ptychadena christyi.*

DESCRIPTION: Males reach a length of 53 mm and females, 57 mm. Each nostril is midway between the tip of the snout and the eye. The tympanum is smaller than the eye, separated from the eye by a distance equal to two or three times the diameter of the tympanum. First and second fingers are of equal length. The tibia is longer than the foot and 60% of the body length. Webbing is well developed, with two phalanges of the fourth toe free of web and the fifth toe webbed to the tip. The inner metatarsal tubercle is one-third to two-thirds the length of the inner toe. There is no tarsal fold. A continuous glandular ridge runs from behind the eye to the leg. Other skin folds are discontinuous to the rear of the body. A glandular ridge starts below the eye and runs to the arm insertion. The underside is smooth except for granulations near the vent. The dorsum is yellow-brown, with a black interorbital bar, and a black band from the snout to behind the tympanum. The dorsolateral fold carries a series of black dashes. The back of the thighs is mustard, marbled with black. The throat is cream in females, and the belly, yellow. The soles of the feet are dark, while the undersides of the hands are light.

HABITAT AND DISTRIBUTION (FIG. 238): This frog is a forest-floor inhabitant and is known from eastern Democratic Republic of Congo and western Uganda.

ADVERTISEMENT CALL: Unknown.

BREEDING: Unknown.

TADPOLES: Unknown.

KEY REFERENCES: Loveridge 1944, Guibé 1966.

Yellow-Bellied Ridged Frog

Ptychadena chrysogaster Laurent, 1954
(Plate 20.8)

The name *chrysogaster* means "golden belly," referring to the color of the ventral skin.
Golden-bellied rocket frog.

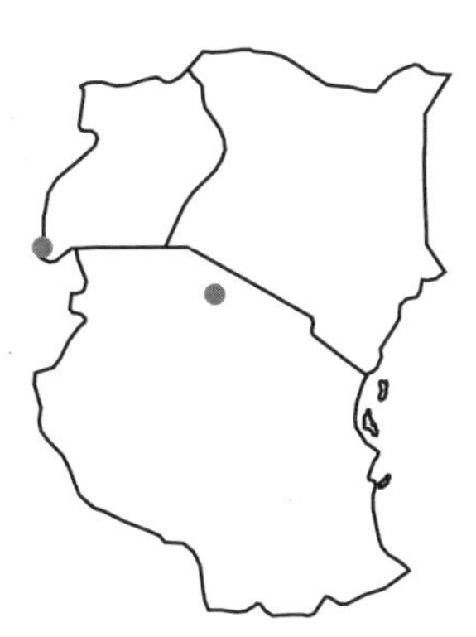

Fig. 239 *Ptychadena chrysogaster.*

DESCRIPTION: The males reach a length of 50 mm and the females, 58 mm. The snout is bluntly pointed, with nostrils slightly closer to the snout tip than the eyes. The distance between the nostrils is equal to the width of the upper eyelid. The tympanum is a little smaller than the eye, but larger than the distance between the eye and the tympanum. Tibia length is 61%–70% of body length. Webbing is moderate, with two and three-fourths to three phalanges of the fourth toe free. The length of the inner metatarsal tubercle is less than half the length of the first toe. Males in breeding condition have smooth skin covered with minute spines. There are 8–10 dorsal skin folds. The underside is smooth with distinct pectoral folds, but is granular around the thighs. Vocal sac openings are in line with the lower margin of the arm. The back is green or brown with small black spots. A pale dorsal stripe is present with irregular indistinct margins. A dark canthal stripe is present. The lower jaw has irregular dark markings that fade into speckling under the throat. There is a pale longitudinal line along the thigh and tibia, with four to seven transverse bars. The posterior surface of the thigh has pale marbling.

HABITAT AND DISTRIBUTION (FIG. 239): This species is known from forest at altitudes above 1800 m, from eastern Democratic Republic of Congo, Rwanda, Burundi, western Uganda, and Tanzania.

ADVERTISEMENT CALL: Unknown.

BREEDING: Unknown

TADPOLES: Unknown.

KEY REFERENCE: Laurent 1954.

Mahnert's Ridged Frog

Ptychadena mahnerti Perret, 1996

This frog was named for Volker Mahnert, director of the Geneva Museum, who took part in the museum's zoological expeditions to Kenya (1974–1977), during which the species was collected.
Highland ridged frog.

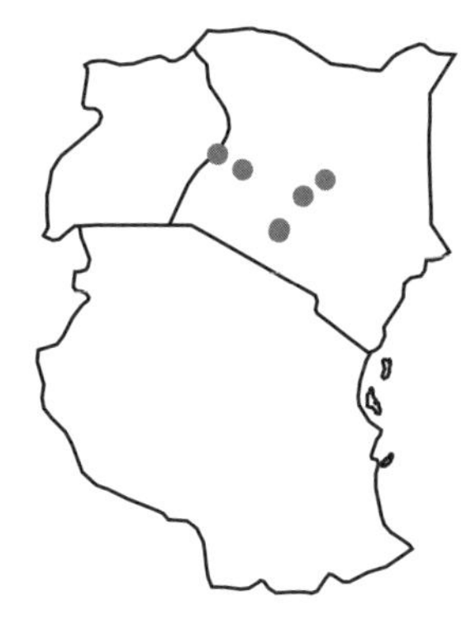

Fig. 240 *Ptychadena mahnerti.*

DESCRIPTION: Males are known up to 42 mm long, while females reach 49 mm. This species can be distinguished by the pattern on the back of the thighs: a marble-patterned area on the outside of each thigh, separated by a uniform inner surface. The head is slightly longer than wide. The snout is longer than the width of the upper eyelid. The nostril is midway between the eye and the snout. The width of the upper eyelid is about equal to the horizontal diameter of the tympanum, which is three-fourths the size of the eye. The tibia is more than half the body length, and nearly as long as the foot. The toes are half-webbed, with three to three and a half phalanges of the fourth toe free of web. The internal metatarsal tubercle is small, positioned at the end of a fine tarsal fold. The dorsal glandular skin ridges are fine and variable in number between individuals.

Dorsal coloration is variable, commonly a dark brown background with a broad paler vertebral band. Black spots are present on the inner skin ridges and on the front of the thigh and tibia. A black band runs from each nostril, backward through the lower part of the eye, the tympanum, and ending at the arm insertion. An alternative pattern is a uniform rich brown back. Males may have a nuptial pad at the base of the thumb, including the second finger and reaching the edge of the third. Paired vocal pouches open obliquely, pointing toward the arm insertion.

HABITAT AND DISTRIBUTION (FIG. 240): This species is known from the moist savanna and high grasslands at altitudes above 2000 m, including the edge of cleared forest. It has been found on the highlands of Kenya, north of Nairobi to the uplands around Mt. Kenya and Mt. Elgon in eastern Uganda.

ADVERTISEMENT CALL: The call is a series of rapid notes that produces a typical trill. The call may consist of 4–11 notes at a rate of 6/s. Each note of 30–110-ms duration consists in turn of 4–13 pulses. The dominant frequency is 3.1 kHz.

BREEDING: Breeding takes place in permanent ponds and flooded vegetation. Details are unknown.

TADPOLES: Unknown.

KEY REFERENCE: Perret 1996.

Mascarene Ridged Frog

Ptychadena mascareniensis (Duméril & Bibron, 1841)
(Plates 21.1, 21.2)

This species was named for the Mascarene Islands.
Mascarene frog, mascarene grass frog, mascarene rocket frog, *Nilfrosch* in German, *makeri* in Kiji, *nanhengo* in Makonde at Kitaya, *nanihengo* in Makonde at Mbanja, *nanmiengo* in Maviha, *marembera* in Lega.

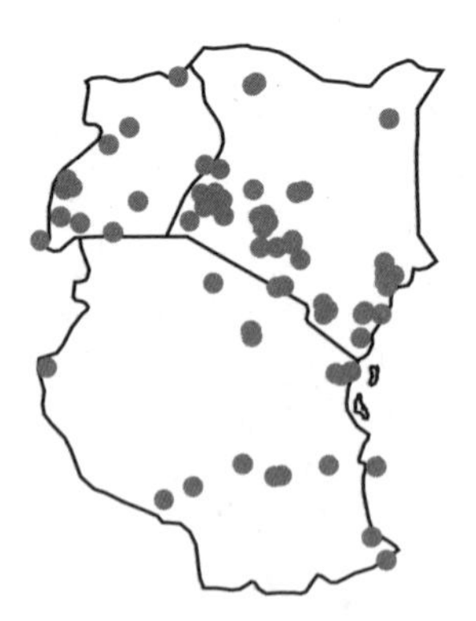

Fig. 241 *Ptychadena mascareniensis.*

DESCRIPTION: Males reach a length of 53 mm and females, 65 mm. The internarial distance is equal to the nostril-snout distance. The nostrils are situated midway between the snout tip and the eyes. The male gular pouch slits end above the level of the arm insertion. The outer metatarsal tubercle is absent. There is no row of tubercles under the fourth metatarsal. Two phalanges of the fourth toe, sometimes two and a half, are free of web. There are eight dorsal folds. The back is generally dark brown with a broad vertebral stripe, which may range from green, through yellow to brown. A light line is present on the upper surface of the tibia. A dark band runs from knee to knee below the vent. The posterior face of the thigh is banded.

HABITAT AND DISTRIBUTION (FIG. 241): This species is found along streams and in temporary and permanent standing water, from small ponds to large lakes. It occurs widely from West Africa to eastern South Africa. This species is interesting as it is also found on Madagascar, Seychelles, and the Mascarene Islands off East Africa (but see Notes below).

ADVERTISEMENT CALL: Males call from the edge of water, sometimes from emergent vegetation. The call is a series of clucking sounds repeated rapidly. Each call is 0.2–0.3 s long, consisting of 16–24 pulses, with dominant harmonics at 0.7–1.0 kHz and a higher harmonic starting at 2.1 kHz, then rising to 3.0 kHz. Males call antiphonally and may form a dense calling group of six to eight individuals in a small area of flooded grass. They use the vocal pouch for the advertisement call, but produce a "chuckle" call, presumably a male-male interaction call, by vibrating the flanks.

BREEDING: Breeding takes place during the long and short rainy seasons. The eggs are small and white with a black hemisphere.

TADPOLES: See the chapter on tadpoles.

NOTES: These frogs eat winged ants, beetles, winged termites, spiders, caterpillars, earthworms, grasshoppers, snails, roaches, bugs, crickets, isopods, wasps, bugs, dragonflies, moths, and flies. They are carnivorous, taking smaller frogs. A large number of predators eat this species, including the cattle egret *Bubulculus ibis*, olive sand snake *Psammophis sibilans*, striped olympic snake *Dromophis lineatus*, white lipped snake *Crotaphopeltis hotamboeia*, rhombic night adder *Causus rhombeatus*, water snake *Grayia tholloni*, brown house snake *Boaedon l. lineatus*, beaked snakes *Rhamphiophis* spp., cormorants *Phalacrocorax* spp., the dimorphic egret *Egretta dimorpha*, and the edible bullfrog *Pyxicephalus edulis*.

This frog escapes from disturbance by jumping away from water. It has been reported that those living near a hot spring escape by briefly entering the water at its hottest point, where it is not possible to immerse a hand for more than a few seconds. In West Africa local people eat these frogs. When captured, they have a "foam and moan" display, where they produce skin secretions and make a moaning call while they adopt a rigid posture. Although this behavior is an attempt to distract predators, they are ready to escape by jumping at every opportunity. Maximum recorded longevity in captivity is 6 years 7 months.

The mascarene ridged frog probably consists of a number of cryptic species.

KEY REFERENCES: Loveridge 1933, 1944; McIntyre & Ramanamanjato 1999.

Mozambique Ridged Frog

Ptychadena mossambica (Peters, 1854)
(Plate 21.3)

The specific name *mossambica* refers to Mozambique, where this species was first found.
Mozambique grass frog.

Fig. 242 *Ptychadena mossambica.*

DESCRIPTION: These are medium-sized frogs, with males ranging up to 44 mm long and females 53 mm. The internarial distance is greater than the distance from the nostril to the tip of the snout. Male gular pouch slits end level with the lower edge of the arm insertion. A row of tubercles is present under the fourth metatarsal. Two and two-thirds to three phalanges are free of web. Paravertebral folds are replaced behind by a pair of para-urostylar folds. This is a variable species in terms of markings. A light line is often present on the tibia, with a characteristic broad light vertebral band. The back is green or brown with dark brown markings.

HABITAT AND DISTRIBUTION (FIG. 242): This species prefers open savanna. It is found from Somalia through East Africa to South Africa.

ADVERTISEMENT CALL: Males call while well concealed. The call is a repeated quacking, 0.1 s long, with dominant energy at 0.8 kHz and a rising harmonic starting at 2.3 kHz and increasing to 3 kHz. The chorus call is a series of discrete clucks at 2.8 kHz. The territorial call is a complex series of notes.

BREEDING: Breeding takes place during the short rains, with some residual activity later in the long rains. No details of eggs or breeding sites have been reported.

TADPOLES: Unknown.

NOTES: This species appears to be very variable, with an east-west difference in size associated with differences in webbing, and calls that are also slightly different, with the western calls being of a higher frequency. It is possible that two related species are being confused under one name.

KEY REFERENCES: Passmore 1977, Poynton & Broadley 1985, Channing 2001.

Sharp-Nosed Ridged Frog

Ptychadena oxyrhynchus (Smith, 1849)
(Plates 21.4, 21.5)

The specific name *oxyrhynchus* means "sharp snout."
Sharp-nosed frog, sharp-nosed grass frog, *kengele* in Taita, *chuachanco* in Pokomo, *nanhengo* in Makonde at Kitaya, *nanihengo* in Makonde at Mbanja, *nanmiengo* in Maviha.

DESCRIPTION: These are medium-sized frogs, with males up to 62 mm and females up to 85 mm in snout-vent length. Internarial distance is less than the distance from the nostril to the snout tip. The tympanum is as large as the eye. Gular pouch slits end level with the lower edge of the arm insertion. There are six or more dorsal folds. No row of

Fig. 243 *Ptychadena oxyrhynchus.*

tubercles exists under the fourth metatarsal. The toes are webbed to the tips of the third and fifth toes, with only one and a half to two phalanges of the fourth toe free of web. The tibia is longer than the foot. There is a distinct pale triangle on the top of the snout, but it does not continue backward as a pale band. A thin pale vertebral stripe may be present. The back is brown, with reddish infusions on the skin ridges. A narrow dark stripe passes backward from the snout through the nostril. It widens behind the eye to include the tympanum, and then continues to the leg as a dark broken band. The posterior face of the thigh is mottled.

HABITAT AND DISTRIBUTION (FIG. 243): This species is widely distributed in savanna from Senegal to southeastern South Africa. In East Africa it is known from sea level to an altitude of 1600 m.

ADVERTISEMENT CALL: Males call from open ground away from water, or along the water's edge. The call consists of a loud trill repeated rapidly. Each trill consists of 10–15 pulses and is frequency modulated. The dominant harmonic starts at 1.0 kHz, rising to 1.5 kHz, before dropping slightly at the end of the call. The duration of each trill is 0.4–0.5 s. A second less-emphasized harmonic rises to 3.0 kHz.

BREEDING: These frogs breed during the height of the rainy seasons. Amplexus is axillary. The eggs are 1 mm in diameter within 3-mm capsules. They are laid on water in short strings 75–200 mm long, which then break up, permitting the eggs to float away. The eggs are whitish with a dark brown pole. One clutch was counted as having 3476 eggs. This species has been recorded breeding in rock pools.

TADPOLES: See the chapter on tadpoles.

NOTES: This species is known to eat grasshoppers, snails, beetles, spiders, crickets, cockroaches, woodlice, crickets, caterpillars, and bugs. It is preyed on by the brown house snake *Lamprophis fuliginosus*, spotted wood snake *Philothamnus semivariegatus*, southeastern green snake *Philothamnus hoplogaster*, Mulanje water snake *Lycodonomorphus mlanjensis*, rufous beaked snake *Rhamphiophis rostratus*, boomslang *Dispholidus typus*, savanna vine snake *Thelotornis capensis*, and Nile monitor *Varanus niloticus*.

Sexual maturity is reached 8–9 months after metamorphosis, when females may reach 59 mm in length. The tadpoles accumulate very high levels of dieldrin when their habitat is sprayed with 2% dieldrin to control mosquitoes! This species is reported to be capable of very long jumps and is still the holder of the Guinness Book of Records long-jump record, obtained in Cape Town, South Africa, in the 1950s. The maximum recorded longevity in captivity is 8 years 11 months. Perhaps two species, one from West Africa and one from eastern and southern Africa, are confused under one name.

KEY REFERENCES: Loveridge 1933, 1944; Inger 1968; Barbault & Trefaut Rodriques 1978.

Grassland Ridged Frog

Ptychadena porosissima (Steindachner, 1867)
(Plate 21.6)

The specific name *porosissima* means "very spotted."

Fig. 244 *Ptychadena porosissima.*

DESCRIPTION: These are medium-sized frogs, with males up to 43 mm long and females up to 49 mm. The internarial distance is equal to the distance from the nostril to the tip of the snout. Gular pouch slits end level with the lower edge of the arm insertion. Three phalanges on the fourth toe are free of web. Dorsal color is brown with darker spots arranged along the dorsal skin ridges. A pale vertebral stripe is present, often with a lateral stripe on each side. The posterior face of the thigh has longitudinal rows of light spots. A pale tibial line is present.

HABITAT AND DISTRIBUTION (FIG. 244): These frogs are found in moist grassland and forest at higher altitudes and are widely distributed from Ethiopia, through East Africa to South Africa. Their distribution is not well known in East Africa.

ADVERTISEMENT CALL: Males call from beneath grass and other vegetation in shallow water. The calls are typically *Ptychadena*-like, a high-pitched short buzz, 0.1–0.2 s long, with dominant harmonics at 2 and 4 kHz. Each call consists of 10–12 pulses.

BREEDING: The eggs are 1 mm in diameter within 5-mm capsules. The eggs float and are laid in shallow grassy pools.

TADPOLES: See the chapter on tadpoles.

NOTES: Recorded food items include earthworms, snails, arthropods, and other frogs. Grassland ridged frogs are eaten by the stripe-backed green snake *Philothamnus ornatus*.

KEY REFERENCES: Inger & Marx 1961, Passmore 1977.

Schilluk Ridged Frog

Ptychadena schillukorum (Werner, 1907)
(Plate 21.7)

This frog is named for the Schilluk region of the Sudan, west of the Nile.

DESCRIPTION: These frogs are small with short legs. Males may reach 53 mm long and females, 49 mm. The internarial distance is greater than the distance from the nostril to the tip of the snout. Gular pouch slits

Fig. 245 *Ptychadena schillukorum.*

end at the middle of the arm insertion. There is no row of tubercles under the fourth metatarsal. Webbing is reduced, with only two to three phalanges of the fourth toe free of web. Dorsal skin folds are not continuous from head to sacrum. The posterior face of the thigh usually has fine vermiculations. The lower jaw is marbled. The tibiotarsal articulation usually reaches the level of the eye.

A thin vertebral stripe is sometimes present.

HABITAT AND DISTRIBUTION (FIG. 245): This species is known from grassland at high and low altitudes, from West Africa to Ethiopia, through Kenya and Tanzania to Mozambique.

ADVERTISEMENT CALL: Males call while floating in a spread-eagled manner, from very dense vegetation at the edge of shallow pools. The call is a series of clicks at 1.7 kHz, the clicks being repeated at a rate of about 8/s.

BREEDING: Unknown.

TADPOLES: Unknown.

NOTES: This species is known to eat spiders, grasshoppers, caterpillars, beetles, ants, and moths.

KEY REFERENCES: Loveridge 1933, 1936.

Narrow-Headed Ridged Frog

Ptychadena stenocephala (Boulenger, 1901)
(Plate 21.8)

The specific name *stenocephala* refers to the relatively narrow head in this species.

Fig. 246 *Ptychadena stenocephala.*

DESCRIPTION: Females are known to be up to 47 mm long, with the males much smaller, at 37 mm. The head is one and a half times as long as wide. The snout projects strongly beyond the mouth. Male vocal pouch slits align below arm. The nostril is midway between the eye and the tip of the snout. The tympanum diameter is two-thirds to three-fourths the eye diameter. The web extends only to the base or middle of the basal phalanx of the toes, and two phalanges of the fifth toe are free of web. The outer metatarsal tubercle is very small or indistinct. The tibia length is about two-thirds of the body length. The back has six to eight glandular ridges, with one running from beneath the eye to the arm insertion. No middorsal skin ridge is present. The top surface is spotted or streaked, with a black stripe from the snout through the eye. A broad orange vertebral stripe is sometimes present. The upper lip and outer dorsal ridge are yellow, and there are no crossbars on the limbs. The ventrum is pale and unmarked.

HABITAT AND DISTRIBUTION (FIG. 246): This species is known in savanna from West Africa to Uganda and northern Tanzania.

ADVERTISEMENT CALL: Unknown.

BREEDING: Unknown.

TADPOLES: Unknown.

KEY REFERENCES: Perret 1979, Lamotte & Ohler 2000, Rödel 2000.

Small Ridged Frog

Ptychadena taenioscelis Laurent, 1954
(Plate 22.1)

The specific name *taenioscelis* means "banded leg."
Spotted-throated ridged frog, dwarf grass frog.

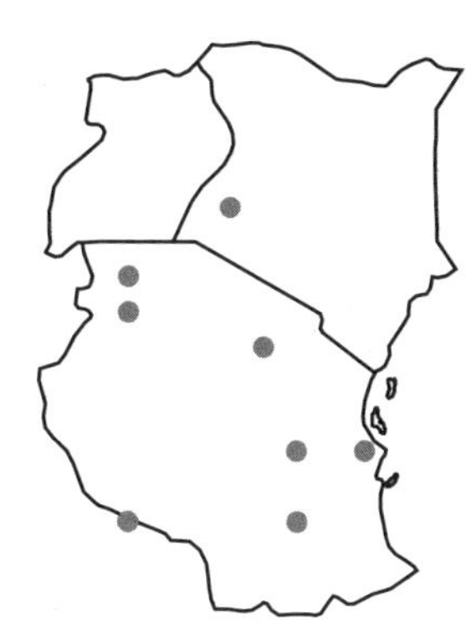

Fig. 247 *Ptychadena taenioscelis.*

DESCRIPTION: These are small, thin frogs, with males growing up to 35 mm and females up to 40 mm. The internarial distance is less than the distance from the nostril to the tip of the snout. Gular pouch openings end at the middle of the arm insertion. There is no row of tubercles under the fourth metatarsal. Three phalanges of the fourth toe are free of webbing. A pair of dorsal skin folds extends onto the snout. A pale ridge runs under the eye to the arm. The tympanum is smaller than the eye. Coloration is brown dorsally, with darker rectangular markings on a lighter background. There are three thin pale lines on the back. The posterior face of the thigh possesses longitudinal bands, one of which runs below the vent from knee to knee.

HABITAT AND DISTRIBUTION (FIG. 247): This grassland species is widely distributed from Kenya through Tanzania to Angola and South Africa.

ADVERTISEMENT CALL: Males call from short flooded grass, while sitting in water. The call is a rapid chirp, 0.1–0.3 s long, with dominant energy at 2.8–3.5 kHz. Each call consists of 22–25 rapid pulses. The pitch rises slightly during the call. Three to eight males congregate in chorus. Apart from the advertisement call, a chorus call (a harsher call probably used to space males) and a territorial call (a clucking uttered when males are within 200 mm of each other) are known.

BREEDING: Males call in the early evening, and amplexus follows a cautious creeping approach by the male. The amplexing pair deposits eggs over a 3-hour period, moving up to 8 m in the process. Two to 10 eggs are deposited at a time, in shallow water where they sink.

TADPOLES: Unknown.

KEY REFERENCES: Laurent 1964, Passmore 1976.

Upemba Ridged Frog

Ptychadena upembae (Schmidt & Inger, 1959)
(Plate 22.2)

The specific name *upembae* refers to the Upemba National Park in the Democratic Republic of Congo, where the species was first discovered.

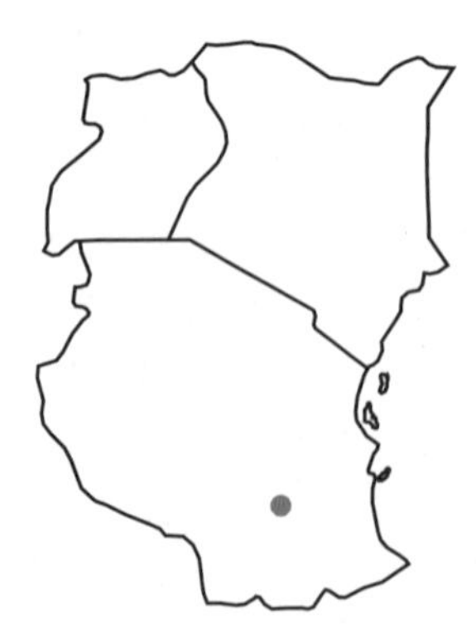

Fig. 248 *Ptychadena upembae.*

DESCRIPTION: These are medium-sized frogs, with males being up to 44 mm long and females up to 51 mm. The internarial distance is equal to the distance from the snout tip to the nostril. Gular pouch openings end level with the lower edge of the arm insertion. A row of tubercles is found under the fourth metatarsal. Three phalanges of the fourth toe are free of web. A light vertebral stripe is present. Darker spots are arranged along the longitudinal skin ridges. A large white dorsolateral fold runs from the tympanum almost to the top of the leg. There is no light line on the upper surface of the tibia. A dark longitudinal band is present on the posterior face of the thigh, at least on one leg.

HABITAT AND DISTRIBUTION (FIG. 248): This species is found in moist wooded savanna. It is known from southern Tanzania, Mozambique, Malawi, Zambia, the Democratic Republic of Congo, and Angola.

ADVERTISEMENT CALL: Males call from the base of grass tussocks in deep flooded areas. The call is a harsh trill, 0.3 s long, with a pulse rate of 40/s. The call rises to a peak with emphasized harmonics at 1.6 and 3.5 kHz, before tapering off.

BREEDING: No details of the breeding biology are available.

TADPOLES: Unknown.

NOTES: These frogs escape disturbance by jumping away from water into vegetation.

KEY REFERENCES: Stewart 1967, Poynton 1977, Channing 2001.

Udzungwa Ridged Frog

Ptychadena uzungwensis (Loveridge, 1932)
(Plate 22.3)

This species is named for the Udzungwa Mountains in Tanzania.

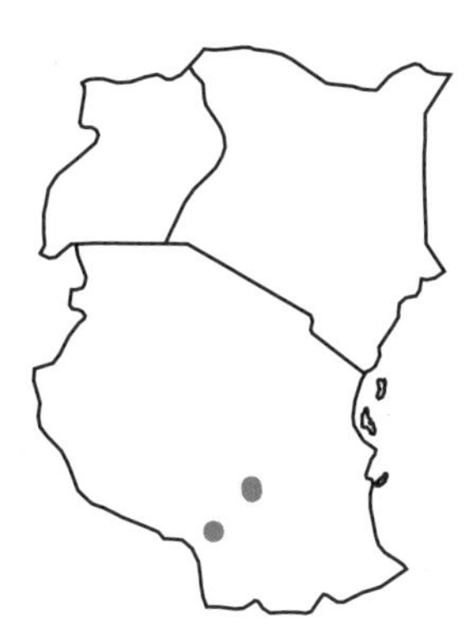

Fig. 249 *Ptychadena uzungwensis.*

DESCRIPTION: These are medium-sized frogs, with males being up to 42 mm long and females up to 48 mm. The internarial distance is less than the distance from the nostril to the tip of the snout. Gular pouch slits end at the middle of the arm insertion. There is a row of tubercles under the fourth metatarsal, and three phalanges of the fourth toe are free of web. A pair of short skin folds run from the upper eyelid toward the snout. A light vertebral stripe is present, and the outer dorsolateral fold is a white ridge. The posterior face of the thigh has a row of light spots, although sometimes this is replaced by a light band. The dorsal pattern consists of dark spots along the skin ridges, which may fuse in some individuals and almost obscure the ground color. The underside is white, while the throat and groin of males are deep yellow.

HABITAT AND DISTRIBUTION (FIG. 249): This species is found in high-altitude grassland, from eastern Democratic Republic of Congo, Rwanda, through southern Tanzania to Angola, South Africa, and Mozambique.

ADVERTISEMENT CALL: Males call from seepages in shallow water. The call is a trill, with males calling antiphonally. A typical trill consists of 7 pulses, produced at a rate of 46/s in 130 ms. Two harmonics are present, showing a slight rise in frequency during each call. The lower harmonic starts at 1.7 kHz, reaching a peak on the fifth pulse of 1.8 kHz. The upper harmonic starts at 3.4, reaching a peak of 3.6 kHz.

BREEDING: Unknown.

TADPOLES: Unknown.

NOTES: These frogs are known to eat snails, arthropods, and other frogs. Individuals may show a defensive behavior when captured, by bowing the body, stiffening and straightening the hind legs, and squeaking.

KEY REFERENCE: Loveridge 1936.

Bullfrogs—Genus *Pyxicephalus*

This group includes the largest frog in East Africa, the African bullfrog. Bullfrogs live in savannas and remain burrowed until heavy rains produce suitable temporary breeding pools. The tadpoles form large schools. Male African bullfrogs exhibit parental care of tadpoles. There are three species in this genus, of which two occur in East Africa. Both of these species are eaten by humans.

KEY TO THE SPECIES

1a. Toothlike projections (odontoids) on lower jaw as long as wide, similar to the middle protuberance *Pyxicephalus edulis*
1b. Toothlike projections on lower jaw (odontoids) longer than wide, larger than middle protuberance (Fig. 250) *Pyxicephalus adspersus*

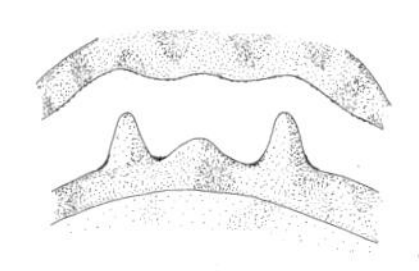

Fig. 250 Toothlike projections on lower jaw.

African Bullfrog

Pyxicephalus adspersus Tschudi, 1838
(Plates 22.4, 22.5)

The specific name refers to the sprinkling of markings on the back. South African speckled frog, giant bullfrog, Tschudi's African bullfrog, giant pyxie, *Grabfrosch*, *Ochsenfrosch*, or *Gespenkelter Grabfrosch* in German.

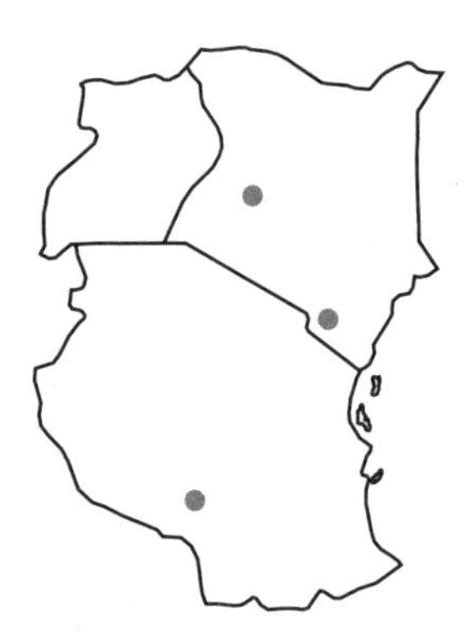

Fig. 251 *Pyxicephalus adspersus*.

DESCRIPTION: This frog is very large and robust, recorded at 1.075 kg. It is unmistakable by its size and the presence of two long, razor-sharp projections on the lower jaw (odontoid processes), one on either side of a bump on the midline. The male may grow to 230 mm long, with a jaw 90 mm wide. The female is smaller than the male. There is no outer metatarsal tubercle, and the inner metatarsal tubercle is spadelike. The distance between the eye and the tympanum is greater than the width of the tympanum. The back has distinct longitudinal skin folds. Coloration is variable, mostly a brown or dark green background, but usually with a bright green component. Even blue patterns are known. The side of the body around the arm is often deep yellow, especially in breeding males. The newly metamorphosed juvenile is bright green with a vertebral stripe.

HABITAT AND DISTRIBUTION (FIG. 251): This species is widely distributed in the drier savannas. It is found in inland Kenya and Tanzania, southward to South Africa and Namibia. This species has been confused with the edible bullfrog in some areas, and museum records should be checked.

ADVERTISEMENT CALL: The male calls from shallow water. The very low-pitched "whoop" call resembles the bellowing of cattle, hence, the name *bullfrog*. The call is quite soft for a large frog, and is up to 2 s long, with an emphasized harmonic between 200 and 250 Hz. The males utter a snorting call when fighting, but this is not always distinct.

BREEDING: Unlike most frogs, this bullfrog will only emerge and start breeding after exceptionally heavy rain, usually well after the start of the season. Breeding is initiated by 65 mm of rain over the previous day or two and takes place during daytime, with spawning happening from 7h00min through to 16h00min. There are two slightly different breeding strategies, depending on the age of the male. Younger males are not as aggressive as the full-grown males. Younger (smaller) males congregate in a small area, perhaps only 1 or 2 m^2, in shallow water. The larger males occupy the center of these breeding arenas or leks and attempt to chase off other males. A female approaches the group of males by swimming along at the surface until she is within 3 m or so of the group of males. The female then dives and reappears on the surface in the middle of the male group. She is soon grasped by one of the large males, and mating follows. Fertilization takes place above water level—the female arches her back to position her vent above the surface—so the eggs are fertilized before they reach the water. The spent female then shakes her head from side to side, and the clasping male releases her. Egg laying occurs along the shallow edge of the pond. Eggs are 1.1–1.3 mm in diameter within 4-mm capsules. All females are attracted to the same area where the males are calling and chasing one another. All egg laying occurs in this small area, and all the tadpoles presumably hatch together, ready to form the characteristic large schools.

Full-grown males fight, causing injury and even killing one another. The dominant male attempts to prevent other males from participating in breeding. Most of the females are mated by the dominant male, in his territory. This dominant male remains in the pool and takes care of the tadpoles (see below).

PARENTAL CARE: The adult male is often found near the eggs, and in or near a school of tadpoles. In one study, investigators built an earth dam around a tadpole school in a pond. Tadpoles isolated at the edges of the drying pond were able to reach the main body of water by following a channel dug by the attendant male. This provides experimental evidence of parental care. Channels over 15 m long are common.

TADPOLES: See the chapter on tadpoles.

NOTES: Newly metamorphosed frogs eat anything that moves, even siblings. It is not uncommon to find two newly emerged bullfrogs swallowing a third, one from each end! The adult is known to eat bullfrog tadpoles and young frogs. Bullfrogs have voracious appetite, with one recorded eating 17 young rinkhals (spitting cobras) and a small chicken. They are known to feed on termites, the red toad *Schismaderma carens*, river frogs *Afrana* spp., snout-burrowers *Hemisus* spp., and toads *Bufo* spp. One 170-mm frog ate a large snake, up to 760 mm long, by holding the snake near the head, centering it with the forelimbs and fingers, and then lunging forward with a wide gape.

Like other frogs that spend the dry season underground, the African bullfrog forms a cocoon around itself of old skin, which serves as waterproofing. Adults have been found during the dry season buried under the sand of a dry riverbed. They reach full size after 28 years, and the maximum longevity for one in captivity was estimated at 45 years. Bullfrogs have long been used as food in Africa, and remains of bullfrogs have been found at Stone Age sites in many parts of the continent.

Birds are major predators of bullfrogs. Records include the pink-backed pelican *Pelecanus rufescens*, the saddle-billed stork *Ephippiorhynchus senegalensis*, white-headed vulture *Trigonoceps occipitalis*, yellow-billed kite *Milvus migrans*, tawny eagle *Aquilla rapax*, African fish eagle

Haliaeetus vocifer, bateleur *Terathopius ecaudatus*, lesser spotted eagle *Aquilla pomarina*, and yellow-billed egret *Mesophoyx intermedia*. The helmeted terrapin *Pelomedusa subrufa* preys on tadpoles, holding them in its jaws and ripping them with clawed forelimbs. Tadpoles are also preyed on by the Nile monitor *Varanus niloticus*, which eats mouthfuls of swarming tadpoles in shallow water.

These large frogs are threatened by the destruction of their breeding habitat as well as the surrounding natural habitat such as forest, thicket, and termitaria in which they may take refuge in the dry season.

KEY REFERENCES: Balinsky & Balinsky 1954, Grobler 1972, Loveridge 1979, Cruz-Uribe & Klein 1983, Paukstis & Reinbold 1984, Kok et al. 1989, Van Wyk et al. 1991, Channing et al. 1994, Channing 2001.

Edible Bullfrog

Pyxicephalus edulis Peters, 1854
(Plate 22.6)

The specific name means "edible," as this species is eaten along the African east coast.
Peter's bullfrog, *liola* in Yao, *kitowa* in Makonde.

Fig. 252 *Pyxicephalus edulis.*

DESCRIPTION: This frog is large and robust, with males being up to 98 mm long and females up to 145 mm. It resembles a small African bull-frog but can be distinguished from it by the narrower head (some females resemble large river frogs *Afrana* spp.). The back does not have continuous elongated longitudinal folds, although it may be rough with short folds or bumps. The tympanum is placed about one eye width behind the eye. The distance from the nostril to the anterior corner of the eye is two-thirds of the distance from the nostril to the snout tip. The tympanum is distinct, round, aid no larger than half the eye width. Three equally sized protuberances in the anterior lower jaw fit into three grooves in the upper jaw. These consist of a tooth (odontoid process) on either side of the median protuberance. The "teeth" on the lower jaw are as long as wide. There is an egg-shaped metacarpal tubercle at the base of the first finger, with a larger, flatter tubercle on the palm. The first finger is longer than the second. Only the proximal phalanges of the toes are webbed. There is a shovel-shaped metatarsal tubercle. The dorsal skin is smooth.

The breeding male is bright green, while the female is duller and brownish. The back has dark spots, often with a pale vertebral stripe.

HABITAT AND DISTRIBUTION (FIG. 252): This species is found in flooded grasslands and is widely distributed on the eastern coastal plain, from Kenya to northern South Africa. It has been confused with the African bullfrog in some areas, and museum records should be checked.

ADVERTISEMENT CALL: The male calls at night after even light rain early in the season, but is known to call during the day under cloudy overcast conditions during the long rains in April. The call sounds like a small dog barking. The male calls from vegetation in the water, and his white vocal pouch flashes with reflected moonlight as he calls. The call is short, 0.19 s, and frequency modulated, with the highest frequencies reached in the middle of the call. The emphasized harmonic reaches 450–600 Hz, measured at the peak, typically with closely spaced lower and higher harmonics present.

BREEDING: Breeding takes place in shallow water early in the rainy season, or once sufficient rain has fallen. These bullfrogs breed at night after rain, and males may be found in amplexus with the smaller females throughout the pond. No leks are formed. As many as 900–3500 eggs are deposited and then scattered.

TADPOLES: Unknown.

NOTES: Recorded food items include beetles, ants, snails, winged termites, grasshoppers, crabs, centipedes and millipedes, tadpoles, young bullfrogs, the mascarene ridged frog *Ptychadena mascareniensis*, reed frogs *Hyperolius* sp., and the southern long-tailed lizard *Latastia johnstoni*. One individual is known to have eaten three scorpions, plus a range of insects. The edible bullfrog is preyed on by the northern stripe-bellied sand snake *Psammophis sudanensis* and black-necked spitting cobra *Naja nigricollis*. It is eaten by various peoples along the east coast of Africa.

KEY REFERENCES: Loveridge 1936, 1942a, 1944; Rödel 2000; Cook et al. 2001.

Stream Frogs—Genus *Strongylopus*

This genus is characterized by extremely long toes and reduced webbing, which permits them to move rapidly through grass. These frogs are found in a range of habitats at all altitudes. There are nine species, of which two are found in the highlands of East Africa. The taxonomy of stream frogs is not yet settled; more species remain to be described.

KEY TO THE SPECIES

1a. Vertical diameter of tympanum less than eye-nostril distance, webbing notch between fourth and fifth toes not reaching the proximal tubercle of fourth toe. *Strongylopus merumontanus*

1b. Vertical diameter of tympanum equal or greater than eye-nostril distance, notch of webbing between the fourth and fifth toes reaching proximal subarticular tubercle of fourth toe (Fig. 253) *Strongylopus kitumbeine*

Fig. 253 Webbing reaching proximal subarticular tubercle.

Kitumbeine Stream Frog

Strongylopus kitumbeine Channing & Davenport, 2002
(Plate 22.7)

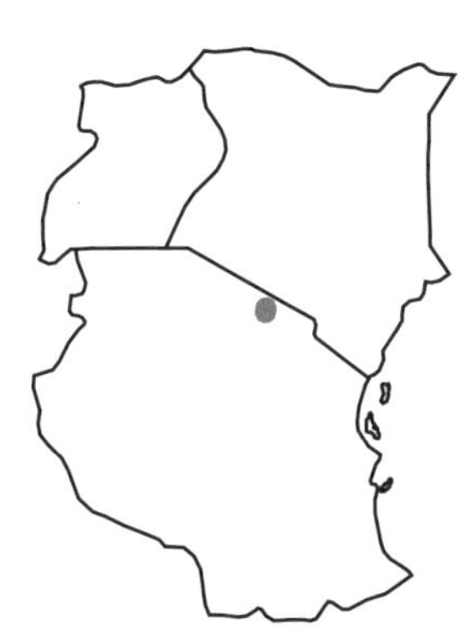

Fig. 254 *Strongylopus kitumbeine.*

DESCRIPTION: Males reach a length of 38 mm and females, 55 mm. The head is bluntly rounded in dorsal view. The nostrils are positioned midway between the snout tip and the anterior corners of the eyes. The tympanum is distinctly edged and subcircular, with the vertical diameter slightly larger than the horizontal. The horizontal diameter of the tympanum is more than half the eye width. The vertical diameter of the tympanum is equal or greater than the eye-nostril distance. The supratympanic ridge is smooth and slightly developed, extending to the arm insertion. The upper lip extends backward as a smooth glandular ridge to the arm insertion. The tibia is 60% of the body length, and the fourth toe is 66% of the body length. The toes are long, with bluntly rounded tips. Webbing is weakly developed, with the main web (the notch between two toes) reaching the proximal subarticular tubercle of the first toe, just reaching the base of the proximal subarticular tubercle of the second toe and reaching the proximal subarticular tubercle of the fourth toe. The back is smooth, with a thin dark band extending from the snout tip through the nostril to the anterior corner of the eye. Behind the eye it continues between the upper lip and the supratympanic ridge, backward to the arm insertion. The tympanum is slightly paler within this dark brown band. The top of the snout is pale, with a dark margin formed by a series of interocular markings.

The dorsum is pale brown, with irregular, small, scattered darker blotches. The sides are light, with a narrow darker irregular line running from above the arm insertion to the groin. A second, less distinct line or series of small irregular mottles is present below this. The front of the thigh has large transverse mottles, with other smaller blotches. The tibia has a series of anterior and posterior transverse markings that do not extend across the limb. The side of the foot and lower arm are likewise blotched.

The ventral surfaces are white, with fine darker speckling under the jaw; these speckles are grouped into indistinct blotches around the arm insertions. The lower belly and underside of the legs and arms are unpigmented.

HABITAT AND DISTRIBUTION (FIG. 254): This species is known from an altitude of 2800 m to the base of Kitumbeine (an extinct volcano in northern Tanzania) in all vegetation types where pools form. It is endemic to Kitumbeine, but it may occur in other springs and water bodies in the arid northern parts of Tanzania.

ADVERTISEMENT CALL: The call consists of one note, alternated often with one male or more to produce a ripple of notes. Each note consists of 1–6 pulses. Pulses are repeated rapidly, although the rate may slow slightly at the end of the call. The midpoint frequency is 2.0 kHz, with the second harmonic of 6.0 kHz emphasized.

BREEDING: Unknown. The males were calling in April.

TADPOLES: Unknown.

KEY REFERENCE: Channing & Davenport 2002.

Mt. Meru Stream Frog

Strongylopus merumontanus (Lönnberg, 1910)
(Plate 22.8)

This species is named for Mt. Meru in Tanzania.
Long-toed frog, *jeraboka* in Hehe, *chula* in Kinga.

DESCRIPTION: Males may reach a length of 43 mm, with females growing up to 50 mm long. The tympanum diameter is three-fourths the eye diameter. Snout length is one and a half times the eye diameter. The nostrils are closer to the eyes than the snout tip. The tympanum is equal to half the eye diameter. There is very little webbing. The inner

Fig. 255 *Strongylopus merumontanus.*

metatarsal tubercle is compressed. There are eight longitudinal dorsal skin folds, all interrupted. The belly is smooth. The dorsum is straw-brown, with a yellowish green vertebral streak bordered by a black raised skin fold extending from the snout to the vent. A deep yellow streak edged with black commences at the eye and merges into the coloring in front of the groin. The lips are white. Young can be bright yellow. An alternative color pattern is a uniform reddish brown back, with a dark band running from the nostrils through the lower half of the eyes, enclosing the tympanum and tapering to the forearm. A dark stripe from knee to ankle is found in both color patterns.

HABITAT AND DISTRIBUTION (FIG. 255): This species is found in high grassland at altitudes up to 3000 m. It is known from eastern Zambia, Malawi, and Tanzania.

ADVERTISEMENT CALL: The males call from flooded grass. The call consists of a single rising note, with dominant energy at 2.7 kHz and a second harmonic at 7.9 kHz. The call is short, only 23 ms. Calls are repeated at 2-intervals, but the call rate increases at the height of the breeding season and as temperatures warm up.

BREEDING: Unknown. On Mt. Meru the frogs breed at the beginning of the year, with juveniles present by April.

TADPOLES: Unknown.

NOTES: This frog runs through grass, leaping only occasionally. It is known to eat caterpillars, beetles, craneflies, grasshoppers, cockroaches, and spiders. This species was previously called *Strongylopus fuelleborni* in the south of Tanzania and Malawi.

KEY REFERENCES: Loveridge 1925, Barbour & Loveridge 1928a (as *Rana fasciata merumontana*), Channing & Davenport 2002.

Sand Frogs—Genus *Tomopterna*

This group of small burrowing frogs is found throughout Africa in sandy areas. Many species have been discovered to be tetraploids—with a double set of chromosomes, compared to the normal diploid species. The tadpoles are robust and develop in temporary pools.

A number of recently discovered species look similar to other species, although the former have distinct advertisement calls and DNA sequences. It is not possible to identify many museum specimens unless calls were recorded when they were collected. On the maps in the individual species accounts, we include only records of positive identifications. A genus-level map shows the distribution of specimens that cannot be positively identified (Fig. 256). There are nine species in Africa, of which four are found in East Africa.

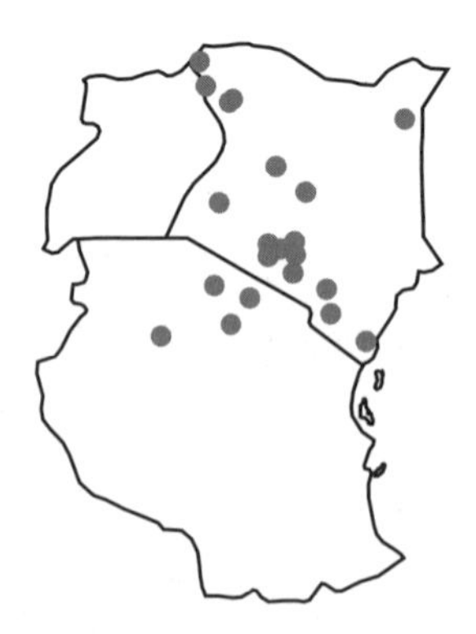

Fig. 256 Combined map of unidentified *Tomopterna* species.

KEY TO THE SPECIES

1a. Outer metatarsal tubercle distinctly elevated — *Tomopterna tandyi*
1b. Outer metatarsal tubercle small or absent — 2

2a. Inner metatarsal tubercle greater than 140% length of second toe *Tomopterna cryptotis*
2b. Inner metatarsal tubercle less than 140% length of second toe 3

3a. Webbing notch between the fourth and fifth toes not extending to distal subarticular tubercle of fifth toe *Tomopterna tuberculosa*
3b. Webbing notch between the fourth and fifth toes extending to distal subarticular tubercle of fifth toe *Tomopterna luganga*

Cryptic Sand Frog

Tomopterna cryptotis (Boulenger, 1907)
(Plate 23.1)

The specific name *cryptotis* refers to the hidden tympanum. Tremolo sand frog, *Kleiner Grabfrosch* in German.

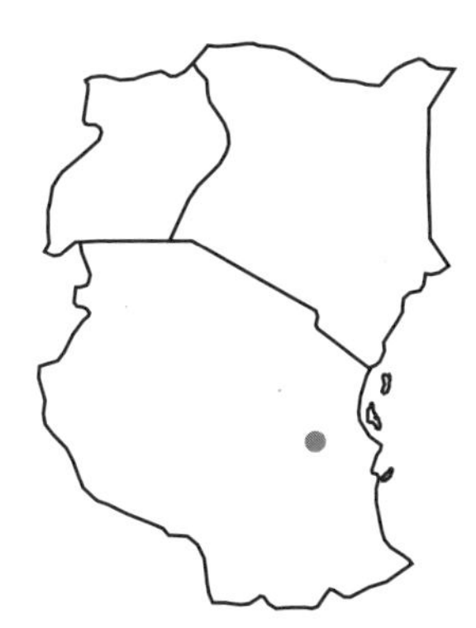

Fig. 257 *Tomopterna cryptotis.*

DESCRIPTION: This frog is small and robust, with males growing up to 45 mm long and females up to 58 mm. The length of the tibia is equal to the head width. A glandular ridge is present below the tympanum. The inner metatarsal tubercle is large and flattened and used for digging. Up to three and a half segments of the fourth toe are free of webbing. The back is variable, with various lighter and darker markings. Grays, reds, and white are common colors. A thin vertebral stripe is sometimes present, and a lighter head patch is common.

HABITAT AND DISTRIBUTION (FIG. 257): This species is associated with sandy soils and is found along drainage lines. It occurs in the central highlands of South Africa through Namibia to East Africa. Due to confusion with *Tomopterna tandyi*, the full extent of its distribution is unknown.

ADVERTISEMENT CALL: Males call from the edge of temporary pools. Often these pools are muddy, and the frogs sit on the mud at the water's edge. The call is a long series of short, high-pitched notes. The emphasized harmonics are 3.2–3.7 kHz. Each note is 0.03 s long, and the call rate is 10–12 notes/s.

BREEDING: The eggs are laid singly in shallow water. Clutch size is 2000–3000. Each egg is 1.5 mm in diameter within a 3-mm capsule.

TADPOLES: See the chapter on tadpoles.

NOTES: It is very difficult to distinguish this species from the other sand frogs in the area unless the calls have been heard. Known predators include the hammerkop *Scopus umbretta* and the barn owl *Tyto alba*. The tadpole is preyed on by fishing spiders. Adults may retreat into termite mounds during the day.

KEY REFERENCES: Channing & Bogart 1996, Dawood et al. 2002.

Red Sand Frog

Tomopterna luganga Channing et al., 2004
(Plate 23.2)

The species name is derived from the Hehe word *luganga* meaning sand. Russet-backed sand frog.

Fig. 258 *Tomopterna luganga.*

DESCRIPTION: This species is large, with females as long as 52 mm and males as long as 47 mm. The body is robust with short stocky limbs. The distance between the anterior corners of the eyes is nearly twice the internostril distance. The nostril is situated closer to the eye than the tip of the snout. The tympanum is vertically elliptical, with a horizontal diameter nearly twice the size of the eye-tympanum distance. The upper eyelid is smooth with small warts posteriorly. The fingers are tapered and without terminal discs. The hind limbs are short and an outer metatarsal tubercle is absent. The dorsal skin is smooth with small flat rounded warts. A narrow glandular ridge runs from the lower edge of the tympanum past the angle of the jaw. The flanks and lower femur are granular.

The back is a reddish orange with dark red warts. A paler occipital blotch behind the eyes is outlined with a fine black line. The eye is gray marbled with black, and a pale golden band runs across the top of the eye, showing darker blood vessels. The pupil is outlined with a pale yellow border, which is notched below. The glands below the eye and the glandular ridge running below the tympanum to the angle of the jaw are orange. The upper surfaces of the fore and hind limbs are reddish orange with dark brown markings. The undersurface is white, with faint darker lines under the jaw, and the flanks are speckled with black on white, which forms a broad band separating the dorsal and ventral colors. Some specimens are marbled in brown and gray.

HABITAT AND DISTRIBUTION (FIG. 258): The red sand frog is known only from Tanzania. The known range of this species is expected to expand once more fieldwork is carried out.

ADVERTISEMENT CALL: Males call from the edge of pools and quiet backwaters of rivers. The call is a series of notes, sometimes uttered singly, but often produced in groups from 2 to 11. The first harmonic is dominant, between 1.05 and 1.17 kHz. The note rate varies from 5.1 to 6.6/s.

BREEDING: Unknown.

TADPOLES: Unknown.

KEY REFERENCE: Channing et al. 2004.

Tandy's Sand Frog

Tomopterna tandyi Channing & Bogart, 1996
(Plate 23.3)

This species was named for Mills Tandy, who was the first collector of this species.

Fig. 259 *Tomopterna tandyi.*

DESCRIPTION: This species is morphologically indistinguishable from Delalande's sand frog and the cryptic sand frog. The arms and legs are short but well built, and the enlarged inner metatarsal tubercle is flattened and used for digging. Three segments of the fourth toe are free of webbing. The row of glands below the tympanum is not fused to form a ridge. The background color is gray, with darker and lighter patches. There is often a light patch in the head region. A pale vertebral stripe is usually present, with thin pale stripes on either side.

HABITAT AND DISTRIBUTION (FIG. 259): This species occurs in sandy flat areas where temporary pans form. Ditches and dams also provide breeding habitat. It is known from eastern South Africa, across to Namibia and northward through Tanzania and Kenya. Due to the lack of recordings of its advertisement call, its distribution is uncertain.

ADVERTISEMENT CALL: The male calls on the muddy edges of newly formed pools. It may also call from below vegetation in flooded areas. The call is a series of notes with emphasized harmonics at 2.6–2.8 kHz. The note repetition rate is temperature-dependent, but at a typical evening temperature of 19 °C the notes are produced at a rate of 8/s.

BREEDING: Unknown.

TADPOLES: See the chapter on tadpoles.

NOTES: Records for *Tomopterna cryptotis* in many countries may actually be *T. tandyi.*

KEY REFERENCES: Channing & Bogart 1996, Dawood et al. 2002.

Rough Sand Frog

Tomopterna tuberculosa (Boulenger, 1882)
(Plate 23.4)

The specific name refers to the rough skin on the back.
Tuberculate sand frog, beaded pyxie, warty frog, beaded dwarf bullfrog.

Fig. 260 *Tomopterna tuberculosa.*

DESCRIPTION: This frog is robust, with the male growing up to 40 mm long and the female 45 mm. The tympanum diameter is half the eye diameter. The snout is as long as the diameter of the orbit. The back has two rows of warts from the snout to the posterior of the body. The nostril is slightly nearer to the eye than the snout tip. Three and a half to four phalanges on the fourth toe are free of webbing. The inner metatarsal tubercle, which is flattened and serves as a digging flange, is shorter than the first toe. Webbing is reduced, just passing the inner metatarsal tubercle of the fourth toe. The width of the head is usually less than the length of the tibia. The back has numerous warty ridges. The coloration of the back is variable, from a uniform brown in western areas to blotched with a light vertebral stripe in eastern areas. A light bar between the eyes separates two darker markings.

HABITAT AND DISTRIBUTION (FIG. 260): This species is known from Namibia and Angola eastward to the Democratic Republic of Congo, Zimbabwe, and Tanzania.

ADVERTISEMENT CALL: The male calls from the edge of water, often camouflaged among pebbles. The call is a continuous fast rattle. The notes are produced at a rate of 13/s, and each note is brief with an emphasized frequency of 2.6 kHz.

BREEDING: Unknown.

TADPOLES: Unknown.

KEY REFERENCES: Schmidt & Inger 1959, Channing 2001.

Foam-Nest Frogs—Family Rhacophoridae

This large family has only one genus in Africa, but many species throughout Asia. Many of these lay eggs in nests made of foam produced by secretions from the female.

Foam-Nest Frogs—Genus *Chiromantis*

This genus consists of four species, all of which occur within the area covered by this book. The hands and feet are webbed, and the fingers of each hand are arranged in two opposing pairs, with which they can grasp thin branches while climbing. All lay eggs in foam nests. The tadpoles develop in seasonal pools. *Chiromantis rufescens* is found in rain forest, while the other species live in drier habitats.

KEY TO THE SPECIES

1a.	One joint or less on fourth finger free of web	2
1b.	More than one joint on fourth finger free of web (Fig. 261)	3

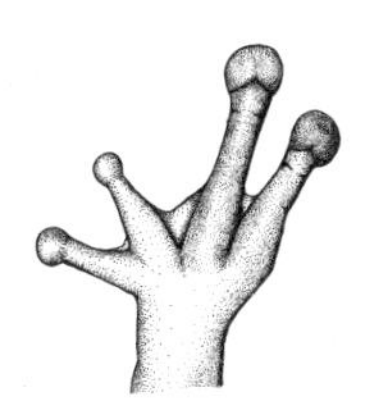

Fig. 261 More than one joint on fourth finger free of web.

2a. Tarsal fold present, less than one joint on fourth finger free of web *Chiromantis rufescens*
2b. No tarsal fold, one joint on fourth finger free of web *Chiromantis xerampelina*

3a. Disc of fourth toe half tympanum diameter, males up to 45 mm long, toe discs about the same size as subarticular tubercles *Chiromantis petersii*
3b. Disc of fourth toe larger than half the tympanum diameter, males up to 56 mm long, toe discs much larger than subarticular tubercles *Chiromantis kelleri*

Keller's Foam-Nest Frog

Chiromantis kelleri Boettger, 1893

This species was named for the collector, C. Keller.
Central foam-nest treefrog.

Fig. 262 *Chiromantis kelleri.*

DESCRIPTION: Males reach a length of 56 mm, with females growing up to 92 mm. Half of the vomerine tooth rows extend behind the posterior level of the internal nostrils. The broad web notch extends between 40% and 70% of the distance between the tubercles of the third finger. The ratio of the width of the disc on the third finger to internarial distance is 44%–54%. The disc of the fourth toe is more than half the tympanum diameter. The side of the body that is covered by the folded arms and legs shows dark freckling. This extends ventrally to become a dark background between lighter-colored granulations of the abdominal skin. Darkening also occurs on the inner side of the femur. Breeding males have gray throats.

HABITAT AND DISTRIBUTION (FIG. 262): This frog is found in dry savanna and is known from southern Tanzania around Lake Rukwa to northern Somalia.

ADVERTISEMENT CALL: The males produce a slow croaking call, with isolated clicks. Each croak has a duration of 0.3 s, with a pulse rate of 25/s, and an emphasized frequency of 200 Hz.

BREEDING: Nests may be laid on mud, grass, or bushes around water holes, and often are attached to grass or rocks overhanging temporary pools.

TADPOLES: Unknown.

NOTES: A group was found together in a bird nest on top of a baobab in the dry season.

KEY REFERENCES: Loveridge 1933, Schiøtz 1999, Poynton 2000b.

Peters' Foam-Nest Frog

Chiromantis petersii Boulenger, 1882
(Plate 23.5)

This species was named for the early fieldworker and herpetologist Wilhelm Peters (1815–1883), who collected in southern and eastern Africa.
Central foam-nest treefrog.

DESCRIPTION: This is the smallest species in the genus, with males reaching a length of 45 mm and females 65 mm. The vomerine tooth rows are placed between the choanae, and at most extend only margin-

Fig. 263 *Chiromantis petersii.*

ally behind the posterior level of the internal nostrils. The notch of broad web extends not more than 30% of the distance between the tubercles of the third finger. The ratio of the width of the disc on the third finger to internarial distance is less than 44%. There are no distinct lateral markings, although sometimes irregular lines are found on the back. The area covered by the folded arms and legs is not darkened; neither is the inner surface of the femur and abdomen. The disc of the fourth toe is about half the diameter of the tympanum. There are small discs on the toes, unlike *Chiromantis kelleri.*

HABITAT AND DISTRIBUTION (FIG. 263): This woodland species is known from southwestern Tanzania through the Serengeti Plain to northern Kenya.

ADVERTISEMENT CALL: Males call from sturdy bushes about a meter above the ground. The call is a series of quiet creaks. Each creak consists of 5 pulses, with a duration of 120 ms and dominant energy between 500 and 1500 Hz.

BREEDING: Nests are constructed on mud, grass, or bushes around water holes. Breeding takes place after the short rains, but may be repeated later during the long rains.

TADPOLES: Unknown.

KEY REFERENCES: Schiøtz 1999, Poynton 2000b.

Western Foam-Nest Frog

Chiromantis rufescens (Günther, 1869)
(Plate 23.6)

The species name *rufescens* means "reddish."

Fig. 264 *Chiromantis rufescens.*

DESCRIPTION: Males reach a length of 49 mm and females, 60 mm. Dorsal surfaces are gray with a greenish tinge. Darker irregular markings are found on the back. The underside is white. The hidden parts of the limbs are blue-green, as is the inside of the mouth. Males have white nuptial pads on the first two fingers. A tarsal fold is present and the feet are fully webbed.

HABITAT AND DISTRIBUTION: This species is known from rain forest, with a distribution including Uganda and the Congo Basin, to Nigeria and Sierra Leone.

ADVERTISEMENT CALL: Males call during the rainy season from vegetation. The call is quiet and consists of a regular knocking, followed by a short creak. The emphasized harmonic of the knocks is at 1800 Hz, with the creak at 1500 Hz. The creak is 0.3 s long, consisting of 10–13 pulses.

BREEDING: Eggs are laid in foam nests overhanging temporary water bodies. The nests may be constructed by a single female and more than one male. The tadpoles develop up to a point in the nest and then drop into the water below.

TADPOLES: See the chapter on tadpoles.

NOTES: Monkeys are the main spawn predators, consuming many of the foam nests.

KEY REFERENCES: Schiøtz 1999, Márquez et al. 2000, Poynton 2000b, Rödel et al. 2002.

Southern Foam-Nest Frog

Chiromantis xerampelina Peters, 1854
(Plate 23.7)

The specific name *xerampelina* means "dry vine leaves," referring to the color of the frog.

Great gray tree frog, African gray tree frog, foam nest frog, gray foam-nest tree frog, southern foam-nest tree-frog, *Grauer Ruderfrosch* in German, *schûre* in Tete and Sena, *zhulankombe* in Kalanga, *chitowa* in Kiyao, *kitowa* in Kimakonde (a general name for frogs).

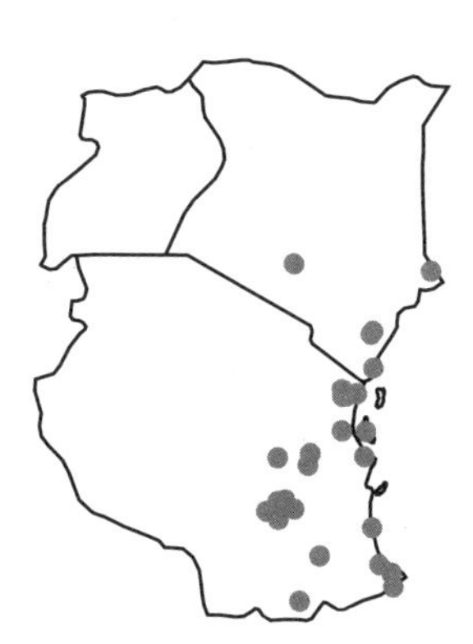

Fig. 265 *Chiromantis xerampelina.*

DESCRIPTION: This frog is large, with males growing up to 75 mm long and females 90 mm. The vomerine tooth rows are placed between the internal nostrils, at most extending only marginally behind the posterior level of the openings. The broad web extends more than 70% of the distance between the tubercles of the third finger. The ratio of the width of the disc on the third finger to internarial distance is 44%–62%. There are no distinct lateral markings. The area covered by the folded arms and legs is not darkened, and neither is the inner surface of the femur and abdomen. The hind legs and groin of breeding frogs are tinged with yellow. The back is brown or gray with darker markings. In sunlight the animal can often be nearly white. The throat may be lightly speckled to heavily marked in black. The skin is roughly textured, and the eyes are large and protruding with horizontal pupils. The fingers and toes have large adhesive discs. Both toes and fingers are webbed.

HABITAT AND DISTRIBUTION (FIG. 265): This species occurs in savanna from northeastern South Africa, across to northwestern Namibia, and northward to Kenya.

ADVERTISEMENT CALL: The males congregate in vegetation and call from above ground level, in grass, low bushes, or trees next to water. The advertisement call is a variety of clicks, croaks, and buzzes. The dominant frequency varies between 1.2 and 2.2 kHz, with 4 or 5 equally spaced pulses in each call. The pulse rate may vary from 20 to 44/s.

BREEDING: The breeding season coincides with the early short rains in October to January, and at the start of the long rains in March or April, and breeding may occur after each rain. The southern foam-nest frog has a remarkable breeding strategy. After the male selects a suitable site on a branch, rock, or vegetation overhanging water, he starts to call. The advertisement call attracts other males, and soon up to eight males can be calling nearby. The female arrives a little later, approaching by moving above ground level where there is adequate cover, sometimes 2–3 m high. A male clasps the female, and she starts to produce a secretion that she beats into a foam by paddling with her legs. She deposits eggs into the foam. The eggs are fertilized by the amplexing male, who places his cloaca next to that of the female as the eggs are released.

Often extra males clasp the amplexing pair. From one to seven extra males gather around the pair and compete with each other and the amplexed male to fertilize the eggs. In a study to test if the peripheral males played a role in fertilization, the primary male was prevented from fertilizing the eggs. Many of the eggs developed, showing that the peripheral males were responsible for fertilizing a proportion of the eggs in the nest.

The female is unable to produce sufficient foam for one nest without rehydrating. After the male releases her, she climbs down to the pool, where she absorbs water through her belly skin. She then returns to continue building the nest. She may then mate again with the same male or another male. On average it takes nearly two and a half sessions to complete a nest.

From 500 to 1226 eggs are deposited in each nest. The female may return to a nest the night after its construction, to add more foam but no new eggs. One pair in captivity produced a new foam nest on three successive nights. The developing eggs rely initially on the bubbles in the foam nest for oxygen.

The outside of the nest dries, and the eggs develop into small black tadpoles within. After 3–5 days the bottom of the nest gives way, probably due to the action of enzymes produced by the developing tadpoles, and the whole mass of tadpoles drops into the water below.

TADPOLES: See the chapter on tadpoles.

NOTES: Recorded food items include grasshoppers, crickets, caterpillars, and beetles. These frogs are preyed on by the boomslang *Dispholidus typus*, the vine snake *Thelotornis capensis*, and the large slit-faced bat *Nycteris grandis*, which will take the frog in captivity. Even the African bullfrog is reported to eat these foam-nest frogs. The gentle monkey *Cercopithecus mitis* will eat the foam nests containing eggs. Perhaps one of the most bizarre egg predators is Fornasini's spiny reed frog *Afrixalus fornasini*, which locates fresh nests before the outside has hardened. It sticks its head deep into the foam nest and takes a mouthful of eggs and foam.

The southern foam-nest frog is a master at surviving the dry hot African savanna. It conserves water by tucking the limbs under the body to reduce the amount of exposed skin. Also, during the dry season it secretes fluid under the outer layer of skin to form a waterproof cocoon, effectively turning the old skin into a protective layer. In addition, it is uricotelic, which means that it excretes uric acid as a mechanism to concentrate wastes and conserve water. Overall, this frog is very efficient at water conservation as it loses water at a slow rate, similar to that in reptiles. It may spend the day exposed to direct sunlight without ill effects. At ambient temperatures between 39 °C and 43 °C, the frog is able to maintain a body temperature 2–4° below ambient by adjusting the rate of evaporative water loss using nervous (sympathetic) control. The blood of the southern foam-nest frog is adapted to carry oxygen at high temperatures. It is able to survive 6 months or more without food or water.

KEY REFERENCES: Loveridge 1936, 1970; Drewes & Altig 1996; Schiøtz 1999; Poynton 2000b.

Caecilians—Order Gymnophiona

Caecilians are limbless amphibians that resemble earthworms. They burrow into soft soil and leaf litter. Their eyes are covered by skin and/or bone. Two families of caecilians occur in Africa, and representatives of both families are known from East Africa.

KEY TO THE FAMILIES

Most readers should be able to identify a caecilian using only the photographs, descriptions, and maps. These keys are only appropriate for caecilians from East Africa.

Biologists who have skeletal material available will be able to use the identification keys provided by Nussbaum and his coworkers. Their works are listed as key references below.

1a. Transverse vent — Caeciliidae
1b. Longitudinal vent — Scolecomorphidae

Caecilians—Family Caeciliidae

This family is widespread, with representatives in India, the Seychelles, Africa, Central America, and South America. Six species in two genera are known from East Africa. Many of the distinguishing features, such as the scales, require a stereomicroscope or hand lens to be seen. We suggest that reference to the photographs and descriptions will be generally sufficient.

KEY TO THE SPECIES

1a. Terminal shield absent, cycloid scales embedded in the skin — *Schistometopum gregorii*
1b. Terminal shield present, no scales embedded in skin — 2

2a. Tentacle midway between nostril and corner of mouth — 3
2b. Tentacle closer to nostril or corner of mouth, 132–148 annuli (body rings) — 4

3a. Blue-gray with small white spots — 5
3b. Bright pink, wormlike, 150–152 annuli — *Boulengerula changamwensis*

4a. Tentacle nearer to corner of mouth than nostril, pink and slender body *Boulengerula uluguruensis*
4b. Tentacle nearer to nostril than corner of mouth, 26–28 upper teeth, glossy black above *Boulengerula taitana*

5a. 125–135 annuli *Boulengerula boulengeri*
5b. 139–150 annuli *Boulengerula* sp.

Caecilians—Genus *Boulengerula*

This genus is known from equatorial Africa south to Malawi. Where known, the species are oviparous with direct development. There are six species, of which five are known from East Africa and the fifth (*Boulengerula fischeri*) is known from Rwanda.

Boulenger's Caecilian

Boulengerula boulengeri Tornier, 1897
(Plate 24.1)

The species name honors G. A. Boulenger, herpetologist of the Natural History Museum in London.
Usambara bluish-gray caecilian, *mcudi* in Shambala.

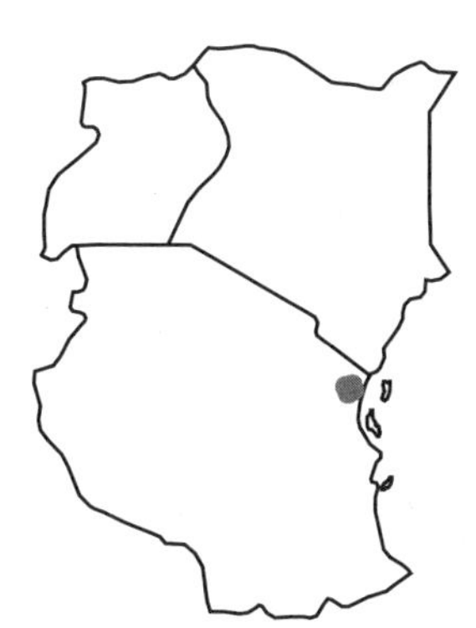

Fig. 266 *Boulengerula boulengeri.*

DESCRIPTION: This species reaches a length of 278 mm, with a diameter of 6.5 mm. There are 125–135 annuli. The front of the tongue is attached to the gum. Adults are pale blue-gray with small white spots and a darker band along the back and paler anteriorly. Below, the color is similar except for a pink throat. The young are more pinkish, especially anteriorly. *Boulengerula boulengeri* is predominantly a burrower in soil. This appears to be correlated with its morphology: a bullet-shaped head with a solid skull and near-cylindrical body. It has reduced eyes covered by bone and lacks the temporal fossae, which are holes in the upper part of the skull.

HABITAT AND DISTRIBUTION (FIG. 266): This species has been found under logs and stones in damp spots, but it mostly lives within the soil. It is known from the East Usambara and Magrotto mountains of Tanzania.

BREEDING: The female lays a small number of eggs in a nest. These develop directly into small versions of the adults, with the female in attendance.

NOTES: Known food is termites. The only known predator is the Usambara garter snake *Elapsoidea nigra*.

KEY REFERENCES: Barbour & Loveridge 1928a, Nussbaum & Hinkel 1994, Gower et al. 2004.

Changamwe Caecilian

Boulengerula changamwensis Loveridge, 1932

This species is named for Changamwe, a village near Mombassa on the coast of Kenya.
Changamwe lowland caecilian.

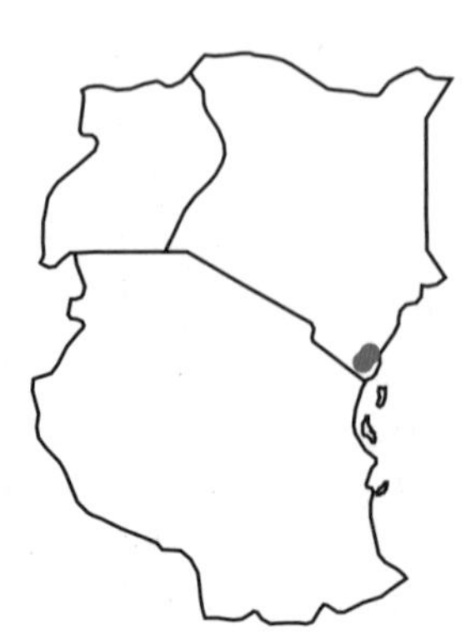

Fig. 267 *Boulengerula changamwensis.*

DESCRIPTION: The longest specimens are 235 mm, with a diameter of 5 mm. The snout projects far beyond the lower jaw. The eye is indistinguishable, and the tentacle is round, surrounded by a faint circular groove, about halfway along the head just above the middle of the upper jaw. There are 18 teeth on the upper jaw. A vertical keel is present on the terminal shield. The body is a livid pink, with 150–152 annuli. The tongue is free, not attached to the gum.

HABITAT AND DISTRIBUTION (FIG. 267): This species is found in leaf litter and loose soil. It is known from southeastern Kenya and the Shire Highlands of southern Malawi.

BREEDING: Unknown. This species probably lays eggs.

NOTES: This species is known to eat termites.

KEY REFERENCES: Loveridge 1936, Nussbaum & Hinkel 1994.

Milky Caecilian

Boulengerula sp.

This species is in the process of being formally named.
Usambara milky-blue caecilian, *mcudi* in Shambala.

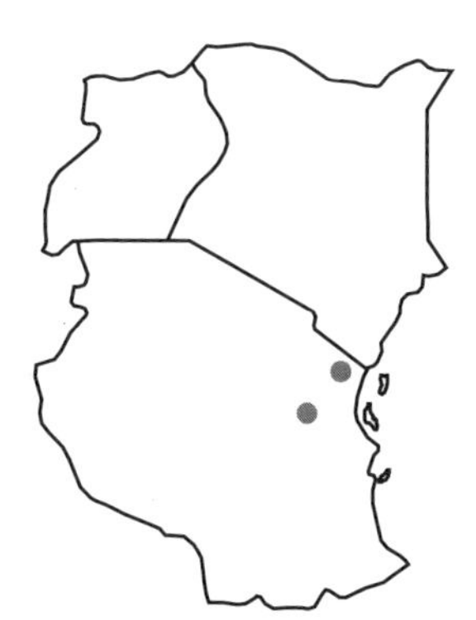

Fig. 268 *Boulengerula* sp.

DESCRIPTION: This species reaches a length of 278 mm, with a diameter of 6.5 mm. There are 139–150 annuli. Adults resemble *Boulengerula boulengeri* with a dark blue-gray middorsal stripe and light blue on the sides and below. The throat is often pinkish brown.

HABITAT AND DISTRIBUTION (FIG. 268). This caecilian has been found under logs and stones in damp spots, and within soil. One was discovered inside a standing, rotten log at Mazumbai. It is known from the West Usambara and Nguru mountains in Tanzania.

BREEDING: Unknown. This species is probably egg laying.

NOTES: In captivity these caecilians take termites and beetle larvae.

KEY REFERENCE: Vestergaard 1994.

Taita Hills Caecilian

Boulengerula taitana Loveridge, 1935
(Plate 24.2)

The species name refers to the Taita Hills in southern Kenya.
Murwe in Taita.

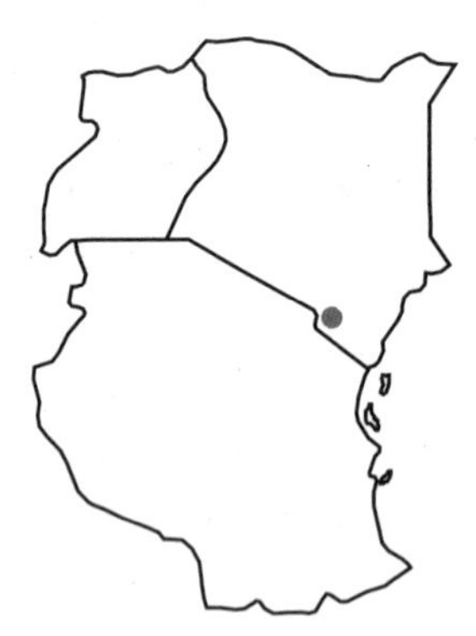

Fig. 269 *Boulengerula taitana.*

DESCRIPTION: This caecilian grows up to 360 mm long and 7 mm wide. There are two rows of teeth in the lower jaw, with 26–28 upper teeth. Along the body there are 136–148 annuli, interrupted along the back except at the nape and the tail. A lateral tentacle is found midway along the upper jaw, nearer the nostril than the corner of the mouth. The body is glossy black above, with each annular groove a paler blue-gray. The undersurface is blue-gray blotched with brown, except for a pink throat.

HABITAT AND DISTRIBUTION (FIG. 269): This caecilian has been found in soil between 4 and 64 cm deep, mostly in fallow agricultural land. It has only been found on the Taita Hills of Kenya, at altitudes between 1200 and 2000 m.

BREEDING: This species lays eggs, which are deposited in nests made just below the surface of moist soil, away from free water. The female remains with the eggs to protect them. The eggs develop directly into miniature versions of the adults.

NOTES: Known food items include very small termites and perhaps detritus. Although this species lives in an area that was once heavily forested, it seems to be very successful when the forest is removed for agriculture.

KEY REFERENCES: Loveridge 1936, Hebrard et al. 1992, Nussbaum & Hinkel 1994.

Uluguru Pink Caecilian

Boulengerula uluguruensis Barbour & Loveridge, 1928

This species is named for the Uluguru Mountains.
Mvuvi in Kami.

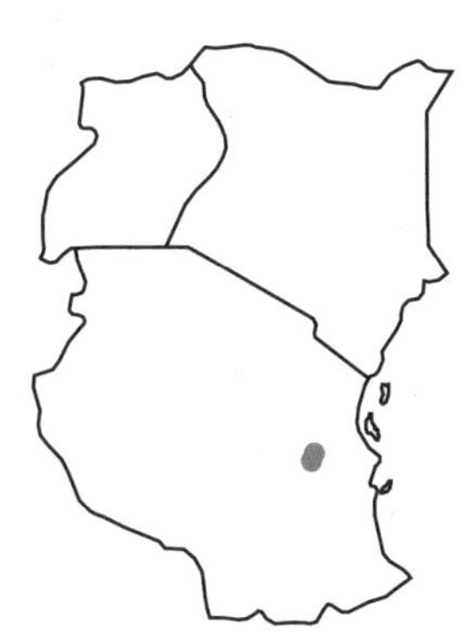

Fig. 270 *Boulengerula uluguruensis.*

DESCRIPTION: This species is very wormlike in appearance, with a length of 272 mm and a diameter of 5 mm. There are 132–148 annuli along the body, which is slender with a pointed snout that projects far beyond the lower jaw. The tentacle is round, surrounded by a circular groove, about halfway along the head just above the middle of the upper jaw. The tongue is not attached to the gum. The skin folds are interrupted dorsally except on the nape and tail. These caecilians are a transparent livid pink above and below.

HABITAT AND DISTRIBUTION (FIG. 270): This species is found in loose soils on the Nguru and Uluguru mountains of Tanzania.

BREEDING: Unknown. This species is probably egg laying.

NOTES: Recorded food items are termites.

KEY REFERENCES: Barbour & Loveridge 1928a, Nussbaum & Hinkel 1994.

Caecilians—Genus *Schistometopum*

This genus is characterized partly by the details of its skull anatomy. There are two species, one in São Tomé and nearby islands, and the other in East Africa.

Flood Plain Caecilian

Schistometopum gregorii (Boulenger, 1894)
(Plate 24.3)

This species was named for J. W. Gregory, a geologist based in Glasgow, after whom the Gregory Rift is also named.
Nyoka mai in Kami, *sango* or *ntzango* in Pokomo.

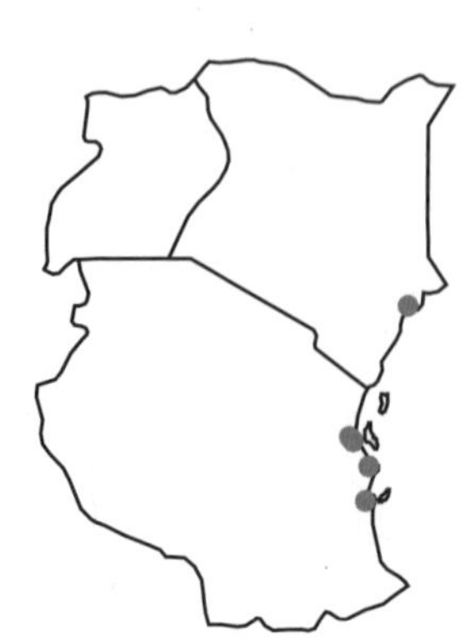

Fig. 271 *Schistometopum gregorii.*

DESCRIPTION: This species reaches a maximum length of 360 mm and a width of 14.5 mm. Females are probably much wider when they breed. The eye is distinct. There are a large number of secondary annuli on the posterior of the body. Small scales are embedded in the skin. The body is a shiny black above, and where the skin stretches over the skull the color is a delicate gray. Below it is slate gray.

HABITAT AND DISTRIBUTION (FIG. 271): Specimens have been found in moist soil and mud of river floodplains. They are known from the Tana River in Kenya, and the Ruvu, Wami, and Rufiji river floodplains in Tanzania.

BREEDING: Unknown, although the other species in this genus is viviparous, and the young are independent immediately after birth.

NOTES: Known food includes termites and earthworms. These caecilians are said to concentrate in damp areas in the dry season, where they are easy to dig up. They may prey on frog eggs under water.

KEY REFERENCES: Loveridge 1936, Nussbaum & Pfrender 1998.

Caecilians—Family Scolecomorphidae

The scolecomorphids spend their life below the surface of forest litter and soft tropical soils. This family of caecilians is restricted to Africa and differs from other caecilians in the details of its skull morphology. They are usually only discovered after very heavy rain, and so very little is known of their biology. There are two genera, *Crotaphatrema* from Cameroon and *Scolecomorphus* from eastern and central Africa.

Caecilians—Genus *Scolecomorphus*

This genus has been found in the forests of eastern and central Africa. Eyespots are visible in young animals but disappear in older individuals as the skull grows. A short tentacle is present on each side of the head and fits into a socket. This genus is viviparous. Three species are known from the area covered here.

KEY TO THE SPECIES

1a. Purple with a broad black dorsal band — *Scolecomorphus vittatus*
1b. Blue, brown, or black above; no bright yellow sides — 2

2a. Dark purple or black above, 131–152 annuli
Scolecomorphus kirkii
2b. Blue to black above with a pink throat, 124–151 annuli
Scolecomorphus uluguruensis

Kirk's Caecilian

Scolecomorphus kirkii Boulenger, 1883
(Plate 24.4)

This species is named for Sir John Kirk (1832–1922), who was vice consul at Zanzibar and a keen naturalist.
Lake Tanganyika caecilian, *timagwini* in Bena, *melawuletzi* in Kinga.

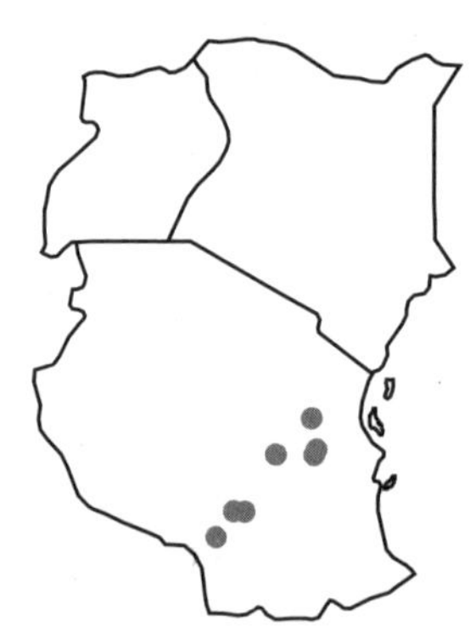

Fig. 272 *Scolecomorphus kirkii.*

DESCRIPTION: This large caecilian is 185–451 mm long with a diameter of 7–20 mm. The body, which is earthworm-like, has 131–152 rings, and the posterior end is blunt. The snout tip projects forward of the mouth. Small nostrils are present at the sides of the snout. A short sensory tentacle, only about 1 mm long, is below each nostril. The small dark eyespots at the base of the tentacle are visible only in young individuals. Above, this caecilian is purple or black, with brown sides. The undersurface is white in front, becoming pink toward the back.

HABITAT AND DISTRIBUTION (FIG. 272): This species is known from forest and cultivated areas. It has been found from southern Malawi through to the Ubena and Mahenge highlands, and the Nguru, Uluguru, and North Pare mountains of Tanzania.

BREEDING: This species is probably live-bearing. A very young specimen showed a number of unusual features arranged on the head around the mouth, including various rows of teeth, and fleshy structures.

NOTES: Many specimens collected so far have emerged after heavy rain. It would be very valuable to discover more of the biology of this animal. This is the first vertebrate found to have highly mobile, protrusible eyes. In members of this family, the eyes are not in sockets but instead are attached to the base of the tentacles. There is a track of translucent skin along the side of the head so that as the tentacles are extended and retracted, the eyes remain exposed to light. In this species, the eyes are carried beyond the skull and exposed to the outside environment when the tentacles are fully extended.

KEY REFERENCES: Loveridge 1933, Nussbaum 1985, O'Reilly et al. 1996, Loader et al. 2003.

Uluguru Black Caecilian

Scolecomorphus uluguruensis Barbour & Loveridge, 1928

The species name refers to the Uluguru Mountains in Tanzania.

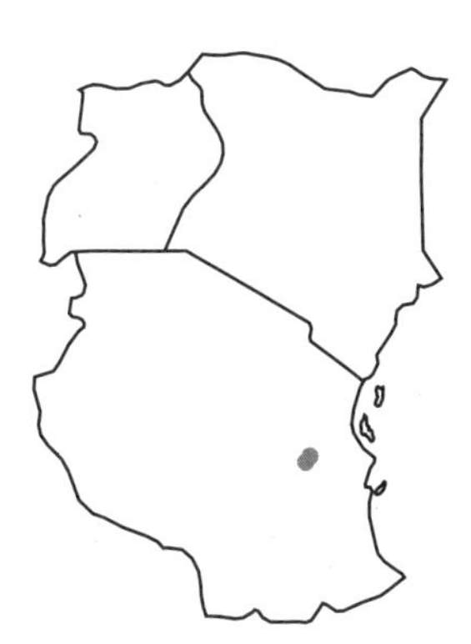

Fig. 273 *Scolecomorphus uluguruensis.*

DESCRIPTION: The head is very small, with a thick body and blunt tail. The snout projects far beyond the lower jaw. There are 16 upper teeth, 12 lower, and an additional 6 recurved palatine teeth. The eye is hidden. The tentacle is found in a horseshoe-shaped groove, opening anteriorly in a line with the nostril and the apex of the lower jaw. Along the body there are 124–151 annuli, with those on the nape pronounced. Behind the first 14 or so, the annuli are interrupted on the vertebral line to the last 20 mm of body and tail. The top is blue to jet black, sometimes glossy; the lower surface is gray; and the throat and anal area are bright pink.

HABITAT AND DISTRIBUTION (FIG. 273): This species is known only from the Uluguru Mountains in Tanzania, at altitudes up to 2500 m. Specimens have been found in forest.

BREEDING: This species is viviparous. Females hold between three and seven fetuses. A large sample showed that all were in the same stage of development, indicating a brief mating season. The fetuses possess three-branched, feathery external gills.

KEY REFERENCES: Barbour & Loveridge 1928a, Parker & Dunn 1964, Nussbaum 1985.

Ribbon Caecilian

Scolecomorphus vittatus (Boulenger, 1895)
(Plate 24.5)

The specific name *vittatus* means "ribbon."

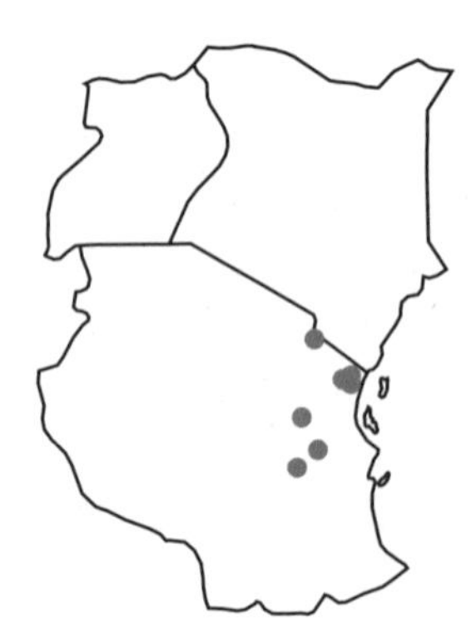

Fig. 274 *Scolecomorphus vittatus.*

DESCRIPTION: These caecilians grow to a length of 300 mm and a width of 9 mm. There are 122–148 annuli. The eye is not visible in large specimens over 200 mm. There are 4–6 teeth on the palate. In the young the tentacle is in front of the apex of the lower jaw, but in large specimens it is on either side of the apex. They are glossy purplish black above and pink or mauve below with a cream tail. This species spends some time on the surface. Its morphology includes a relatively well-developed eye, although covered by bone. The eye moves forward with the tentacle during development, so that in adults it lies in a groove at the base of the tentacle, where it is covered by translucent skin.

HABITAT AND DISTRIBUTION (FIG. 274): This species is known from leaf mold beneath rotting logs in damp rain forest, or in similar situations near water. It has been found on the Usambara, Uluguru, and North Pare mountains of Tanzania.

BREEDING: This species is live-bearing. Females give birth to young that have scraping teeth in the mouth and others that lie outside the mouth, whose function is unclear.

NOTES: When picked up, a specimen exuded a sticky mucus. Food items include earthworms and caterpillars.

KEY REFERENCES: Barbour & Loveridge 1928a, Nussbaum 1985, Loader et al. 2003, Gower et al. 2004.

Tadpoles

Most frogs lead a double life: As tadpoles they live in water, eat vegetation, and have a blood system that allows oxygen to be taken up from the watery environment. They later metamorphose into adults that eat largely insects, live on land, and require hemoglobin to take up oxygen from the air. This strategy prevents direct competition between the adults and tadpoles. Not all species have free-living tadpoles, and not all tadpoles are the same. Many filter out small algae from the water, some chew the stems and leaves of water plants, while a few are carnivorous, feeding on other tadpoles.

Only a proportion of the tadpoles of East Africa are known. Very valuable observations can be made by collecting some eggs from known parent frogs and allowing these to develop. Some should be allowed to develop into froglets, to check their identification. A few should be preserved in 5% formalin at the stage where the hind limbs are just visible. The preserved specimens should be clearly labeled with as much information as possible, and can be deposited in a natural history museum or sent to a specialist at a university.

Tadpoles are well regarded by local people. In Luo a tadpole is known as *oluk*, while in Nandi the word *kibilbiliot* is used, which describes the swimming motion of the larvae.

Identification of Tadpoles

Tadpoles are very easy to identify to the level of family, and often to genus. Unlike the adults, which are normally secretive during the day, tadpoles can be found and collected from most water bodies. The identification relies on anatomical features of the mouth, vent, spiracle, and certain body proportions. These are best seen through a stereomicroscope, although a good hand lens can be used for all but the smallest tadpoles. Many species of frogs are direct developers: the embryos pass through a tadpole stage within the egg and hatch as small frogs. Some forest toads in the genus *Nectophrynoides* are ovoviviparous: the eggs are retained within the oviducts, where the tadpoles develop through to small frogs before emerging from the cloaca of the female. There is thus no free-swimming stage in both of these cases.

Characters Used for Identification

Tadpoles should be carefully examined under a microscope or hand lens to check the features listed below. Pale structures like papillae around the mouth, and the spiracle opening are best seen if the tadpole is lightly stained—food coloring works well. Tadpole identification may be supported by behavioral data. Some species of tadpoles school, some live in mid-water, while many are bottom dwellers. Most frogs will only lay their eggs in particular habitats, so the tadpoles will only be found in those places. In the key and notes that follow, an indication is given of the likely habitat and behavior of the tadpoles. The tadpoles of many species are unknown or undescribed. We have indicated the species for which no information is available in the species accounts. Check the distribution of these species to be aware of the possible presence of tadpoles when attempting to identify specimens. Tadpoles are frequently very similar within a genus. For this reason an illustration of only a typical tadpole for each genus has been provided.

Tadpole identification in the field is usually not a problem, as for each area the species maps will often quickly establish a list of expected tadpoles, which accounts for only a small subset of the species described in this book. Exceptions are the large numbers of reed frogs *Hyperolius* spp. or tree frogs *Leptopelis* spp. that may breed in the same pond, sometimes at the same time or a few days or weeks apart. They have similar tadpoles that cannot effectively be identified to the species level. Tadpoles remain a very interesting part of the biology of frogs, and hopefully this book will stimulate the discovery and description of tadpoles for all the remaining species.

Finally, the identification of a tadpole can only be established without doubt if either the tadpole is reared from the eggs laid by known parents, or the tadpole is permitted to grow and metamorphose into a frog that can be identified.

Terminology Used in the Keys and Descriptions of Tadpoles (Figs. 275, 276)

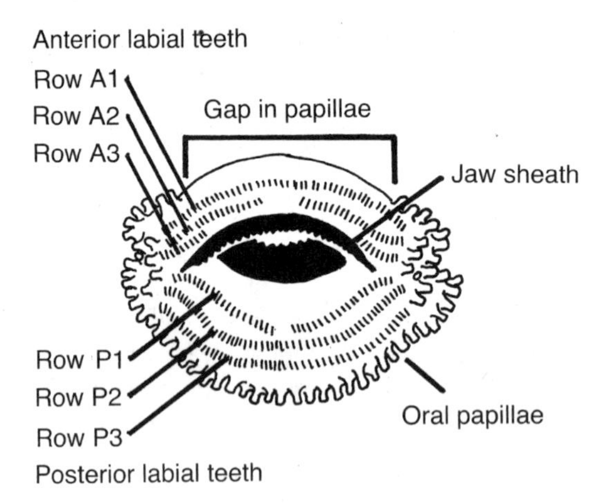

Fig. 275 Tadpole mouthparts. This example of *Strongylopus* has an upper gap in the papillae, and three anterior rows of labial teeth, with only the most anterior row (A1) complete. There are three posterior rows, with only the first row (P1) incomplete. The labial tooth row formula is 3(2–3)/3(1). The dark biting mouthparts are the keratinized jaw sheaths.

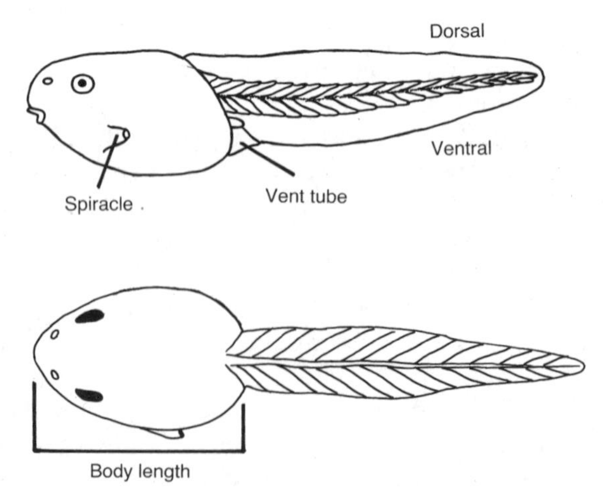

Fig. 276 Tadpole body parts. The body is measured from the snout to the start of the caudal muscles. The tubelike structure on the left side of the body is the spiracle.

Body length. A straight line measured laterally from the tip of the snout to the junction of the tail muscles.

Dorsal. Pertaining to the upper surface of the body.

Jaw sheaths. Serrated keratinized sheaths overlying upper and lower cartilages that serve as cutting or abrasive structures.

Labial tooth row. A line of small labial teeth embedded in a tooth ridge.

Labial tooth row formula. A notation for designating the number of tooth rows on the upper and lower labia, and the position of rows with medial gaps. Upper rows are numbered from the lip to the mouth, and lower rows are numbered from the mouth posteriorly. The numbers in parentheses indicate rows with medial gaps. A formula of 3(2–3)/3(1) indicates that there are three upper rows, with the two rows closest to the mouth having medial gaps, and there are three posterior rows, of which the row closest to the mouth has a medial gap.

Medial gap. A break in the middle of a labial tooth row.

Oral angle. The sides of the mouth.

Oral disc. The structures around the mouth, including the upper and lower labia, usually with transverse rows of labial teeth, and papillae on the face and margin of the disc.

Oral papilla. A fleshy projection on the face or margin of the oral disc.

Spiracle. One or two openings of different shapes and positions for the exit of water pumped through the mouth and internal gills for respiration and feeding.

Tail length. A distance measured laterally from the junction of the tail muscles with the body, to the tip of the tail.

Tentacle. A long sensory structure originating below the eye.

Vent. The posterior opening of the intestine, which may end in a vent tube.

Vent tube. A tube that projects the vent away from the body. It can be in line with the ventral fin (medial) or lie to the right of the fin (dextral).

Ventral. Pertaining to the lower surface of the body.

Key and Brief Descriptions of Known East African Tadpoles

Tadpoles may change quite markedly during their development. The key is based on fully grown tadpoles, at the stage where five toes are just visible, or older. It only covers free-swimming tadpoles. Less than half of the East African tadpoles are known.

KEY TO THE FAMILIES

1a. Oral papillae present, and usually also labial tooth rows and keratinized jaw sheaths — 3
1b. No papillae, keratinized jaw sheaths, or labial tooth rows — 2

2a. Mouth with a fold in the lower lip, and one spiracle opening situated midventrally — Microhylidae
2b. Mouth a horizontal slit, tentacles present at oral angle — Pipidae

3a. A broad lower gap in the oral papillae present, vent median in the fin margin — Bufonidae
3b. A lower gap absent, or very narrow if present; vent on the right of the fin margin — 4

4a. A narrow lower gap in the oral papillae present — Rhacophoridae
4b. Lower gap absent — 5

5a. Anterior third of the tail muscle obscured by opaque connective tissue, which extends ventrally and dorsally beyond the muscles, giving the appearance of a thicker muscle region — Hemisotidae
5b. Tail muscle not obscured anteriorly — 6

6a. One anterior row of labial teeth, or no anterior tooth rows — Hyperoliidae (part)
6b. Two or more rows of anterior labial teeth — 7

7a. Nostrils small and far apart (distance between nostrils greater than 10 times nostril width), tadpoles typically found in thin mud — Hyperoliidae (part)
7b. Not as above — Ranidae

Family Bufonidae

Genus *Bufo*

The tadpoles are typically small and dark and found in shallow water. Small differences in pigmentation patterns are useful distinguishing characters. The tadpole of *B. gutturalis* is illustrated (Fig. 277).

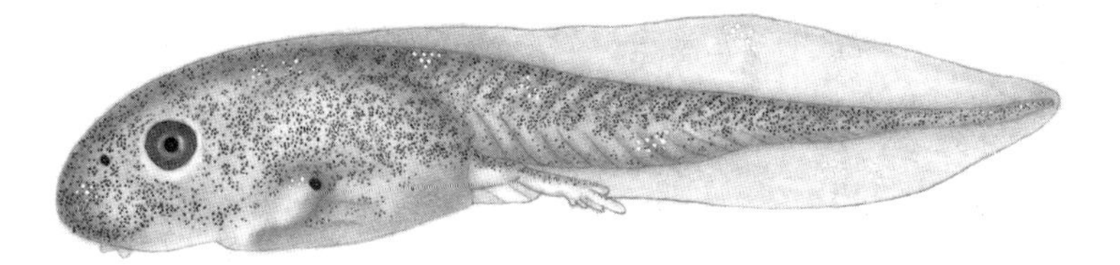

Fig. 277 *Bufo gutturalis.*

Bufo garmani

The tadpoles have been observed feeding upside down at the surface. They assume a lighter or darker coloration depending on the background, with lighter tadpoles being found on pale backgrounds. Metamorphosis takes from 64 to 91 days.

Bufo gutturalis

The tadpoles are small, up to 25 mm long, and black with iridescent spots. There is no pigmentation across the anterior throat region, and the belly midline has little pigmentation. Tadpoles are free-swimming within 2–3 days after the eggs have been laid. It takes a further 74 or 75 days to metamorphosis. Tadpoles may be found aggregated on sand at the water's edge.

Bufo kisoloensis

Full-grown tadpoles are about 20 mm long. Overall the tadpole is brown, rather than black as is common in many species of *Bufo*. The upper pigmentation is a uniform dark brown. The lower pigmentation is a uniform lighter brown. The tail is evenly pigmented almost to the lower surface over the front two-thirds. The dorsal fin is completely pigmented, while the ventral fin is only pigmented anteriorly. Some tadpoles have a light mottling on the ventral fin.

Bufo maculatus

The tadpole at the five-toed stage is 14–17 mm long. The oral formula is 2(1)/3. The tail accounts for 44% of the total length. Pigmentation consists of small dark brown stipples, uniformly covering the back and extending onto the sides, with a wide unpigmented belly stripe. The tail muscles are countershaded, with heavy upper pigment, a middle mottled area, and a lower unpigmented region. Tadpoles sometimes aggregate, which reduces predation.

Bufo xeros

The tadpole is large for the genus, reaching 34 mm.

Genus *Mertensophryne*

The tadpoles develop in very small pockets of water. They are similar in ecology and form to those of *Stephopaedes*.

Mertensophryne micranotis

The tadpole reaches a length of 15 mm. A characteristic ring of tissue surrounds the eyes and nostrils. A pale line runs from the eye to the nostril. The labial tooth formula is 1/2. The dorsal fin is higher at the back of the tail (Fig. 278). The tadpoles develop in small pockets of water in snail shells, empty coconuts, or small rock crevices.

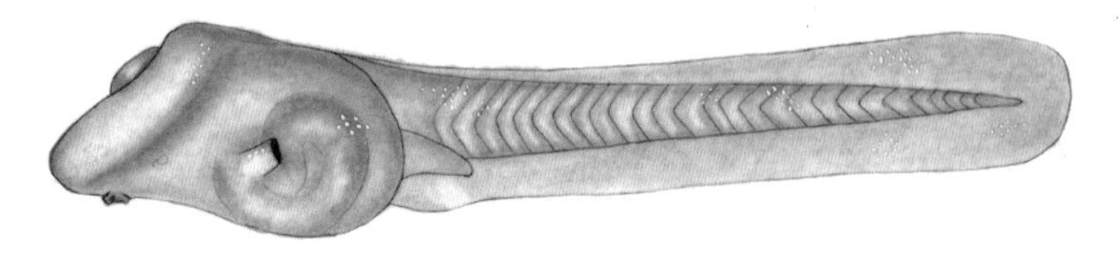

Fig. 278 *Mertensophryne micranotis.*

Genus *Schismaderma*

The tadpoles swarm together in a compact ball. This ball of tadpoles appears to roll slowly through the water, with its center below water level, but floating well clear of the bottom. The ball slowly rises and sinks. Small groups of tadpoles are sometimes seen on the bottom of the pond, all facing inward, strongly attracted to each other.

They have been found in a mixed swarm with African bullfrog *Pyxicephalus adspersus* tadpoles.

Schismaderma carens

The tadpoles are large, up to 35 mm long, and black. The oral formula is 2/3. The tadpole is quite different from other toad tadpoles in this region on two counts: a peculiar flap of skin on the head and their gregarious behavior. The horseshoe-shaped flap of skin extends from the eyes to midway along the body (Fig. 279). When the water is fetid, the tadpoles swim very near the surface, with the flap of skin serving as a float. The flap is well supplied with capillaries and functions as a respiratory organ. In an experiment, tadpoles grown in polluted water developed head flaps twice as large as those of tadpoles grown in clean, aerated water.

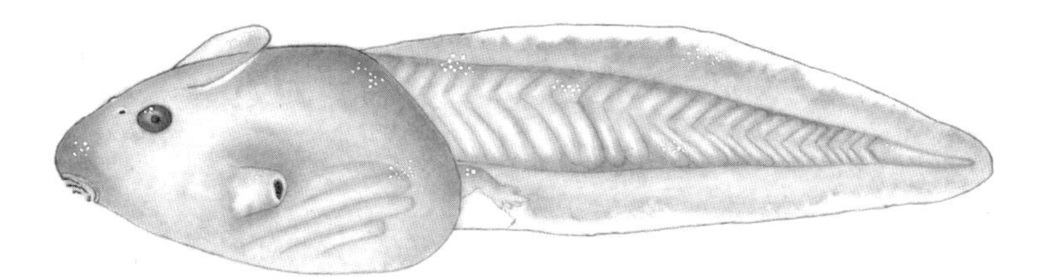

Fig. 279 *Schismaderma carens.*

Genus *Stephopaedes*

No tadpoles are known of the East African species, but the southern form in Zimbabwe, *S. anotis*, has small tadpoles that have a ring of tissue surrounding the eyes and nostrils, similar to *Mertensophryne micranotis*.

Family Hemisotidae

Genus *Hemisus*

The tadpoles have a characteristic thickened sheath covering the front of the tail muscles.

Hemisus marmoratus

The tadpole reaches a length of 55 mm (Fig. 280). The mouthparts are characterized by a few long lower papillae. Variation in labial tooth row formula is from 5(2–5)/4(1) to 5(3–5)/4(1). The eggs hatch into tadpoles after 8 days. The tadpole leaves the burrow by swimming out when it floods, or can be carried by the female to a suitable pond. The female may also dig a slide about 1 cm deep in the mud surface, along which the tadpoles follow her to open water. The tadpoles spend the day resting on the bottom of muddy pools. They commonly occur together with sand frog tadpoles. They have been found in flooded grassy depressions or in shallow pans. The tadpoles are slow moving and can be caught by hand, although they can swim very fast when the need arises!

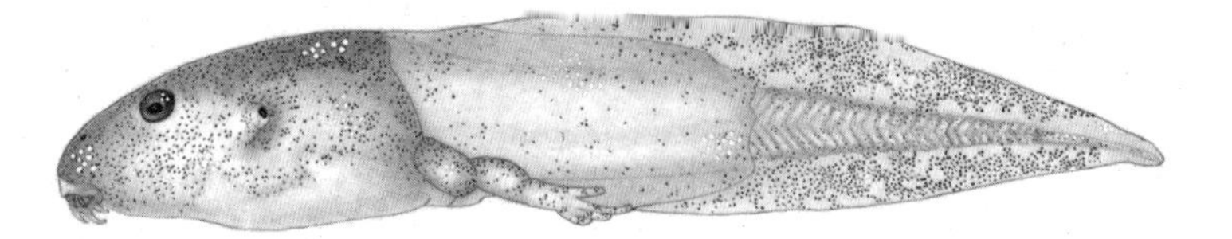

Fig. 280 *Hemisus marmoratus.*

Family Hyperoliidae

Genus *Afrixalus*

The tadpoles in this genus are long and thin, with no anterior labial tooth rows. The eyes are positioned on the side of the head, with the mouth at the front of the body. Many species are reported to have carnivorous tadpoles. The tadpole of *A. uluguruensis* is illustrated (Fig. 281).

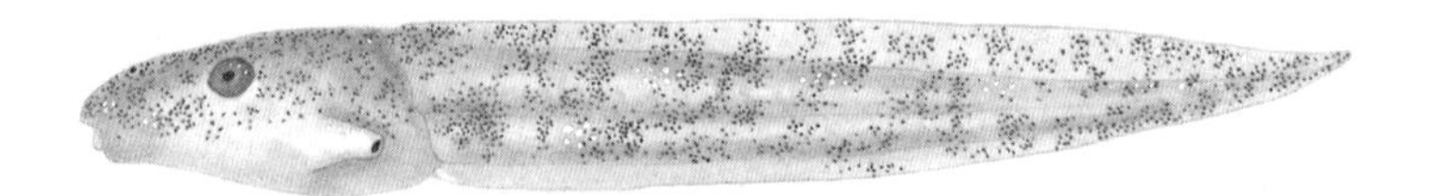

Fig. 281 *Afrixalus uluguruensis.*

Afrixalus fornasini

The tadpole reaches a length of 65 mm. It is brown with darker markings. The tail tip is sharp. There is a triple row of papillae below the mouth. The labial tooth row formula is 0/1. The tadpole drops into the water 5–10 days after the eggs are deposited on vegetation. Metamorphosis takes place after 10–12 weeks.

Afrixalus laevis

The tadpole has a narrow fin and a tooth formula of 0/0.

Afrixalus uluguruensis

The tadpole reaches a length of 37 mm. The head is rounded when viewed from above. The eyes are orientated upward, and the nostrils are very small. The tail has a sharp tip, and the ventral fin is unpigmented anteriorly. The labial tooth formula is 0/0. The vent is difficult to see. The tadpoles have been found in slow-flowing streams.

Genus *Hyperolius*

The tadpoles in this very widespread genus are not well known. The eyes are situated on the side of the head. The labial tooth row formula is 1/3. The details for the better-known species are listed below. The tadpole of *H. pictus* is illustrated (Fig. 282).

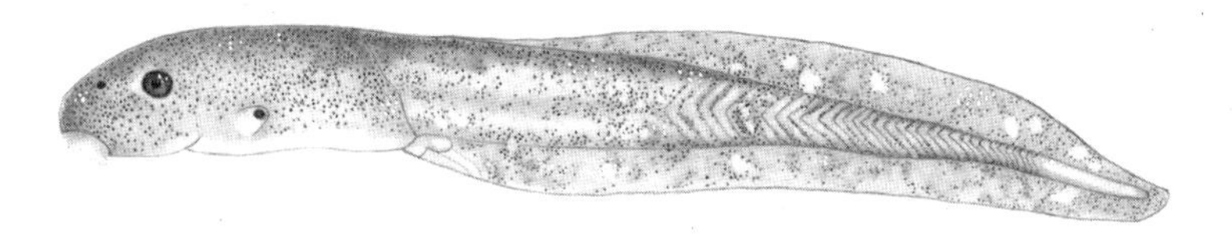

Fig. 282 *Hyperolius pictus.*

Hyperolius acuticeps

The tadpole reaches a length of 33 mm. It is brown with mottled fins, which are sharply pointed with a dark tip. The most posterior row of labial teeth (P3) is half the length of the first row (P1). There is a double row of papillae below the mouth. The tadpole is known from deep pools fringed with reeds.

Hyperolius argus

The tadpole reaches a length of 48 mm. It is a light brown with a pale underside and mottled fins. The most posterior row of labial teeth is short. The papillae are in a double row below the oral disc with a narrow gap in the middle. The keratinized jaw sheaths are heavy. The tadpoles are known from pools with dense vegetation.

Hyperolius mitchelli

The tadpole can be distinguished from other similar reed frog tadpoles by the papillae on the front of the head, which are continuous with the oral papillae. Another unusual feature is the complex of papillae that form a dense filter at the front of the mouth. The tadpole was described from specimens collected in the East Usambaras in Tanzania.

Hyperolius pusillus

The tadpole reaches a length of 35 mm. It is brown with a greenish tinge, but white below. The fins have a dark edge that fades out toward the tail. The jaw sheaths are delicate, and a double row of papillae is found below the mouth. Metamorphosis takes place after 5–6 weeks.

Hyperolius spinigularis

The tadpole has a characteristic black V on the snout tip. The labial tooth row formula is 1/3(1). The tadpole hatches after 5 days and wriggles from the leaf into the water.

Hyperolius tuberilinguis

The tadpole reaches a length of 46 mm. It is brown with pale undersides, speckled dorsally with golden pigment cells called iridiophores. The jaw sheaths are delicate, and the most posterior tooth row is much shorter than the other two.

Genus *Kassina*

The tadpoles have very high fins and are found in standing water where there is much vegetation. They feed on plants and have heavy mouthparts with extra accessory plates to enable them to deal with stems and other tough parts. The tadpoles are large and often brightly patterned in stripes or mottles. The eyes are on the side of the head, and the mouth is at the front. The tadpole of *K. senegalensis* is illustrated (Fig. 283).

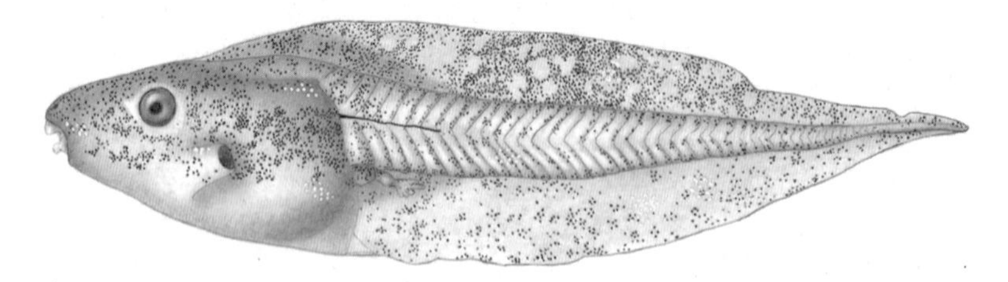

Fig. 283 *Kassina senegalensis.*

Kassina maculata

The tadpole is large, up to 130 mm long. The upper fin is high, starting at the back of the head. The tadpole has contrasting markings. The mouthparts are adapted for dealing with tough food: the labial tooth row formula is 1/3, the jaw sheaths are very heavy. In addition, the tadpoles have a hardened plate on each side of the jaw sheaths. A gap in the lower papillae is present. The tadpole can consume large amounts of plant matter, chewing the stems and leaves. The tadpoles may remain in the pool for 8–10 months.

Kassina senegalensis

The tadpole has very high fins, with the upper fin originating at the level of the eyes. It is large, up to 80 mm long. The tadpole is dark, with red or golden markings on the fins. The labial tooth row formula is 1/2(1), and there are heavy jaw sheaths and additional hardened plates at the side of the sheaths. The tail tip is thin and vibrates, to maintain the position of the tadpole in the pond. The tadpole can bite through stems of waterweeds and is also known to eat mosquito eggs. It tends to be a mid-water dweller, often hiding under the leaves of water plants.

Genus *Leptopelis*

The tadpoles have elongated tails with which they wriggle to the water. The labial tooth row formula of many species is 4(2–4)/3. The tadpole of *L. argenteus* is illustrated (Fig. 284).

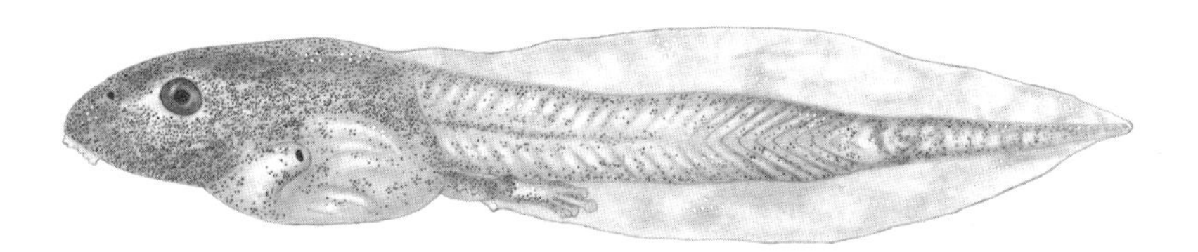

Fig. 284 *Leptopelis argenteus.*

Leptopelis argenteus

The tadpoles are found in muddy pools. The tadpole is brown, elongated, and has the typical labial tooth formula.

Leptopelis barbouri

The tadpoles are long with low fins, found in shallow pools in forest. They are remarkable for color, being bright yellow to blue-green.

Leptopelis oryi

The tadpoles are elongated with a tooth formula of 3(2–3)/3(1). The underside of the head is darkly pigmented.

Leptopelis vermiculatus

Tadpoles reach a length of 34 mm or more. The head slopes to the mouth, when viewed from the side. The tail is pointed, with parallel dorsal and lateral fins. The nostrils are surrounded by a pale area. The body and tail are heavily speckled, but pale below. The labial tooth row formula is 5(2–5)/3. The papillae at the sides of the oral disc are elongated.

Genus *Phlyctimantis*

The tadpoles are found in slow-flowing water and have very high fins.

Phlyctimantis verrucosus

The tail is very high, and the tooth formula is 1/3 or 1/3(2) (Fig. 285).

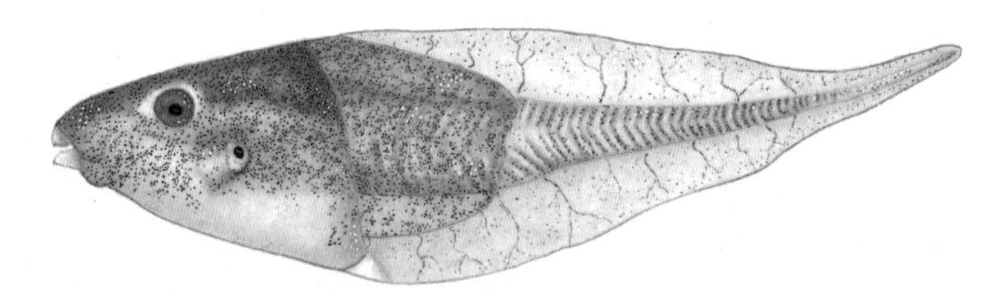

Fig. 285 *Phlyctimantis verrucosus.*

Family Microhylidae

Genus *Hoplophryne*

The tadpoles develop in small pockets of water trapped in fallen hollow bamboo stems or leaf axils of wild banana plants.

Hoplophryne rogersi

The tadpoles reach 28 mm in length. They are similar to the tadpoles of *H. uluguruensis* but have a series of distinct papillae on the snout and upper lips. The tadpole has no labial teeth or jaw sheaths. There are no external nares. The spiracle is median, ventral, and located about one-third along the body from the vent. The toes are webbed during development, but the webbing is lost at metamorphosis. The tadpole uses lungs and skin, not external gills, for respiration. Characteristic, almost triangular, skin flaps are present on either side of thc throat (Fig. 286). These are rough and assist in locomotion. The tadpoles feed on animal and plant tissue, including ants and frog eggs. They also feed on eggs of the same species.

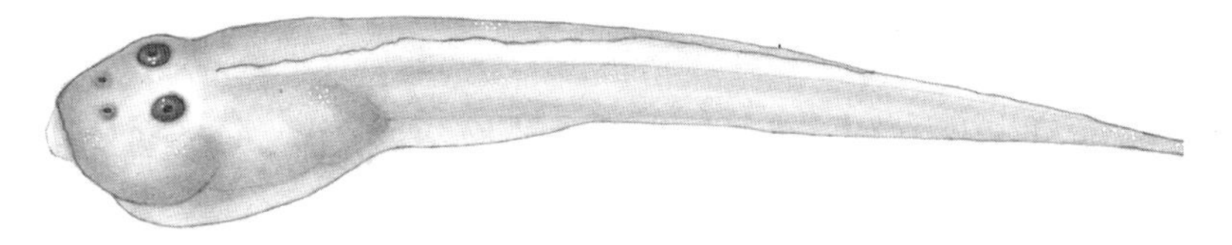

Fig. 286 *Hoplophryne rogersi.*

Hoplophryne uluguruensis

The eggs develop to an advanced stage before hatching, with the hind limb buds present. The body length is one-fifth the tail length. The mouth opening is at the front of the head, with a folded lower lip. The lip has a median groove and two conspicuous lateral flaps. A flap covers the gill region on each side. The vent is median. The tail is three and a half times as long as deep, and is bluntly pointed. The lower surface is lightly reticulated with pigment. The tadpole is toothless and without jaw sheaths. There are no external nostrils. The lungs are functional at the time of hatching, and there are no external gills at any stage, an unusual situation for frog larvae. The tadpole therefore uses lungs and skin, not external gills, for respiration.

Genus *Phrynomantis*

The tadpoles form schools and while feeding move slowly through the water. The eyes are situated on the side of the head. The mouth is a wide slit, with a V-shaped lower lip. There are no hard mouthparts. The spiracle is wide and flat, situated on the midline near the posterior end of the body.

Phrynomantis bifasciatus

The tadpole has a wide head, which is sharp when viewed from the side. It may reach 37 mm long. It is transparent, with darker pigmentation and many iridiophores (shiny pigment cells) along the midline. The tail possesses wide black longitudinal bands, sometimes also narrow red bands. The tail tip is thin, extending beyond the pigmented region of the fins (Fig. 287). Usually the tadpoles are gregarious, motionless, and remain about 1 cm below the surface. The tadpole eats mostly floating unicellular algae, along with desmids and diatoms. They have been found in muddy water pools at 40 °C. Metamorphosis takes 90 days in captivity.

Fig. 287 *Phrynomantis bifasciatus.*

Family Pipidae

Genus *Xenopus*

The tadpoles of *Xenopus* are transparent and without hard mouthparts. Many species school in the water column, vibrating the tail tip to remain in position. Tadpoles can breathe air or water. However, being filter feeders they use the same gill filter surfaces for food trapping and aquatic respiration. When there is a dense food suspension, the tadpoles breathe air more frequently.

Tadpoles living in algae-rich ponds may become quite green. The larger tadpoles prefer deeper water, while the smaller tadpoles are often found around vegetation.

The tentacles appear to compensate for poor maneuverability and an anterior blind spot, by preventing the tadpole from becoming trapped in vegetation. When the tadpole needs to come to the surface to breathe, it does so in unison with other tadpoles, which may offer some advantage against predators. The tadpole of *X. borealis* is illustrated (Fig. 288).

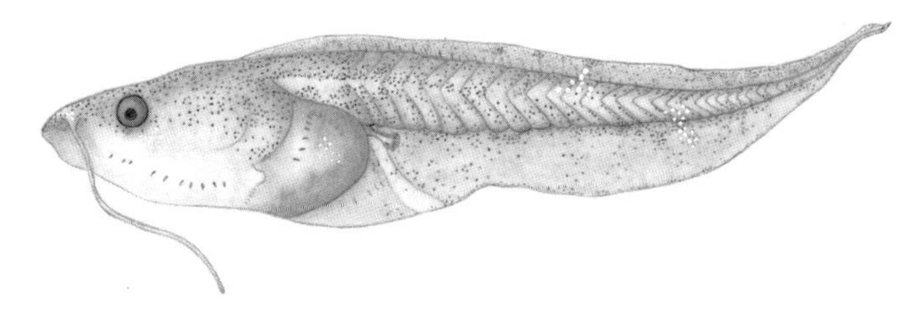

Fig. 288 *Xenopus borealis.*

Xenopus borealis

The body is heavy, and the tail is broad with a sharply upturned tip. These tadpoles do not school, although they may be found together in shallow water in groups usually not larger than 10. Most are found alone in the pool. They feed along the muddy bottom of the pond, sucking up food with the mouth just touching the surface. The internal organs and heart are sheathed in a white membrane. When disturbed, the tadpole races away for about 40 cm while twisting the body to show flashes of the white undersurface.

Xenopus muelleri

The body is more rounded than the tadpole of the common clawed frog *Xenopus victorianus*. The tentacles are long, accounting for more than 25% of the tadpole length. Tadpoles school 50–70 mm apart, with their bodies horizontal in mid-water. The tadpoles feed on particles of vegetation and unicellular algae.

Xenopus victorianus

The tadpole may become fairly large, reaching 80 mm. It has a long tail and high fin. The head is flattened and often translucent. The tadpole is transparent with various degrees of pigmentation, and possesses long tentacles. Eggs hatch rapidly into tadpoles, taking only 24 hours in the warmer areas. The tail tip vibrates constantly, maintaining the tadpole in a head-down position. The tadpole maintains its height in the water visually, changing only the amplitude of the tail tip oscillations to adjust position, and schools in parallel with others. It uses the lateral line sensory organs to maintain position in the dark, and they eyes to assist during daytime. The gregarious nature of a typical tadpole is shown by the fact that it tolerates other tadpoles well, and will only move if another individual comes within two body lengths during daytime or within one body length at night. The tadpole is pale in light, becoming more pigmented in the dark. Pale tadpoles choose pale backgrounds, while dark tadpoles choose dark backgrounds.

Family Ranidae

Genus *Afrana*

The tadpoles are large with many labial tooth rows. The tadpole of *A. angolensis* is illustrated (Fig. 289).

Fig. 289 *Afrana angolensis.*

Afrana angolensis

The tadpole grows to a large size, over 80 mm long. Overall the color is brown, with flecks. The mouthparts are distinct with a labial tooth row formula of 4(2–4)/3(1–2). The tadpole often lies quietly in sunny parts of the water.

Afrana ruwenzorica

Tadpoles with high numbers of labial tooth rows have been collected, one with 9/10, one with 9/6. It was thought that these might be *A. ruwenzorica*, but the tadpoles were collected from two localities where no adults were collected. Fieldwork is needed to confirm the identification.

Genus *Amnirana*

The tadpoles have small glandular patches on the body that are believed to function as a chemical defense against predators. The tadpole of *A. albolabris* is illustrated (Fig. 290).

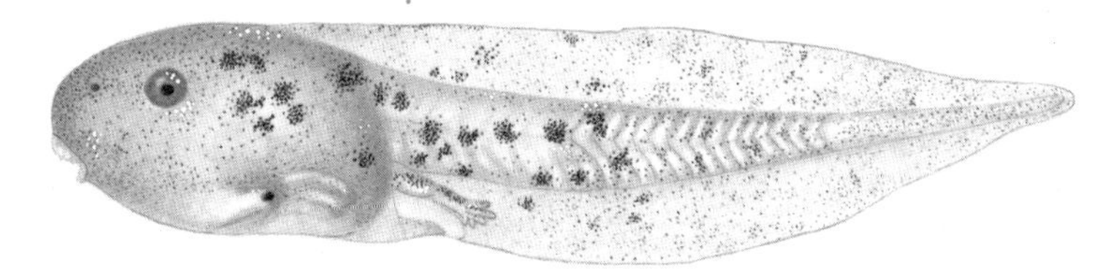

Fig. 290 *Amnirana albolabris.*

Amnirana albolabris

The tadpoles reach 60 mm in length by the time the hind legs start to develop. The body is salmon pink to red with darker spots. Mouthparts consist of a few elongated lower papillae, with an oral formula of 4(2–4)/3(1). Some tadpoles have five anterior rows of labial teeth. Characteristic skin glands are found arranged in three pairs—one pair behind the eyes, one in the midline near the tail, one either side of the vent—with a single median gland on top of the body. These are believed to secrete protective substances against predators such as catfish. These tadpoles can be seen during the day in flowing water.

Amnirana galamensis

The tadpoles reach a length of 60 mm. They are darkly spotted, with a tooth formula of 2/1, or 1/2.

Genus *Arthroleptides*

Tadpoles develop on wet rock faces, apparently feeding on algae growing on the rock. They have very specialized mouthparts. The hind limbs develop early, compared to other ranid tadpoles, as they use the limbs and tail muscle to move on the wet rock. The tadpoles complete development in 2–3 months, all the time staying on the home rock. Juveniles disperse and feed in wet vegetation along streams and on the leaf litter of the forest floor. The tadpole of *A. yakusini* is illustrated (Fig. 291).

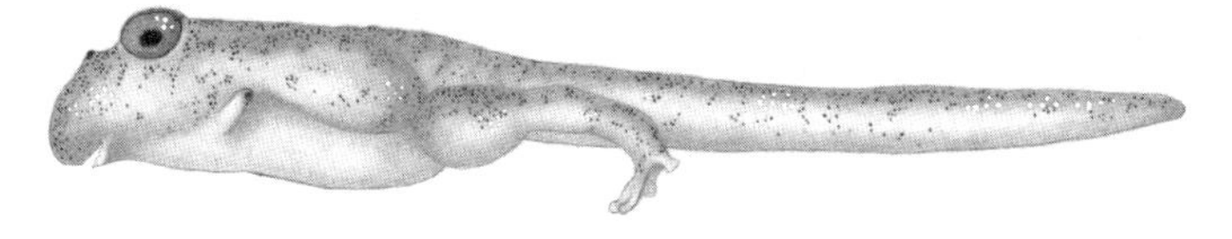

Fig. 291 *Arthroleptides yakusini.*

Arthroleptides martiensseni

The tadpole develops on wet rock or in shallow streams. The fins are very narrow, and the hind limbs develop early so that the tadpole can move around on the rock. The tooth formula is 3(1–3)/3(1). The eyes protrude on top of the head. The upper jaw sheath is strongly curved, resembling the beak of a parrot. The lower jaw sheath is acutely curved, fitting inside the upper sheath when the jaws close. The upper labium (lip) covers the mouth when closed. Saddle-like markings are present along the back and tail.

Arthroleptides yakusini

The eyes protrude dorsally, with a labial tooth formula of 3(1–3)/3(1). The upper jaw sheath is strongly curved, resembling the beak of a parrot. The lower jaw sheath is acutely curved, fitting inside the upper sheath when the jaws close. The posterior edge of the oral disc has a double row of marginal papillae, becoming single toward the angle of the jaw. The papillae are short and rounded. When the disc is closed, the upper lip covers the mouth. The dorsal fin is very reduced, being merely a ridge along the tail. The ventral fin is absent. Femoral glands extend from near the knee for one-third along the back of the thigh. The toes are separate with distinct discs. The bifid dorsal disc scutes are not yet developed. Nine light dorsal patches on the tail extend onto the sides. These are saddle-like in living specimens. Tadpoles hatch from the eggs and remain on the wet rock faces. They may occur at high densities on rock faces in a film of water, with 22 counted in 1 m^2 near the Uluguru North Forest Reserve, grazing on algae. They complete development in the film of water, taking 8–10 weeks. After the tail is resorbed, juveniles leave the nursery rock and can be found feeding in vegetation along streams and in leaf litter in forest.

Genus *Cacosternum*

Cacosternum sp.

The tadpole is small and moderately pigmented (Fig. 292). The mouthparts are variable, usually with a labial tooth row formula of 4(2–4)/3 or 4(2–4)/4.

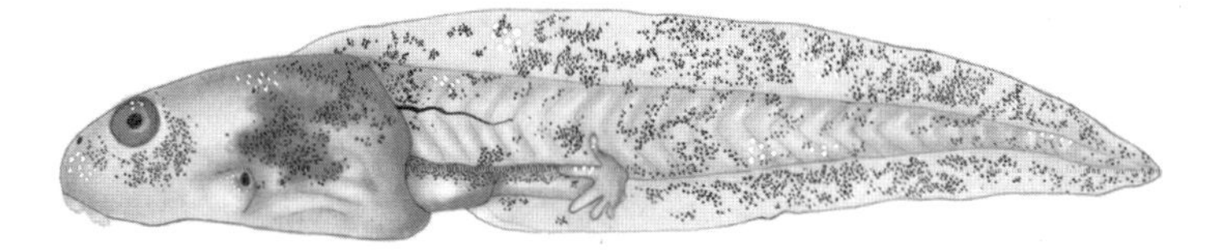

Fig. 292 *Cacosternum* sp.

Genus *Hildebrandtia*

The tadpoles are large and carnivorous. The tadpole of *H. ornata* is illustrated (Fig. 293).

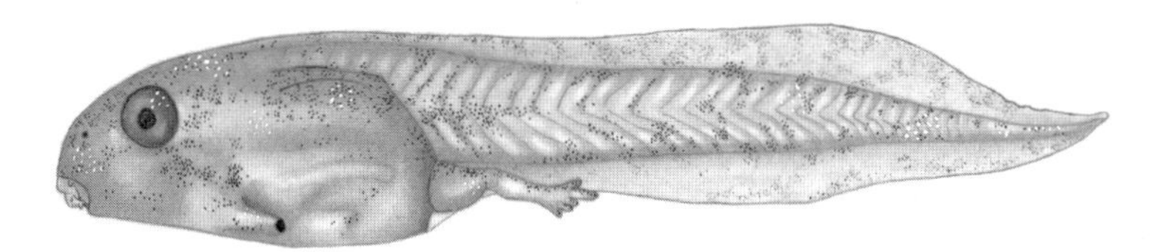

Fig. 293 *Hildebrandtia ornata.*

Hildebrandtia macrotympanum

The tadpole is large, up to 95 mm long, and heavily built. The body is dark above and pale below, with a metallic green sheen. The tadpole has a mouth at the front of the head. The labial tooth row formula is 1/2. Papillae are present only on the posterior lip. Jaw sheaths are large, keratinized, and serrated. It hunts tadpoles of other species but will also scavenge.

Hildebrandtia ornata

The tadpole is large, up to 95 mm long, and heavily built. The body is dark above and pale below, with a metallic green sheen. Specimens in muddy pools may be pale with a silver belly. The labial tooth formula is 0/2, and the jaw sheaths are massive. The mouth is situated at the front of the head. It hunts tadpoles of other species but will scavenge.

Genus *Hoplobatrachus*

These tadpoles are carnivorous, known to hunt tadpoles of ridged frogs *Ptychadena* spp.

Hoplobatrachus occipitalis

The tadpole is reddish gray above with many darker spots, but pale gray below. The tail is yellowish with brown markings. The jaw sheaths are well developed, with a typical "tooth" on the upper sheath that fits between two "teeth" on the lower sheath. The labial tooth formula is 4(3–4)/4(1–2) (Fig. 294).

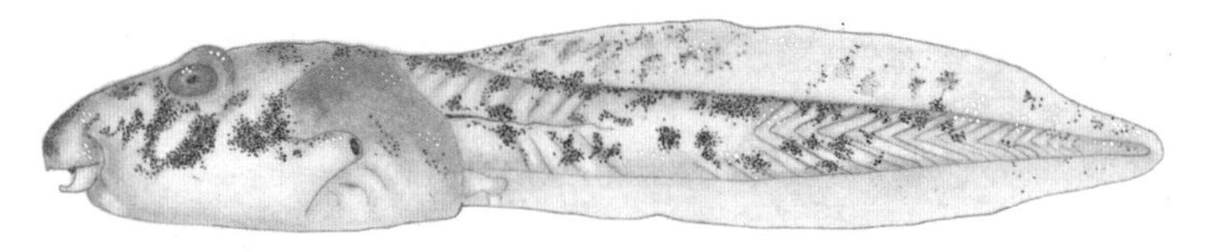

Fig. 294 *Hoplobatrachus occipitalis.*

Genus *Phrynobatrachus*

The tadpoles are dark and are found in almost all slow-flowing stream or pond habitats. At least one species in West Africa has nonfeeding, nonhatching tadpoles. The tadpole of *P. natalensis* is illustrated (Fig. 295).

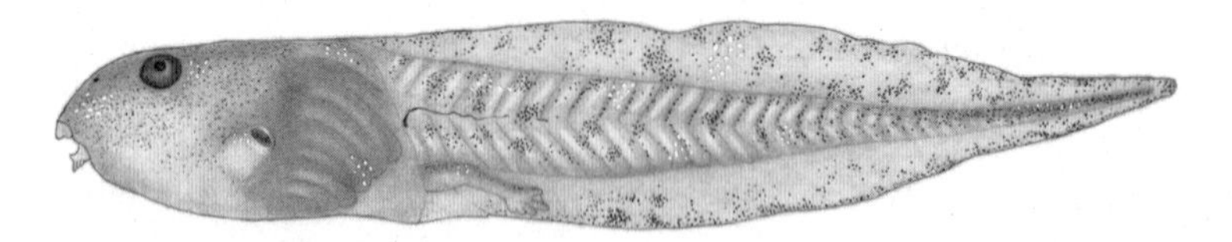

Fig. 295 *Phrynobatrachus natalensis.*

Phrynobatrachus kreffti

The tail is speckled and twice as long as the body. The nostrils are nearer the snout tip than the eyes, and the fins are narrow. The labial tooth formula is 2/3.

Phrynobatrachus mababiensis

The tadpoles are small, reaching 18 mm long. They are brown with dark markings on the fins. The labial tooth formula is 1/2.

Phrynobatrachus natalensis

The tadpole reaches a length of 35 mm. The labial tooth row formula is 2(2)/2 or 1/2. It is brown with clear fins. It is often difficult to find in muddy pools.

Genus *Ptychadena*

The tadpoles vary from generalized pond types to specialized rock-dwelling tadpoles with a very little fin. The nostrils are relatively large and close together. The tadpole of *P. mascareniensis* is illustrated (Fig. 296).

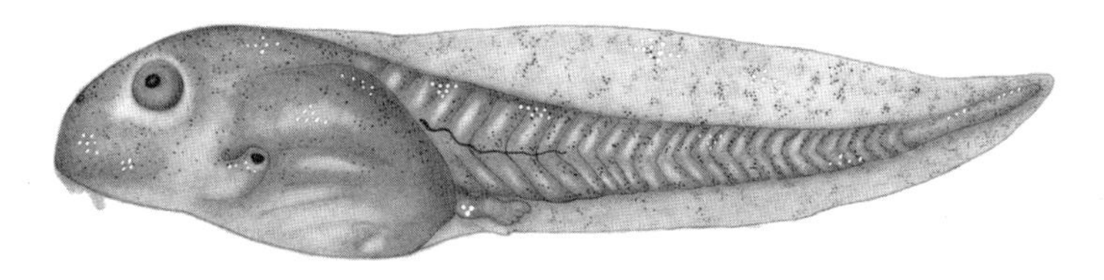

Fig. 296 *Ptychadena mascareniensis.*

Ptychadena anchietae

The tadpole reaches a length of 45 mm. The upper body is gray to brown, with clear fins, but pale below. The tadpole is part of a complex food web involving algae and mineral nutrients in the bottom mud. It takes in bottom mud, grazes on the surface algae, or feeds off the surface film. This promotes the growth of the algae it feeds on by transferring nutrients from the sediment of the pond. The tadpole can withstand temperatures of up to 40 °C in shallow water.

Ptychadena mascareniensis

The tadpoles are small and have a spiracle opening against the body. Besides the standard arrangement of labial tooth rows, other reported labial tooth row formulas are 3(2–3)/2 and 4(2–4)/2. The tadpoles are found in temporary pools. Metamorphosis occurs after 9 weeks.

Ptychadena oxyrhynchus

The tadpoles reach 54 mm in length. Labial tooth row formulas are 1/2, or sometimes 3(2–3)/2. Three rows of papillae are present. The tadpole is gray but pale over the head. The throat has no chromatophores, and the fin is clear with spots.

Ptychadena porosissima

The brown tadpoles reach a length of 41 mm. The labial tooth row formula is 1/2.

Genus *Pyxicephalus*

The tadpoles form large schools. They are dark, with eyes set close together on top of the head (Fig. 297).

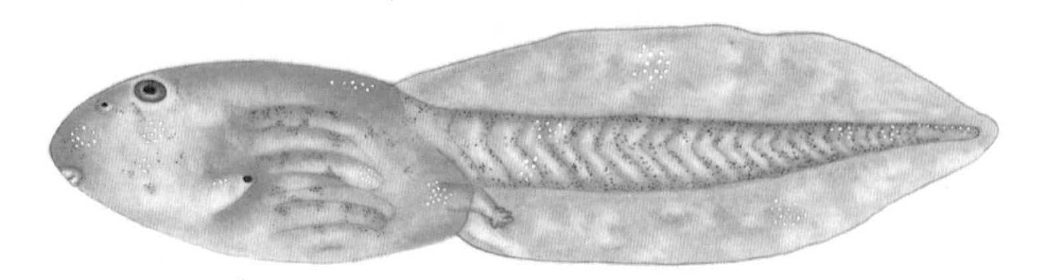

Fig. 297 *Pyxicephalus adspersus.*

Pyxicephalus adspersus

The black tadpoles become large, up to 71 mm long, and robust. They assume a gray color at about 60 mm. The labial tooth row formula is 4(3–4)/3. The tadpoles are gregarious, and up to 3000 have been estimated to form one school. Large schools like these consist of cohorts of different ages. The tadpoles remain in shallow water, even if deeper water is available. The school may form a ball, raft, or ring. The tadpoles feed on algae, and the schools move to deeper water at night. The individuals in a school dive at 45° to the bottom and return to the surface, while the rest swim over, so that the school appears to roll along. A school can move as fast as 46 m/hour. Sometimes the schools consist of two species; the African bullfrog *Pyxicephalus adspersus* and the red toad *Schismaderma carens* have been found schooling together.

Metamorphosis in warm shallow pools may take as little as 18–33 days or up to 47 days in captivity. Many of the shallow pools in which the tadpole develops become very hot during the day.

Genus *Strongylopus*

The tadpoles are robust, with upper jaw sheaths pigmented to the base (Fig. 298).

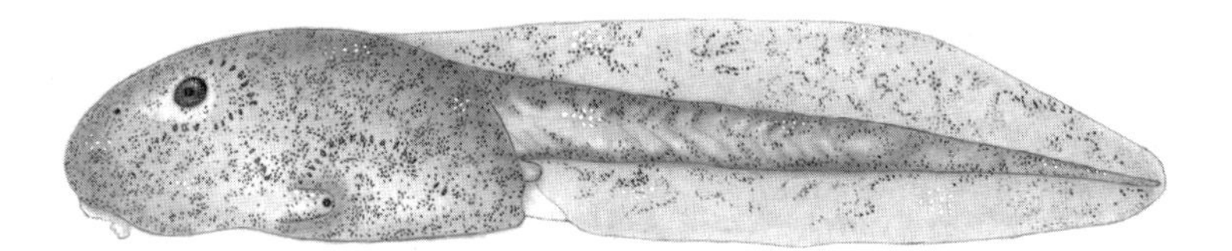

Fig. 298 *Strongylopus* sp.

Genus *Tomopterna*

The tadpoles are heavy-bodied and found in temporary pools. The tadpole of *T. cryptotis* is illustrated (Fig. 299).

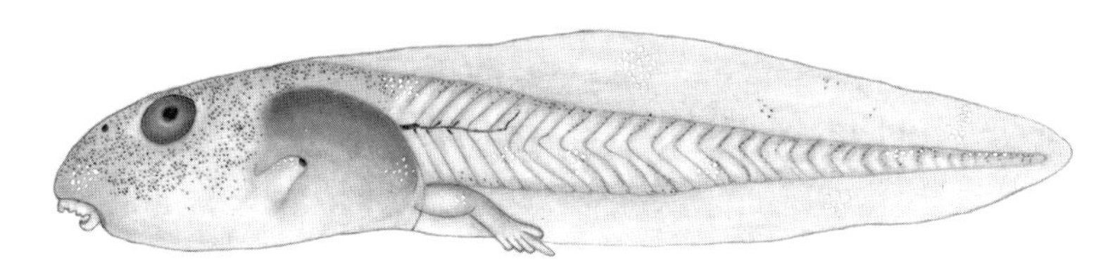

Fig. 299 *Tomopterna cryptotis.*

Tomopterna cryptotis

The tadpole is large-bodied, reaching 39 mm long. It is brown, sometimes with darker markings. It spends much time resting on the bottom and, when disturbed, will dive under the mud or swim to the deeper middle. The tadpole occurs together with terrapins but spends most of the time in the shallows, presumably where the terrapins are less likely to be feeding. The tadpole often feeds upside down on the surface film. Metamorphosis may take 5 weeks.

Tomopterna tandyi

The labial tooth row formula is 3(2–3)/3. The color pattern varies with the turbidity of the pool: In a clear pool the tadpoles were black. They were mottled in a slightly muddy pool, yet striped in a very turbid pool.

Family Rhacophoridae

Genus *Chiromantis*

The tadpoles are developed before they leave the foam nest and drop or wriggle to the water. The tadpole of *C. rufescens* is illustrated (Fig. 300).

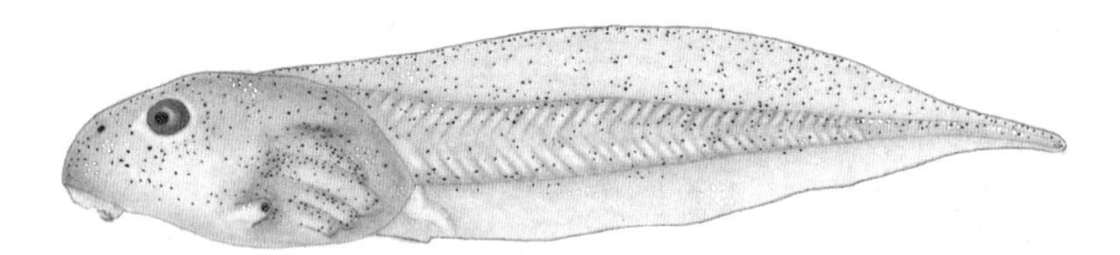

Fig. 300 *Chiromantis rufescens.*

Chiromantis rufescens

The tadpoles have an oral formula of 5(2–5)/3(1).

Chiromantis xerampelina

The tadpole has an oval body and is small, only 18 mm long. The labial tooth row formula is 3(2–3)/3. The lower gap in the papillae is about 60% of the width of the oral disc. The tadpoles are found in temporary pools in tropical savanna.

Kwa Wasomaji wa Kiswahili (For the Swahili Reader)

Vyura wana sehemu muhimu katika mazingira. Wanakula idadi kubwa ya wadudu, ikiwa ni pamoja na wadudu waharibifu na wanaoeneza magonjwa ya binadamu. Aidha, chura wanaliwa na wanyama wengi hasa wale ambao ni wakubwa kuliko wao, kwa mfano mamalia wadogo, ndege wakubwa, nyoka na aina nyingine za vyura. Baadhi ya watu katika bara la Afrika wanakula "bullfrogs," ingawa kwa sasa vyura hawa wamo hatarini kutoweka na hivyo wanastahili kulindwa.

Kuna takribani spishi (aina) 200 wa amfibia katika Afrika Mashariki, yaani Tanzania, Kenya na Uganda. Idadi kubwa ya amfibia hawa ni vyura lakini kuna takribani dazani moja ya caecilians ambao hawana miguu, wanafanana na minyoo na wanaishi chini ya ardhi. Vyura wanajilinda dhidi ya magonjwa kwa kutoa majimaji kutoka kwenye ngozi na maji haya yana vitu ambavyo vinaua vidubini vinavyoleta magonjwa. Baadhi ya vitu hivi vinavyotolewa kwenye ngozi vina manufaa kwa binadamu; mfano wake ni dawa za kutuliza maumivu na viuavijasumu (antibiotics) vinavyopambana kikamilifu na baadhi ya aina ya vijimea vilivyoko kwenye ngozi ambavyo ni vigumu kutibu. Baada ya kugundua "compound" hizi kutoka kwenye majimaji ya ngozi, zinaweza kutengenezwa kwenye maabara na vyura hawadhuriki kwa namna yoyote ile. Kwa hiyo ni muhimu kuwafahamu vyura walioko kwenye mazingira yetu.

Katika sehemu hii, tunaeleza jinsi ya kutumia kitabu hiki na kuorodhesha majina ya makundi makuu ya vyura. Sehemu juu ya viluwiluwi mwishoni mwa kitabu hiki inatoa taarifa juu ya viluwiluwi ambao wako kwenye hatua ya kuogelea na ambao wanakula mimea ya majini mpaka wanapofikia umri wa kubadilika (metamorphose) kuwa chura na kuishi kwenye nchi kavu. Sehemu ya marejeo ni mwanzo wa kutafuta taarifa zaidi za kila mnyama. Maktaba yoyote ile itakusaidia kupata rejea za mwanzo kabisa.

Taarifa za kila spishi ya chura zimewekwa chini ya familia za vyura, kwa mfano Arthroleptidae au Bufonidae. Taarifa za kila spishi inaanza na orodha ya majina ya asilia, kama yapo. Aidha kuna maelezo mafupi juu ya taarifa muhimu zinazosaidia kumtambua kila mnyama. Pia kuna taarifa zinazoonyesha kila spishi inapatikana wapi. Aidha kuna ramani inayoonyesha mahali spishi ilipoonekana na mahali pengine ambapo inaweza kupatikana. Idadi kubwa ya vyura wana miito ya kipekee ambayo wanatumia kuwavutia vyura wa kike; mtu akijua miito hiyo anaweza kuwatambua vyura mbalimbali. Picha za vyura wengi zinapatikana katika kitabu hiki; mtu akiangalia ramani, picha ya chura na taarifa zilizotolewa kwenye makala hii, anaweza kutambua spishi nyingi za vyura.

Inabidi kukumbuka kwamba bado hatuwafahamu vyura wote na "caecilians" wa Afrika Mashariki, wala mahali wanapopatikana. Taarifa za spishi mpya na mahali zinapopatikana zipelekwe kwa wataalamu wa makambusho au chuo kikuu au zitumwe kwa mhariri wa jarida la African Herpetological Association. Anwani zinaweza kubadilika lakini zinaweza kupatikana kwa kutumia Internet, au kupitia makumbusho na chuo kikuu.

Vyura wako katika kundi la amfibia ambao ni wanyama wenye uti wa mgongo na ngozi laini. Kila mtu anawafahamu vyura lakini kuna aina nyingine ya amfibia inayojulikana kama "caecilians," amfibia minyoo. Amfibia hawa wana kiwiliwili kama cha mnyoo na nyoka na hawana mikono wala miguu; wana macho madogo sana na wanaishi katika udongo wenye unyevunyevu.

Ingawa kuna karibu ya spishi 200 za vyura, na vyura wanapatikana karibu kila mahali katika Afrika Mashariki, spishi nyingi hazina jina maalumu katika Kiswahili au lugha nyingine za asili. Kutokana na hali hii, tumejaribu kueleza maana ya jina la Kiingereza la kila familia na jenasi na kutoa taarifa chache, badala ya kuunda majina ya Kiswahili kwa spishi hizo.

Orodha Ya Jenasi Za Amfibia Wa Afrika Mashariki

Kabila Anura—Vyura

Familia Arthroleptidae—Vyura Filimbi "Squeakers"

Vyura hawa ni wadogo wanaoishi kwenye majani ya miti yaliyonyauka na kuanguka ardhini msituni. Jina la vyura filimbi "squeaker" linatokana na sauti ya juu na kali ya madume, inayofanana na sauti za baadhi ya wadudu au filimbi. Vyura hawa hawatagi mayai kwenye maji ila kwenye udongo wenye unyevunyevu na kwenye majani yaliyoanguka ardhini.

Jenasi *Arthroleptis*—Vyura Filimbi "Squeakers"

Vyura katika jenasi hili ni wakubwa kuliko vyura wengine katika kundi la vyura filimbi. Kuna spishi tano za vyura hawa Afrika Mashariki.

Jenasi *Schoutedenella*—Vyura Filimbi "Squeakers"

Kwa kawaida, vyura hawa ni wadogo kuliko wengine; kuna spishi tisa za vyura hawa Afrika Mashariki na sauti zao zinafanana na sauti ya wadudu au filimbi.

Familia Bufonidae—Vyura Matomvu "Toads"

Tofauti na vyura wengine, vyura hawa wana ngozi isiyo laini na ina chunjua. Ngozi yao ina mabaka na hutoa matomvu yenye sumu ili kujilinda wasiliwe na nyoka au wanyama wengine kama mbwa na paka. Matomvu hayo yanaweza kumfanya mwanadamu asumbuliwe na macho iwapo atamgusa chura huyo halafu akaupitisha mkono kwenye macho yake bila kunawa kwanza.

Jenasi *Bufo*—Vyura Matomvu "Toads"

Jenasi hii ina vyura wakubwa na wadogo na wanapatikana kwenye mbuga za nyasi, mashambani, porini na misituni. Kuna spishi 23 katika Afrika Mashariki.

Jenasi *Churamiti* "Tree Toads"

Jenasi hii ina spishi moja tu inayoitwa *"Churamiti,"* kwa sababu chura huyu alipatikana akipanda miti kwenye milima ya Ukaguru, mahali pekee ambapo chura huyu ameonekana.

Jenasi *Mertensophryne* "Woodland Toad"

Jenasi hii ina spishi moja tu inayopatikana katika misitu na mapori ya miombo, hasa katika ukanda wa pwani. Vyura hawa hawajapatikana kwenye misitu ya milimani.

Jenasi *Nectophrynoides* "Forest Toads"

Vyura hawa ni baadhi ya vyura wanaozaa vitoto badala ya kutaga mayai majini kama vyura wengine. Aidha vyura hawa wanapatikana katika misitu ya milimani tu.

Jenasi *Schismaderma*—Chura Matomvu Mwekundu "Red Toad"

Kuna spishi moja tu katika jenasi hii. Jina la Kiingereza la chura huyu linatokana na kwamba ana ngozi ambayo ni nyekundu kidogo. Chura huyu anapatikana kwenye mapori.

Jenasi *Stephopaedes* "Forest Toads"

Kuna spishi tatu tu za jenasi hii; mbili zinaishi kwenye misitu ya pwani na moja kwenye miombo.

Familia Hemisotidae "Snout-Burrowers"

Ni rahisi kuwatambua vyura hawa kwa vile wana pua iliyochongoka na bapa kama chepe, wanayoitumia kuchimba ardhini; aidha wana mwili mviringo. Jina la Kiingereza la "snout-burrower" linatokana na kutumia pua wakati wa kuchimba ardhini.

Jenasi *Hemisus* "Snout-Burrowers"

Kuna spishi tatu za jenasi hii na zote zinapatikana kwenye mapori.

Familia Hyperoliidae—Vyura Miti "Tree Frogs"

Vyura wa familia hii wana vidole vyenye ncha za mviringo zinazowasaidia kupanda kwenye sehemu laini ya miti, majani na nyasi. Wanaitwa vyura wa miti (tree frogs) kwa sababu baadhi yao wanakwea miti lakini wengine wanapatikana juu ya majani kwenye aina zingine za uoto.

Jenasi *Afrixalus* "Spiny Reed Frogs"

Vyura hawa ni wadogo na wana miiba midogo kwenye ngozi. Aidha vyura hawa wanapatikana kwenye mbuga za nyasi na misitu. Spishi 12 za vyura hawa zimeonekana Afrika Mashariki. Jina la "spiny reed frogs" linatokana na miiba iliyopo kwenye ngozi na kwa vile vyura hawa wanapatikana mara nyingi kwenye matete.

Jenasi *Hyperolius* "Reed Frogs"

Kuna spishi 31 za vyura hawa Afrika Mashariki; vyura hawa wana ngozi laini na wanapatikana kwenye mbuga za nyasi na misitu.

Jenasi *Kassina* "Kassinas"

Kuna spishi tatu za vyura hawa Afrika Mashariki na wanapatikana kwenye mbuga za nyasi au misitu. Madume wana kipande cha ngozi pana kwenye koo.

Jenasi *Leptopelis*—Vyura Miti "Tree Frogs"

Vyura hawa wanaoishi kwenye miti wana macho makubwa. Spishi 15 zinapatikana Afrika Mashariki kwenye mbuga za nyasi na misitu.

Jenasi *Phlyctimantis* "Wot-Wots"

Kuna spishi mbili tu katika jenasi hii na zote zinaishi msituni. Jina la Kiingereza la "wot-wot" linatokana na sauti ya madume.

Familia Microhylidae "Narrow-Mouthed Frogs"

Vyura wa familia hii wana mdomo mwembamba na mfupi na mwili mkubwa wa mviringo.

Jenasi *Breviceps*—Vyura Waitamvua "Rain Frogs"

Vyura hawa wanaitwa "rain frogs" kwa sababu watu wanaamini kwamba madume yanalia mwanzoni mwa msimu wa mvua. Spishi mbili tu zinapatikana Afrika Mashariki, kwenye mbuga za nyasi.

Jenasi *Callulina* "Warty Frogs"

Vyura hawa wanaishi msituni tu. Ngozi yao inanata sana. Spishi mbili zinapatikana Afrika Mashariki.

Jenasi *Hoplophryne*—Vyura Wenye Vidole Vitatu "Three-Fingered Frogs"

Vyura hawa wanaitwa "three-fingered frogs" kwa sababu wana vidole vitatu tu katika kila mkono. Aina moja huishi misituni tu katika milima ya Uluguru na Nguru, na aina nyingine huishi misituni katika milima ya Usambara.

Jenasi *Parhoplophryne* "Black-Banded Frog"

Kuna spishi moja tu katika jenasi hii, na hupatikana katika misitu ya milima ya Usambara. Vyura hawa wanaitwa "black-banded" kutokana na mstari mweusi unaopita kati ya pua na jicho.

Jenasi *Phrynomantis*—Vyura Mpira "Rubber Frogs"

Kuna spishi moja tu katika jenasi hii. Chura huyu anaitwa "rubber frog" kwa sababu ngozi yake ni laini na inang'aa kama mpira. Aidha chura huyu ana rangi za pekee, nyeusi na mistari miwili myekundu pembeni. Chura huyu anaishi kwenye mapori, kwenye mashamba na karibu na makazi ya watu lakini haishi msituni.

Jenasi *Probreviceps*—Vyura Wa Misituni "Forest Frogs"

Kuna spishi nne katika jenasi hii na vyura hawa wanapatikana tu kwenye misitu ya Safu ya Milima ya Mashariki na misitu jirani. Vyura hawa hutoa majimaji yanayonata na kuteleza.

Jenasi *Spelaeophryne* "Scarlet-Snouted Frog"

Kuna spishi moja tu katika jenasi hii na ni rahisi kumtambua chura huyu kwa sababu ni mweusi na ana alama ya "V" kwenye pua yake.

Familia Pipidae—Vyura Wenye Kucha "Clawed Frogs"

Vyura hawa wanaishi majini. Jina la Kiingereza linatokana na makucha meusi kwenye vidole vya mguu.

Jenasi *Xenopus*—Vyura Wenye Kucha "Clawed Frogs"

Kuna spishi sita Afrika Mashariki na chura hawa wanaishi majini.

Familia Ranidae—Vyura Wa Kawaida "Common Frogs"

Vyura wa familia hii wanaitwa "common frogs" kwa sababu wanaishi mahali pengi. Hata hivyo wengi wao wanaishi majini au karibu na maji.

Jenasi *Afrana*—Vyura Mito "River Frogs"

Vyura hawa wanaitwa "river frogs" kwa sababu wanapatikana mtoni mara nyingi. Kuna spishi tatu Afrika Mashariki.

Jenasi *Amnirana* "White-Lipped Frogs"

Vyura hawa wanaitwa "white-lipped frogs" kwa sababu wana mstari mweupe juu ya mdomo. Kuna spishi mbili na wanaishi msituni na kwenye mbuga za nyasi.

Jenasi *Arthroleptides*—Vyura Maporomoko "Torrent Frogs"

Jina la vyura hawa linatokana na kwamba wanaishi kwenye mito yenye mwendo kasi na yenye maporomoko kwenye Safu za Milima za Mashariki. Ni rahisi kuwatambua kwa sababu ncha za vidole ni bapa ili kuwasaidia kupita kwenye mawe ya mtoni yaliyolowana.

Jenasi *Cacosternum* "Dainty Frog"

Jina la chura huyu "dainty frog" linatokana na umbo lake dogo na miguu midogo. Kuna spishi moja tu inayopatikana Afrika Mashariki kwenye mbuga za nyasi.

Jenasi *Hildebrandtia*—Vyura Maridadi "Ornate Frogs"

Spishi mbili za chura hawa zinaitwa "ornate frog" kwa sababu vyura hawa wana urembo mzuri mgongoni. Vyura hawa wanaishi kwenye mapori.

Jenasi *Hoplobatrachus* "Groove-Crowned Bullfrog"

Jina la Kingereza la chura huyu linatokana na mstari uliopo nyuma ya kichwa. Chura huyu anaishi kwenye mito, mabwawa na maziwa.

Jenasi *Phrynobatrachus*—Vyura Madimbwi "Puddle Frogs"

Vyura hawa wanaitwa "puddle frogs" kwa sababu mara nyingi wanaishi kwenye madimbwi madogo ya maji. Baadhi yao wanaishi msituni na wengine kwenye mbuga za nyasi. Spishi 20 zinapatikana Afrika Mashariki.

Jenasi *Ptychadena* "Ridged Frogs"

Vyura hawa wanaitwa "ridged frogs" kwa sababu wana ngozi ya mgongoni iliyokunjamana na kufanya mistari kutoka kichwani kulekea nyuma.
Mara nyingi vyura hawa wanaishi kwenye mbuga za nyasi. Kuna spishi 13 za vyura hawa Afrika Mashariki.

Jenasi *Pyxicephalus* "Bullfrogs"

Spishi mbili zinapatikana Afrika Mashariki. Madume yana sauti kubwa na nene ambayo katika Kiingereza inafananishwa na ya fahari la ng'ombe. Vyura hawa wanaishi kwenye mapori na mbugani.

Jenasi *Strongylopus* "Stream Frogs"

Ni rahisi kuwatambua vyura hawa kwa sababu wana vidole virefu na vyembamba sana; aidha wanaishi kwenye savanna karibu na mito. Kuna spishi mbili Afrika Mashariki.

Jenasi *Tomopterna*—Vyura Mchanga "Sand Frogs"

Kuna spishi nne za vyura hawa wanaoishi kwenye udongo wa mchanga mchanga katika mapori.

Familia Rhacophoridae—Vyura Mapovu "Foam-Nest Frogs"

Vyura wa familia hii wanaitwa "foam-nest frogs" kwa sababu majike yanataga mayai na pamoja na madume kadhaa, wanatengeneza viota vya mapovu kwenye miti juu ya maji. Viluwiluwi vinapojitokeza, vinaangukia kwenye maji. Kuna spishi nne Afrika Mashariki.

Jenasi *Chiromantis*—Vyura Mapovu "Foam-Nest Frogs"

Kuna spishi nne katika jenasi hii; moja inaishi msituni na zingine tatu kwenye mapori.

Kabila Gymnophiona—Amfibia Minyoo

"Caecilians" hawa ni amfibia wasiokuwa na miguu. Wanaishi mashimoni kwenye udongo wenye unyevunyevu.

Familia Caeciliidae—Amfibia Minyoo "caecilians"

Jenasi *Boulengerula*—Amfibia Minyoo "caecilians"

Kuna spishi tano katika jenasi hii ambayo haina jina la Kiingereza. Amfibia minyoo hawa wanaishi msituni na mashambani karibu na misitu.

Jenasi *Schistometopum*—Amfibia Minyoo "mud caecilian"

Kuna spishi moja tu inayoishi kwenye matope karibu na mito ya Tana, Ruvu, Rufiji na Wami na kwenye mito midogo. Ina rangi nyeusi inayong'aa.

Familia Scolecomorphidae—Amfibia Minyoo "caecilians"

Amfibia katika famili hii hawana miguu wala mikono na hawana jina la Kiingereza.

Jenasi *Scolecomorphus*—Amfibia Minyoo "caecilians"

Spishi tatu za jenasi hii zinapatikana kwenye mashamba na misitu ya Safu za Milima ya Mashariki.

Bibliography

Any library should be able to obtain copies of the following publications.

Ahl, E. 1931. Amphibia: Anura III. Polypedatidae. In F. E. Schulze and W. Kükenthal (Eds.), *Das Tierreich 55*. Berlin: Lieferung, Walter de Gruyter.

Akizawa, T., K. Yamazaki, T. Nakajima, M. Roseghini, G. F. Erspamer, and V. Erspamer. 1982. Trypargine, a new tetrahydro-β-carboline of animal origin: Isolation and chemical characterization from the skin of the African rhacophorid frog, *Kassina senegalensis*. *Biomedi. Res.* 3: 232–4.

Amiet, J.-L. 1974. Voix d'amphibiens camerounais. IV—Raninae: Genres *Ptychadena, Hildebrandtia* et *Dicroglossus*. *Ann. Fac. Sci. Cameroun* 1974: 108–28.

——. 1975. Écologie et distribution des amphibiens anoures de la région de Nkongsamba (Cameroun). *Ann. Fac. Sci. Yaoundé* 20: 33–107.

——. 1991. Images d'Amphiens camerounais. III. Le comportement de garde des oeufs. *Alytes* 9: 15–22.

Amiet, J.-L., and J.-L. Perret. 1969. Contributions a la fauna de la région de Yaonde (Cameroun) Amphibiens Anoures. *Ann. Fac. Sci. Cameroun* 3: 117–37.

Andersson, L. G. 1911. Reptiles, batrachians and fishes collected by the Swedish zoological expedition to British East Africa. 1911. 2. Batrachians. *Svenska Vetensk.—Akad. Handl.* 47: 25–42.

Andrews, P. J., G. E. Meyer, D. R. Pilbeam, J. A. Van Couvering, and J. A. H. Van Couvering. 1981. The Miocene fossil beds of Maboko island, Kenya: Geology, age, taphonomy and paleoecology. *J. Human Evol.* 10: 35–48.

Angel, M. F. 1924a. Description de deux Batraciens nouveaux, d'Afrique Orientale anglaise, appartenant au genre *Phrynobatrachus* (Mission Alluaud et Jeannel, 1911–1912). *Bull. Mus. Natl. Hist. Nat. Paris* 30: 130–2.

——. 1924b. Note préliminaire sur deux Batraciens nouveaux, des genres *Rappia* et *Bufo*, provenant d'Afrique Orientale anglaise (Mission Alluaud et Jeannel, 1911–1912). *Bull. Mus. Natl. Hist. Nat. Paris* 30: 269–76.

Angel, F. 1925. *Voyage de Ch. Alluaud et R. Jeannel en Afrique Orientale (1911–1912). Résults scientifiques. Reptiles et Batraciens.* Paris: P. Lechevalier.

Bachmann, K., H. Hemmer, A. Konrad, and L. R. Maxson. 1980. Molecular evolution and the phylogenetic relationships of the African toad, *Bufo daniellae*, Perret, 1977 (Salientia: Bufonidae). *Amphibia-Reptilia* 1: 173–83.

Backwell, P. R. Y. 1988. Functional partitioning in the two-part call of the leaf-folding frog, *Afrixalus brachycnemis*. *Herpetologica* 44: 1–7.

Baird, T. A. 1983. Influence of social and predatory stimuli on the airbreathing behavior of the African clawed frog, *Xenopus laevis*. *Copeia* 1983: 411–20.

Balinsky, B. I. 1969. The reproductive ecology of amphibians of the Transvaal highveld. *Zool. Afr.* 4: 37–93.

Balinsky, B. I., and J. B. Balinsky. 1954. On the breeding habits of the south African bullfrog *Pyxicephalus adspersus. S. Afr. J. Sci.* 51: 55–8.

Balletto, E., M. A. Cherchi, and B. Lanza. 1978. On some amphibians collected by the late Prof Guiseppe Scotecci in Somalia. *Monit. Zool. Ital.* 9: 221–43.

Barbault, R., and M. Trefaut Rodriques. 1978. Observations sur la reproduction et la dynamique des populations de quelque anoures tropicaux. I. *Ptychadena maccarthyensis* et *Ptychadena oxyrhynchus. Terre Vie* 32: 441–52.

——. 1979. Observations sur la reproduction et la dynamique des populations de quelque anoures tropicaux. III. *Arthroleptis poecilonotus. Trop. Ecol.* 20: 64–77.

Barbour, T., and A. Loveridge. 1928a. A comparative study of the herpetological faunae of the Uluguru and Usambara mountains, Tanganyika territory with descriptions of new species. *Mem. Mus. Comp. Zool.* 50: 87–261.

——. 1928b. New frogs of the genus *Phrynobatrachus* from the Congo and Kenya colony. *Proc. New Engl. Zool.* 10: 87–90.

Bles, E. J. 1907. Notes on Anuran development *Paludicola, Hemisus* and *Phyllomedusa. Budgett Memorial Volume,* 443–58, pls 22–7. Cambridge: Cambridge University Press.

Boettger, O. 1913. *Reise in Ostafrika von A. Voelzkow.* Vol 3, Part 4. Schweizerbart'sche Verlagsbuchhandlung.

Bogart, J. P. 1972. Karyotypes. In W. F. Blair (Ed.), *Evolution in the Genus Bufo,* 171–95. Austin: University of Texas Press.

Bogart, J. P., and M. Tandy. 1976. Polyploid amphibians. Three more diploid-tetraploid cryptic species of frogs. *Science* 193: 334–5.

——. 1981. Chromosome lineages in African ranoid frogs. *Monit. Zool. Ital. (N. S.) Suppl.* 15: 55–91.

Böhme, W. 1994. Amphibien und Reptilien aus dem tropischen Afrika. *DATZ* 47: 240–3.

Boulenger, G. A. 1895. An account of the reptiles and batrachians collected by Dr. A. Donaldson Smith in western Somaliland and the Galla country. *Proc. Zool. Soc. Lond.* 1894: 530–40, pls 29, 30.

——. 1896. Descriptions of two new frogs from Lake Tanganyika presented to the British Museum by Mr W. H. Nutt. *Ann. Mag. Nat. Hist.* 18: 467–8.

——. 1900. A list of the batracians and reptiles of the Gaboon (French Congo), with descriptions of new genera and species. *Proc. Zool. Soc. Lond.* 1900: 433–56, pls 1–6.

——. 1901a. Descriptions of a new frog from British East Africa. *Ann. Mag. Nat. Hist.* 8: 515–16.

——. 1901b. Materiaux pour la faune du Congo. Batraciens et Reptiles. Batraciens nouveaux. *Ann. Mus. Cong. Zool.* 2: 1–14, pls I–VI.

——. 1906. Additions to the herpetology of British East Africa. *Proc. Zool. Soc. Lond.* 1906: 570–3.

——. 1919. Batraciens et Reptiles recueillis par le Dr C. Christy au Congo Belge dans les districts de Stanleyville, Haut-Uelé et Ituri en 1912–1914. *Rev. Zool. Afr.* 7: 1–45.

Bowker, R. G, and M. H. Bowker. 1979. Abundance and distribution of anurans in a Kenya pond. *Copeia* 1979: 278–85.

Broadley, D. G. 1971. The reptiles and amphibians of Zambia. *Puku* 6: 1–143.

Buxton, D. R. 1936. A natural history of the Turkana fauna. *J. E. Afr. Nat. Hist. Soc.* 13: 85–104.

Castanet, J., S. Pinto, M.-M. Loth, and M. Lamotte. 2000. Âge individuel, longévité et dynamique de croissance osseuse chez un amphibien vivipare, *Nectophrynoides occidentalis* (Anoure, Bufonidé). *Ann. Sci. Nat.* 21: 11–17.

Channing, A. 1991. The distribution of *Bufo poweri* in southern Africa. *S. Afr. J. Zool.* 26: 81–4.

——. 2001. *Amphibians of Central and Southern Africa.* Ithaca: Cornell University Press.

Channing, A., and J. P. Bogart. 1996. Description of a tetraploid *Tomopterna* (Anura: Ranidae) from South Africa. *S. Afr. J. Zool.* 31: 80–5.

Channing, A., and M.-R. Capon de Caprona. 1987. The tadpole of *Hyperolius mitchelli* (Anura: Hyperoliidae). *S. Afr. J. Zool.* 22: 235–7.

Channing, A., and T. R. B. Davenport. 2002. A new stream frog from Tanzania (Anura: Ranidae: *Strongylopus*). *Afr. J. Herpetol.* 51: 135–42.

Channing, A., and R. C. Drewes. 1997. Description of the tadpole of *Bufo kisoloensis*. *Alytes* 15: 13–18.

Channing, A., and K. M. Howell. 2003. *Phlyctimantis keithae* defensive behavior. *Herpetol Rev.* 34: 51–2.

Channing, A., and L. Minter. 2004. A new rain frog from Tanzania (Microhylidae: *Breviceps*) *Afr. J. Herpetol.* 53: 147–54.

Channing, A., D. Moyer, and M. Burger. 2002a. Cryptic species of sharp-nosed reed frogs in the *Hyperolius nasutus* complex: Advertisement call differences. *Afr. Zool.* 37: 91–9.

Channing, A., D. C. Moyer, and A. Dawood. 2004. A new sand frog from central Tanzania (Anura: Ranidae: *Tomopterna*). *Afr. J. Herpetol.* 53: 21–8.

Channing, A., D. Moyer, and K. M. Howell. 2002b. Description of a new torrent frog in the genus *Arthroleptides* from Tanzania (Ranidae). *Alytes* 20: 13–27.

Channing, A., N. I. Passmore, and L. du Preez. 1994. Status, vocalizations and breeding of two species of African bullfrogs. *J. Zool. (Lond).* 234: 141–8.

Channing, A., A. R. E. Sinclair, S. A. R. Mduma, D. Moyer, and D. A. Kreulen. 2004. Serengeti amphibians: Distribution and monitoring baseline. *Afr. J. Herpetol.* 53: 163–81.

Channing, A., and W. T. Stanley. 2002. A new tree toad from the Ukaguru Mountains, Tanzania. *Afr. J. Herpetol.* 51: 121–7.

Chapman, B. M., and R. F. Chapman. 1958. A field study of a population of leopard toads (*Bufo regularis regularis*). *J. Anim. Ecol.* 27: 165–286.

Charter, R. R., and J. B. C. MacMurray. 1939. On the "frilled" tadpole of *Bufo carens* Smith. *S. Afr. J. Sci.* 36: 386–9.

Cherry, M. I., M. Stander, and J. C. Poynton. 1998. The amphibians of the Mkomazi Game Reserve. In M. J. Coe, N. C. McWilliam, Y. N. Stone, and M. J. Packer, (Eds.), *Mkomazi: The Ecology, Biodiversity and Conservation of a Tanzanian Savanna*, 405–9. London: Royal Geographical Society.

Chipman, A. D., A. Haas, and O. Khaner. 1999. Variations in anuran embryogenesis: Yolk-rich embryos of *Hyperolius puncticulatus* (Hyperoliidae). *Evol. Dev.* 1: 49–61.

Clarke, B. T. 1988. Real vs apparent distributions of dwarf amphibians: *Bufo lindneri* Mertens 1955 (Anura: Bufonidae)—a case in point. *Amphibia-Reptilia* 10: 297–306.

——. 1989. The amphibian fauna of the East African rainforests, including the description of a new species of toad, genus *Nectophrynoides* Noble 1926 (Anura Bufonidae). *Trop. Zool.* 1: 169–77.

——. 1997. The natural history of amphibian skin secretions, their normal functioning and potential medical application. *Biol. Rev.* 72: 365–80.

Coe, M. J. 1974. Observations of the ecology and breeding biology of the genus *Chiromantis* (Amphibia: Rhacophoridae). *J. Zool. (Lond.)* 172: 12–34.

Colley, B. 1987. *Phrynomerus bifasciatus*—an unpleasant experience. *Herptile* 12: 43.

Cook, C. L., J. W. H. Ferguson, and S. R. Telford. 2001. Adaptive male parental care in the giant bullfrog, *Pyxicephalus adspersus*. *J. Herpetol.* 35: 310–15.

Crutsinger, G., M. Pickersgill, A. Channing, and D. Moyer. 2004. A new species of *Phrynobatrachus* (Anura: Ranidae) from Tanzania. *Afr. Zool.* 39: 19–23.

Cruz-Uribe, K., and R. G. Klein. 1982/83. Faunal remains from some Middle and Later Stone Age archeological sites in South West Africa. *J. S. W. Afr. Sci. Soc.* 36/37: 91–114.

Curry-Lindahl, K. 1956. Ecological studies on mammals, birds, reptiles and amphibians in the eastern Belgian Congo. Part I. *Ann. Mus. R. Congo Belge Tervuren Ser. 8° Sci. Zool.* 42: 5–78, pls I–VII.

Dawood, A., A. Channing, and J. P. Bogart. 2002. A molecular phylogeny of the frog genus *Tomopterna* in southern Africa: Examining species boundaries with mitochondrial 12S rRNA sequence data. *Mol. Phyl. Ecol.* 22: 407–13.

Dawson, P., and P. J. Bishop. 1987. The painted reed-frog (*Hyperolius marmoratus*) aspects of life history, experimental use and husbandry. *Anim. Techn.* 38: 81–6.

De Fonesca, P. H., and D. Mertens. 1979. De gouden zeggekikker *Hyperolius puncticulatus* in de natuur en in het terrarium. *Lacerta* 37: 147–51.

De Witte, G.-F. 1932. Description d'un Batracien nouveau du Katanga. *Rev. Zool. Bot. Afr.* 22: 1.

De Sá, R. O., S. P. Loader, and A. Channing. 2004. A new species of *Callulina* (Anura: Microhylidae) from the West Usambara Mountains, Tanzania. *J. Herpetol.* 38: 219–24.

De Witte, G. F. 1941. Batraciens et reptiles. *Explor. Parc Nat. Albert Miss. De Witte* 33: 1–261.

——. 1952. Amphibiens et reptiles. Results scientifiques de l'exploration hydrobiologique du Lac Tanganika (1946–1947). *Instit. R. Sci. Nat. Belg.* 3: 1–22.

Drewes, R. C. 1972. Report on a collection of reptiles and amphibians from the Ilemi Triangle, southwestern Sudan. *Occ. Pap. Calif. Acad. Sci.* 100: 1–14.

——. 1984. A phylogenetic analysis of the Hyperoliidae (Anura): Treefrogs of Africa, Madagascar, and the Seychelles Islands. *Occ. Pap. Calif. Acad. Sci.* 139: 1–70.

——. 1997. A new species of treefrog from the Serengeti National Park, Tanzania (Anura: Hyperoliidae, *Hyperolius*). *Proc. Calif. Acad. Sci.* 49: 439–46.

Drewes, R. C., and R. Altig. 1996. Anuran egg predation and heterocannibalism in a breeding community of East African frogs. *Trop. Zool.* 9: 333–47.

Drewes, R. C., R. Altig., and K. M. Howell. 1989. Tadpoles of three frog species endemic to the forests of the Eastern Arc Mountains, Tanzania. *Amphibia-Reptilia* 10: 435–43.

Drewes, R. C., S. S. Hillman, R. W. Putnam, and O. M. Sokol. 1977. Water, nitrogen and ion balance in the African tree frog *Chiromantis petersi* Boulenger (Anura: Rhacophoridae), with comments on the structure of the integument. *J. Comp. Physiol. [B]* 116: 257–67.

Drewes, R. C., and J.-L. Perret. 2000. A new species of giant, montane *Phrynobatrachus* (Anura: Ranidae) from the central mountains of Kenya. *Proc. Calif. Acad. Sci.* 52: 55–64.

Drewes, R. C., and J. V. Vindum. 1994. Amphibians of the Impenetrable Forest, southwest Uganda. *J. Afr. Zool.* 108: 55–70.

Dubois, A. 1980. Deux noms d'espèces préoccupés dans le genre *Rana* (Amphibiens, Anoures). *Bull. Mus. Natl. Hist. Nat. Paris* 4° sér. 2: 927–31.

——. 1981. Liste des genres et sous-genres nominaux de Ranoidea (Amphibiens Anoures) du monde, avec identification de leurs espèces-types: Conséquences nomenclaturales. *Monit. Zool. Ital. (N. S.) Suppl.* 15: 225–84.

——. 1988. Miscellanea nomenclatorica batrachologica (XVII). *Alytes* 7: 1–5.

——. 1992. Notes sur la classification des Ranidae (Amphibiens Anoures). *Bull. Mens. Soc. Linn. Lyon.* 61: 305–52.

Dudley, C. O. 1978. The herpetofauna of the Lake Chilwa basin. *Nyala* 4: 87–99.

Dudley, C. O., D. E. Stead, and G. G. M. Schulten. 1979. Amphibians, reptiles, mammals and birds of Chilwa. In M. Kalk, A. J. McLachlan, and C. Howard-Williams (Eds.), *Studies of Change in a Tropical Ecosystem. Monogr. Biol.* 35: 247–443.

Duff-MacKay, A., and A. Schiøtz. 1971. A new *Hyperolius* from Kenya. *J. E. Afr. Nat. Hist. Soc. Nat. Mus.* 19 (128): 1–3.

Emmrich, D. 1994. Herpetological results of some expeditions to the Nguru Mountains, Tanzania. *Mitt. Zool. Mus. Bull.* 70: 281–300.

Fischberg M., and H. R. Kobel. 1978. Two new polyploid *Xenopus* species from western Uganda. *Experientia* 34: 1012–14.

Fischer, E., and H. Hinkel. 1992. *Natur und Umwelt Ruandas. Materialien zur Partnerschaft Rheinland-Pfalz/Ruanda.* Rheinland-Pfalz, Mainz: Ministerium des Innern und für Sport.

FitzSimons, V., and G. Van Dam. 1929. Some observations on the breeding habits of *Breviceps. Ann. Transv. Mus.* 13: 152–3.

Fleischack, P. C., and C. P. Small. 1979. The vocalizations and breeding behaviour of *Kassina* (Anura; Rhacophoridae) in summer breeding aggregations. *Koedoe* 21: 91–9.

Ford, L. S., and D. C. Cannatella. 1993. The major clades of frogs. *Herpetol. Monogr.* 7: 94–117.

Frost, D. R. 2002. Amphibian species of the world: An online reference. V2.21 (15 July 2002). Electronic database available at http://research.amnh.org/herpetology/amphibia/index.html.

Gower, D. J., S. P. Loader, C. B. Moncrieff, and M. Wilkinson. 2004. Niche separation and comparative abundance of *Boulengerula boulengeri* and *Scolecomorphus vittatus* (Amphibia: Gymnophiona) in an east Usambara Forest, Tanzania. *Afr. J. Herpetol.* 53: 183–90.

Grafe, T. U. 2000. *Leptopelis viridis* (West African tree frog) cocoon formation. *Herpetol. Rev.* 31: 100–1.

Grandison, A. G. C. 1972. The status and relationships of some East African earless toads (Anura, Bufonidae) with a description of a new species. *Zool. Meded.* 47: 30–48, pls 1–4.

——. 1980. Aspects of breeding morphology in *Mertensophryne micranotis* (Anura: Bufonidae): Secondary sexual characters, eggs and tadpole. *Bull. Br. Mus. Nat. Hist. (Zool.)* 39: 299–304.

——. 1983. A new species of *Arthroleptis* (Anura: Ranidae) from the West Usambara Mountains, Tanzania. *Bull. Br. Mus. Nat. Hist. (Zool.)* 45: 77–84.

Grandison, A. G. C., and S. Ashe. 1983. The distribution, behavioural ecology and breeding strategy of the pygmy toad, *Mertensophryne micranotis* (Lov.). *Bull. Br. Mus. Nat. Hist. (Zool.)* 45: 85–93.

Grandison, A. G. C., and K. M. Howell. 1983. A new forest species of *Phrynobatrachus* (Anura: Ranidae) from Morogoro region, Tanzania. *Amphibia-Reptilia* 4: 117–24.

Griffiths, J. F. 1958. Climatic zones of East Africa. *E. Afr. Agric. J.* 1958: 179–85.

Grobler, J. H. 1972. Observations on the amphibian *Pyxicephalus adspersus* Tschudi in Rhodesia. *Arnoldia (Rhod.) Misc. Publ.* 6: 1–4.

Guibé, J. 1966. *Ptychadena* (Amphibia Salientia). *Explor. Parc Natl. Albert Ser. 2.* 18: 47–65.

Guibé, J., and M. Lamotte. 1958. La réserve naturelle intégrale du Mont Nimba. Fasc. IV. XII Batraciens (sauf *Arthroleptis, Phrynobatrachus* et *Hyperolius*). *Mem. Inst. Fr. Afr. Noire* 53: 241–73, pls I–XI.

Hamilton, A. C. 1982. *Environmental History of Africa: A Study of the Quartenary.* London: Academic Press.

Hayes, T. B., and K. P. Mendez. 1999. The effect of sex steroids on primary and secondary sex differentiation in the sexually dichromatic reedfrog (*Hyperolius argus*: Hyperoliidae) from the Arabuko Sokoke forest of Kenya. *Gen. Comp. Endocrin.* 115: 188–99.

Hebrard, J. J., G. M. O. Maloiy, and D. M. I. Alliangana. 1992. Notes on the habitat and diet of *Afrocaecilia taitana* (Amphibia: Gymnophiona). *J. Herpetol.* 26: 513–55.

Heyer, W. R., M. A. Donnelly, R. W. McDiarmid, L.-A. C. Hayek, and M. S. Foster (Eds.). 1994. *Measuring and Monitoring Biological Diversity. Standard Methods for Amphibians.* Washington, D.C.: Smithsonian Institution Press.

Howell, K. M. 1978. Ocular envenomation by a toad in the *Bufo regularis* species group: Effects and first aid. *Bull. E. Afr. Nat. Hist. Soc.* 1978 (July/August): 82–7.

——. 1981. The female of *Leptopelis argenteus argenteus* (Pfeffer) (Anura: Rhacophoridae). *J. Herpetol.* 15: 113–4.

——. 1993. The herpetofauna of the eastern forests of Africa. In J. Lovett, and S. Wasser (Eds.), *The Biogeography and Ecology of the Forests of Eastern Africa*, 173–201. Cambridge: Cambridge University Press.

——. 2002. Amphibians and reptiles: The herptiles. In G. Davis (Ed.), *African Forest Biodiversity. A Field Survey Manual for Vertebrates*, 17–44. Oxford: Earthwatch Institute (Europe).

Inger, R. F. 1968. Mission H. de Saeger. *Amphibia. Explor. Parc. Nat. Garamba* (Fasc. 52): 1–190.

Inger, R., and H. Marx. 1961. The food of amphibians. *Explor. Parc. Nat. Upemba* 64: 1–86.

IUCN, Conservation International, and NatureServe. 2004. *Global Amphibian Assessment.* <www.globalamphibians.org>. Accessed on 19 February 2005.

Jackson, J. A., R. C. Tinsley, and S. Kigoolo. 1998. Polyploidy and parasitic infection in *Xenopus* species from western Uganda. *Herpetol. J.* 8: 19–22.

Jaeger, R. G. 1971. Toxic reaction to skin secretions of the frog *Phrynomerus bifasciatus. Copeia* 1971: 160–1.

Jennions, M. D., P. R. Y. Backwell, and N. I. Passmore. 1992. Breeding behaviour of the African frog, *Chiromantis xerampelina*: Multiple spawning and polyandry. *Anim. Behav.* 44: 1091–100.

Jennions, M. D., and N. I. Passmore. 1993. Sperm competition in frogs—testis size and a sterile male experiment on *Chiromantis xerampelina* (Rhacophoridae). *Biol. J. Linn. Soc.* 50: 211–20.

Johansen, K., G. Lykkeboe, S. Kornerup, and G. M. O. Maloiy. 1980. Temperature insensitive O_2 in blood of the tree frog *Chiromantis petersi. J. Comp. Physiol.* 136: 71–6.

Kaminsky, S. K., K. E. Linsenmair, and T. U. Grafe. 1999. Reproductive timing, nest construction and tadpole guidance in the African pig-nosed frog, *Hemisus marmoratus. J. Herpetol.* 33: 119–23.

Karplus, I., D. Algom, and D. Samuel. 1981. Acquisition and retention of dark avoidance by the toad, *Xenopus laevis* (Daudin). *Anim. Learning Behav.* 9: 45–9.

Katz, L. C., M. J. Potel, and R. J. Wassersug. 1981. Structure and mechanisms of schooling in tadpoles of the clawed toad, *Xenopus laevis*. *Anim. Behav.* 29: 20–33.

Kaul, R., and V. H. Shoemaker. 1989. Control of thermoregulatory evaporation in the waterproof treefrog *Chiromantis xerampelina*. *J. Comp. Physiol. [B].* 158: 643–9.

Keith, R. 1968. A new species of *Bufo* from Africa, with comments on the toads of the *Bufo regularis* complex. *Am. Mus. Novit.* 2345: 1–22.

Kingdon, J. 1997. *The Kingdon Field Guide to African Mammals.* London: Academic Press.

Klemens, M. W. 1998. The male nuptial characteristics of *Arthroleptides martiennseni* Neiden, an endemic torrent frog from Tanzania's Eastern Arc Mountains. *Herpetol. J.* 8: 35–40.

Kobel, H. R., C. Loumont, and R. C. Tinsley. 1996. The extant species. In R. C. Tinsley and H. R. Kobel (Eds.), *The Biology of Xenopus*, 9–33. Zoological Society of London. Oxford: Clarendon Press.

Kobelt, F., and K. E. Linsenmair. 1986. Adaptations of the reed frog *Hyperolius viridiflavus* (Amphibia, Anura, Hyperoliidae) to its arid environment. *Oecologia (Berlin)* 68: 533–41.

Kok, D., L. H. du Preez, and A. Channing. 1989. Channel construction by the African bullfrog: Another anuran parental care strategy. *J. Herpetol.* 23: 435–7.

Kosuch, J., M. Vences, A. Dubois, A. Ohler, and W. Böhme. 2001. Out of Asia: Mitochondrial DNA evidence for an Oriental origin of tiger frogs, genus *Hoplobatrachus*. *Mol. Phyl. Evol.* 21: 398–407.

Kühn, E. R., H. Gevaerts, G. Jacobs, and G. Vandorpe. 1987. Reproductive cycle, thyroxine and cortisone in females of the giant swamp frog *Dicroglossus occipitalis* at the equator. *Gen. Comp. Endocr.* 66: 137–44.

Lamotte, M., and A. Ohler. 2000. Révision des espèces du groupe de *Ptychadena stenocephala* (Amphibia, Anura). *Zoosystema* 22: 569–683.

Lamotte, M., and J.-L. Perret. 1961a. Contribution à l' étude des Batraciens de l' Ouest Africain. XI—Les formes larvaires de trois espèces de *Ptychadena*: *Pt. maccarthyensis* And., *Pt. perreti* G. et L. et *Pt. mascareniensis* D. et B. *Bull. Inst. Fr. Afr. Noire* 23A: 192–210.

——. 1961b. Contribution à l' étude des Batraciens de l' Ouest Africain. XIII—Les formes larvaires de quelques espèces de *Leptopelis*: *L. aubryi, L. viridis, L. anchietae, L. ocellatus* et *L. calcaratus*. *Bull. Inst. Fr. Afr. Noire* 23A: 855–85.

——. 1963. Contribution à l'étude des Batraciens de l'Ouest Africain. XV—Le développement direct de l'èspece *Arthroleptis poecilonotus* Peters. *Bull. Inst. Fr. Afr. Noire* 25: 277–84.

Lamotte, M., and F. Xavier. 1972. Les amphibiens anoures a developpment direct d'Afrique. Observations sur la biologie de *Nectophrynoides tornieri* (Roux). *Bull. Soc. Zool. Fr.* 97: 413–28.

——. 1980. Amphibiens. In J. R. Durand and C. Lévêque (Eds.), *Flore et Faune Aquatiques de l'Afrique Sahel-Soudanienne*, 773–816. Paris: Office de la Recherche Scientifigue et Technique Outre-Mer.

Lamotte, M., and M. Zuber-Vogeli. 1954. Contribution à l' étude des Batraciens de l' Ouest Africain. II—Le développement larvaire de *Bufo regularis* Reuss, de *Rana occidentalis* Günther et de *Rana crassipes* (Buch. et Peters). *Bull. Inst. Fr. Afr. Noire* 16A: 940–54.

Largen, M. J., M. Tandy, and J. Tandy. 1978. A new species of toad from the Rift Valley of Ethiopia, with observations on the other species of *Bufo* (Amphibia Anura Bufonidae) recorded from this country. *Monit. Zool. Ital. (N. S.) Suppl.* 10: 1–41.

Laurent, R. 1940a. Description d'un Rhacophoride nouveau du Congo Belge (Batracien). *Rev. Zool. Bot. Afr.* 33: 313–16.

——. 1940b. Contribution à l'osteologie et à la systématique des Ranides Africaines. Première note. *Rev. Zool. Bot. Afr.* 34: 74–97, pls III–V.

——. 1941a. Contribution à l'osteologie et à la systématique des Rhacophorides Africaines. Première note. *Rev. Zool. Bot. Afr.* 35: 85–111.

Laurent, R. F. 1954. Reptiles et Batraciens de la région de Dundo (Angola) (Deuxième note). *Publ. Cult. Comp. Diamant. Angola* 23 (9) 35: 70–84.

——. 1957. Genres *Afrixalus* et *Hyperolius* (Amphibia Salientia). *Explor. Parc. Nat. Upemba* 42: 1–47.

——. 1963. Three new species of the genus *Hemisus*. *Copeia* 1963: 395–9.

——. 1964. Reptiles et Amphibiens de l'Angola (Troisième contribution). *Publ. Cult. Comp. Diamant. Angola* 67: 11–165.

——. 1972a. Tentative revision of the genus *Hemisus* Günther. *Ann. Mus. R. Afr. Centr. Tervuren Ser. 8° Sci. Zool.* 194: 1–67.

——. 1972b. Amphibiens. *Explor. Parc. Nat. Virunga* 22: 1–125, pls I–XI.

Lescure, J. 1981. L'alimentation du crapaud *Bufo regularis* Reuss et de la grenouille *Dicroglossus occipitalis* (Günther) au Sénégal. *Bull. Inst. Fr. Afr. Noire* 33A: 446–66.

Liem, S. S. 1970. The morphology, systematics and evolution of the old world treefrogs (Rhacophoridae and Hyperoliidae). *Fieldiana Zool.* 57: i–vii, 1–145.

Linden, I. 1971. Development of *Leptopelis viridis cinnamomeus* (Bocage) with notes on its systematic position. *Zool. Afr.* 6: 237–42.

Loader, S. P., D. J. Gower, K. M. Howell, N. Doggart, M.-O. Rödel, B. Y. Clarke, R. O. de Sá, B. L. Cohen, and M. Wilkinson. 2004. Phylogenetic relationships of African microhylid frogs inferred from DNA sequences of mitochondrial 12s and 16s rRNA genes. *Org. Div. Evol.* 4: 227–35.

Loader, S. P., J. P. Poynton, and J. Mariaux. 2004. Herpetofauna of Mahenge Mountain, Tanzania: A window on African biogeography. *Afr. Zool.* 39: 71–6.

Loader, S. P., M. Wilkinson, D. J. Gower, and C. A. Msuya. 2003. A remarkable young *Scolecomorphus vittatus* (Amphibia: Gymnophiona: Scolecomorphidae) from the North Pare Mountains, Tanzania. *J. Zool. (Lond).* 259: 93–101.

Lönnberg, E. 1907. Reptilia and Batrachia. Wissenschaftliche Ergebnisse der schwedischen zoologischen Expedition nach dem Kilimandjaro, dem Meru und den umgebenden Massaisteppen 1905–1906 unter Leitung von Prof Dr Yngve Sjöstedt, 20–8, pl 1. Royal Swedish Academy of Sciences. Uppsala: Almquist & Wiksells.

Loumont, C. 1983. Deux espèces nouvelles de *Xenopus* du Cameroun (Amphibia, Pipidae). *Rev. Suisse Zool.* 90: 169–77.

Loveridge, A. 1925. Notes on East African Batrachians, collected 1920–1923, with the description of four new species. *Proc. Zool. Soc. Lond.* 1925: 763–91, pls I, II.

——. 1928. Field notes on vertebrates collected by the Smithsonian-Chrysler East African expedition of 1926. *Proc. U. S. Nat. Mus.* 73: 1–69.

——. 1929. East African reptiles and amphibians in the United States National Museum. *U. S. Nat. Mus. Bull.* 151: 1–135.

——. 1930a. A scientist's expedition in Africa. *Harv. Alumni Bull.* 32 (October 30): 3–10, map.

——. 1930b. A list of the amphibia of the British Territories in East Africa (Uganda, Kenya colony, Tanganyika territory, and Zanzibar), together with keys for the diagnosis of the species. *Proc. Zool. Soc. Lond.* 1930: 7–32.

——. 1932a. New reptiles and amphibians from Tanganyika territory and Kenya colony. *Bull. Mus. Comp. Zool.* 72: 373–87.

——. 1932b. Eight new toads of the genus *Bufo* from East and Central Africa. *Occ. Pap. Boston Soc. Nat. Hist.* 8: 43–53.
——. 1932c. New frogs of the genera *Arthroleptis* and *Hyperolius* from Tanganyika territory. *Proc. Biol. Soc. Wash.* 45: 61–3.
——. 1932d. New races of a skink (*Siaphos*) and a frog (*Xenopus*) from the Uganda protectorate. *Proc. Biol. Soc. Wash.* 45: 113–15.
——. 1933. Reports on the scientific results of an expedition to the southwestern highlands of Tanganyika territory. VII. Herpetology. *Bull. Mus. Comp. Zool. Harv.* 74(7): 197–416, pls 1–3.
——. 1935. Scientific results of an expedition to rain forest regions in eastern Africa. I. New reptiles and amphibians from East Africa. *Bull. Mus. Comp. Zool. Harv.* 79: 3–19.
——. 1936. Reports on the scientific results of an expedition to rain forest regions in eastern Africa. VII. Amphibians. *Bull. Mus. Comp. Zool.* 79: 369–430, pls 1–3.
——. 1941. New geckos (*Phelsuma* and *Lygodactylus*), snake (*Leptotyphlops*) and frog (*Phrynobatrachus*) from Pemba Island, East Africa. *Proc. Biol. Soc. Wash.* 54: 175–8.
——. 1942a. Comments on the reptiles and amphibians of Lindi. *Tanganyika Notes Rec.* 14: 38–51.
——. 1942b. Scientific results of a fourth expedition to forested areas in East and Central Africa. V. Amphibians. *Bull. Mus. Comp. Zool.* 91: 377–436.
——. 1942C. Comments on the reptiles and amphibians of Lindi. *Tanganyika Notes and Records* 14: 38–51.
——. 1944. Scientific results of a fourth expedition to forested areas in east and central Africa. VI. Itinerary and comments. *Bull. Mus. Comp. Zool. Harv.* 94: 191–214, pls 1–4.
——. 1953. Zoological results of a fifth expedition to East Africa. IV. Amphibians from Nyasaland and Tete. *Bull. Mus. Comp. Zool.* 110: 325–406.
——. 1955. On a second collection of reptiles and amphibians taken in Tanganyika Territory by C. J. P. Ionides. *J. East Afr. Nat. Hist. Soc.* 22: 168–98.
——. 1957. Check list of the reptiles and amphibians of East Africa (Uganda; Kenya; Tanganyika; Zanzibar). *Bull. Mus. Comp. Zool.* 117: 153–362.
Loveridge, J. P. 1970. Observations on nitrogenous excretion and water relations of *Chiromantis xerampelina* (Amphibia, Anura). *Arnoldia (Rhod.)* 5: 1–6.
——. 1979. Cocoon formation in two species of southern African frogs. *S. Afr. J. Sci.* 75: 18–20.
Márquez, R., I. de la Riva, and J. Bosch. 2000. Advertisement calls of *Bufo camerunensis, Chiromantis rufescens, Dimorphognathus africanus* and *Phrynobatrachus auritus*, from Equatorial Guinea (Central Africa). *Herpetol. J.* 10: 41–4.
Mattison, C. 1986. Repeated spawnings in *Hyperolius marmoratus*. *Bull. Br. Herpetol. Soc.* 5: 6–8.
Maxson, L. R. 1981. Albumin evolution and its phylogenetic implications in African toads of the genus *Bufo*. *Herpetologica* 37: 96–104.
McDiarmid, R. W., and R. Altig (Eds.). 1999. *Tadpoles. The Biology of Anuran Larvae.* Chicago: University of Chicago Press.
McIntyre, P. 1999. *Hylarana albolabris*. Predation. *Herpetol. Rev.* 30: 223.
McIntyre, P., and J.-B. Ramanamanjato. 1999. *Ptychadena mascareniensis mascareniensis. Diet. Herpetol. Rev.* 30: 223.
McLachlan, A. 1981. Interaction between insect larvae and tadpoles in tropical rain pools. *Ecol. Entomol.* 6: 175–82.

Menegon, M., and S. Salvidio. 2000. *Nectophrynoides viviparus. Diet. Herpetol. Rev.* 31: 41.

Menegon, M., S. Salvidio, and S. P. Loader. 2004. Five new species of *Nectophrynoides* Noble 1926 (Amphibia Anura Bufonidae) from the Eastern Arc Mountains, Tanzania. *Trop. Zool.* 17: 97–121.

Micha, J.-C. 1975. Quelques données ecologiques sur la grenouille Africaine *Dicroglossus occipitalis* (Günther). *Terre Vie* 29: 307–27.

Miller, K. 1982. Effect of termperature on sprint performance in the frog *Xenopus laevis* and the salamander *Necturus maculosus. Copeia* 1982: 695–8.

Mkonyi, F. J., W. Ngalason, C. A. Msuya, K. M. Howell, and A. Channing. 2004. *Probreviceps loveridgei, Probreviceps uluguruensis* and *Probreviceps macrodactylus* (Loveridge's forest frog, Uluguru forest frog, long-fingered forest frog) advertisement calls. *Herpetol. Rev.* 35: 261–2.

Moreau, R. E., and R. H. W. Pakenham. 1941. The land vertebrates of Pemba, Zanzibar and Mafia: A zoogeographical study. *Proc. Zool. Soc. Lond. [A]* 110: 97–128.

Morescalchi, A. 1967. Note cariologiche su *Phrynomerus* (Amphibia Salientia). *Boll. Zool.* 34: 144.

——. 1968. Initial cytotaxonomic data on certain families of amphibious Anura (Diplasiocoela, after Noble). *Experientia* 24: 280–3.

——. 1973. Amphibia. In A. B. Chiarelli and E. Capanna (Eds.), *Cytotaxonomy and Vertebrate Evolution*, 233–348. London: Academic Press.

Morescalchi, A., G. Gorgiulo, and E. Olmo. 1970. Notes on the chromosomes of some amphibia. *J. Herpetol.* 4: 77–9.

Nakajima, T. 1981. Active peptides in amphibian skin. *Trends Pharmacol. Sci.* 2: 202–5.

Newmark, W. D. 2001. *Conserving Biodiversity in East African Forests. A Study of the Eastern Arc Mountains.* Ecological Studies, Vol. 55. Berlin: Springer-Verlag.

Nieden, F. 1911a. Neue ostafrikanische Frösche. *Sitzsungsber. Gesellsch. Naturf. Freunde Berlin* 436–41.

——. 1911b. Verzeichnis der bei Amani in Deutschostafrika vorkommenden Reptilien und Amphibien. *Sitzsungsber. Gesellsch. Naturf. Freunde Berlin* 10: 441–52.

——. 1912. *Amphibia. Wissenschaftliche Ergebnisse der Deutschen Zentral-Afrika Expedition 1907–1908*, Vol. 4, 165–95. Leipzig: Klinkhardt & Biermann.

Noble, G. K. 1924. Contributions to the Herpetology of the Belgian Congo based on the collection of the American Museum Congo expedition, 1909–1915, Part III. Amphibia. *Bull. Am. Mus. Nat. Hist.* 49: 147–303, pls XXII–XLII.

——. 1929. The adaptive modifications of the arboreal tadpoles of *Hoplophryne* and the torrent tadpoles of *Staurois. Bull. Am. Mus. Nat. Hist.* 58: 291–334, pls 15, 16.

Nussbaum, R. 1985. Systematics of caecilians (Amphibia: Gymnophiona) of the family Scolecomorphidae. *Occ. Pap. Mus. Zool. Univ. Mich.* 713: 1–49.

Nussbaum, R. A., and H. Hinkel. 1994. Revision of East African caecilians of the genera *Afrocaecilia* Taylor and *Boulengerula* Tornier (Amphibia, Gymnophiona, Caecilidae). *Copeia* 1994: 750–60.

Nussbaum, R. A., and M. E. Pfrender. 1998. Revision of the African caecilian genus *Schistometopum* Parker (Amphibia: Gymnophiona: Caeciliidae). *Misc. Publ. Mus. Zool. Univ. Mich.* 187: i–iv, 1–32.

O'Reilly, J. C., R. A. Nussbaum, and D. Boone. 1996. Vertebrate with protrusible eyes. *Nature* 382: 33.

Osborne, P. L., and A. J. McLachlan. 1985. The effects of tadpoles on algal growth in temporary rain-filled rock pools. *Freshwater Biol.* 15: 77–87.

Pakenham, R. H. W. 1983. The reptiles and amphibians of Zanzibar and Pemba Islands (with a note on the freshwater fishes). *J. E. Afr. Nat. Hist. Soc. Nat. Mus.* 177: 1–40.
Pallett, J. R., and N. I. Passmore. 1988. The significance of multi-note advertisement calls in a reed frog *Hyperolius tuberilinguis*. *Bioacoustics* 1: 13–23.
Parker, H. W. 1931. Some brevicipitid frogs from Tanganyika Territory. *Ann. Mag. Nat. Hist.* 8: 261–4.
——. 1936. Reptiles and amphibians collected by the Lake Rudolf Rift Valley Expedition, 1934. *Ann. Mag. Nat. Hist.* 18 (10): 594–609.
Parker, H. W., and E. R. Dunn. 1964. Dentitional metamorphosis in the Amphibia. *Copeia* 1964: 75–86.
Passmore, N. I. 1976. Vocalizations and breeding behaviour of *Ptychadena taenioscelis* (Anura: Ranidae). *Zool. Afr.* 11: 339–47.
——. 1977. Mating calls and other vocalizations of five species of *Ptychadena* (Anura: Ranidae). *S. Afr. J. Sci.* 73: 212–14.
Passmore, N. I., and V. C. Carruthers. 1995. *South African Frogs*, 2nd ed. Johannesburg: Southern Book Publishers and Witwatersrand University Press.
Paukstis, G. L., and S. L. Reinbold. 1984. Observations of snake-feeding by captive African bullfrogs (*Pyxicephalus adspersus*). *Br. Herpetol. Soc. Bull.* 10: 52–3.
Perret, J.-L. 1958. Observations sur des rainettes Africaines du genre *Leptopelis* Günther. *Rev. Suisse Zool.* 65: 259–75.
——. 1966. Les Amphibiens du Cameroun. *Zool. Jahrb. Syst.* 93: 289–464.
——. 1971. Les espèces du genre *Nectophrynoides* d'Afrique (Batraciens Bufonidés). *Ann. Fac. Sci. Cameroun* 6: 99–111.
——. 1972. Les espèces des genres *Wolterstorffina* et *Nectophrynoides* d'Afrique (Amphibia Bufonidae). *Ann. Fac. Sci. Cameroun* 11: 93–119.
——. 1977. Les *Hylarana* (Amphibia, Ranidae) du Cameroun. *Rev. Suisse Zool.* 84: 841–68.
——. 1979. Remarques et mise au point sur quelques espèces de *Ptychadena* (Amphibia, Ranidae). *Bull. Soc. Neuchâtel. Sci. Nat.* 102: 5–21.
——. 1987. A propos de *Ptychadena schillukorum* (Werner, 1907) (Anura, Ranidae). *Bull. Soc. Neuchât. Sci. Nat.* 110: 63–70.
——. 1988. Sur quelques genres d'Hyperoliidae (Anura) restes en question. *Bull. Soc. Neuchât. Sci. Nat.* 111: 35–48.
——. 1996. Une nouvelle espèce du genre *Ptychadena* (Anura, Ranidae) du Kenya. *Rev. Suisse Zool.* 103: 757–66.
Perret, J.-L., and J.-L. Amiet. 1971. Remarques sur les *Bufo* (Amphibiens Anoures) du Cameroun. *Ann. Fac. Sci. Cameroun* 5: 47–55.
Pfeffer, G. 1893. Ostafrikanische Reptilien und Amphibien, gesammelt von Herrn Dr. F. Stuhlmann im Jahre 1988 und 1889. *Jahrb. Hamburg Wissen. Anat.* 10: 71–105, pls 1, 2.
Pickersgill, M. 1992. A new species of *Afrixalus* (Amphibia, Anura, Hyperoliidae) from eastern Africa. *Steenstrupia* 18: 145–8.
Power, J. H. 1935. A contribution to the herpetology of Pondoland. *Proc. Zool. Soc. Lond.* 1935: 333–46, pl I.
Poynton, J. C. 1977. A new *Bufo* and associated amphibia from southern Tanzania. *Ann. Natal Mus.* 23: 37–41.
——. 1991. Amphibians of southeastern Tanzania, with special reference to *Stephopaedes* and *Mertensophryne* Bufonidae. *Bull. Mus. Comp. Zool.* 152: 451–73.
——. 1997. On *Bufo nyikae* Loveridge and the *B. lonnbergi* complex of the East African highlands (Anura: Bufonidae). *Afr. J. Herpetol.* 46: 98–102.

——. 1999. Distribution of amphibians in sub-Saharan Africa, Madagascar and Seychelles. In W. E. Duellman (Ed.), *Patterns of Distribution of Amphibians: A Global Perspective,* 483–539. Baltimore: John Hopkins University Press.

——. 2000a. Amphibians. In N. D. Burgess and G. P. Clarke (Eds.), *Coastal Forests of Eastern Africa,* 201–9, Appendix 7, 411–12. Gland, Switzerland: IUCN—World Conservation Union.

——. 2000b. Foam-nest treefrogs in eastern Africa (Anura Rhacophoridae *Chiromantis*): Taxonomic complexities. *Afr. J. Herpetol.* 49: 111–28.

——. 2003a. *Arthroleptis troglodytes* and the content of *Schoutedenella* (Amphibia: Anura: Arthroleptidae). *Afr. J. Herpetol.* 52: 49–51.

——. 2003b. A new giant species of *Arthroleptis* (Amphibia: Anura) from the Rubeho Mountains, Tanzania. *Afr. J. Herpetol.* 52: 107–12.

——. 2004. Stream frogs in Tanzania (Ranidae: Strongylopus): the case of *S. merumontanus* and *S. fuelleborni. Afr. J. Herpetol.* 53: 29–34.

Poynton, J. C., and D. G. Broadley. 1985. Amphibia Zambesiaca 2. Ranidae. *Ann. Natal Mus.* 27: 115–81.

——. 1987. Amphibia Zambesiaca 3. Rhacophoridae and Hyperoliidae. *Ann. Natal Mus.* 28(1): 161–229.

Poynton, J. C., and B. T. Clarke. 1999. Two new species of *Stephopaedes* (Anura: Bufonidae) from Tanzania, with a review of the genus. *Afr. J. Herpetol.* 48: 1–14.

Poynton, J. C., K. M. Howell, B. T. Clarke, and J. C. Lovett. 1998. A critically endangered new species of *Nectophrynoides* (Anura: Bufonidae) from the Kihansi Gorge, Udzungwa Mountains, Tanzania. *Afr. J. Herpetol.* 47: 59–67.

Quang Trong, Y. le. 1976. Étude de la peau et des glandes cutanées de quelques Amphibiens de la famille des Rhacophoridae. *Bull. Inst. Fond. Afr. Noire* 38A: 166–87.

Razzetti, E., and C. A. Msuya. 2002. *Field Guide to the Amphibians and Reptiles of Arusha National Park (Tanzania).* ARUSHA, TANZANIA: Tanzania National Parks.

Richards, C. M. 1976. The development of color dimorphism in *Hyperolius v. viridiflavus,* a reed frog from Kenya. *Copeia* 1976: 65–70.

——. 1977. Reproductive potential under laboratory conditions of *Hyperolius viridiflavus* (Amphibia: Anura: Hyperoliidae), a Kenyan reed frog. *J. Herpetol.* 11: 426–8.

——. 1981. A new colour pattern variant and its inheritance in some members of the super-species *Hyperolius viridiflavus* (Dumeril and Bibron) (Amphibia: Anura). *Monit. Zool. Ital. (N. S.) Suppl.* 15: 337–51.

——. 1982. The alteration of chromatophore expression by sex hormones in the Kenyan reed frog, *Hyperolius viridiflavus. Gen. Comp. Endocr.* 46: 59–67.

Richards, C. M., and A. Schiøtz. 1977. A new species of reed frog *Hyperolius cystocandicans,* from montane Kenya. *Copeia* 1977: 285–94.

Rödel, M.-O. 1998. Kaulquappengesellschaften ephemerer Savannengewässer in Westafrika. Frankfurt am Main: Edition Chimaira.

——. 2000. *Herpetofauna of West Africa.* Vol. 1. *Amphibians of the West African Savanna.* Frankfort am Main: Edition Chimaira.

——. 2003. The amphibians of Mont Sangbé National Park, Ivory Coast. *Salamandra* 39: 91–110.

Rödel, M.-O., and R. Ernst. 2001. Description of the tadpole of *Kassina lamottei* Schiøtz, 1967. *J. Herpetol.* 35: 678–81.

——. 2002. A new reproductive mode for the genus *Phrynobatrachus: Phrynobatrachus alticola* has nonfeeding, nonhatching tadpoles. *J. Herpetol.* 36: 121–5.

Rödel, M.-O., F. Range, J. Séppânén, and R. Noë. 2002. Caviar in the rain forest: Monkeys as frog spawn predators in Taï National Park, Ivory Coast. *J. Trop. Ecol.* 18: 289–94.

Rödel, M.-O., M. Spieler, K. Grabow, and C. Bockheler. 1995. *Hemisus marmoratus* (Peters, 1854), (Anura: Hemisotidae), Fortpflanzungsstrategien eines Savannenfrosches. *Bonn. Zool. Beitr.* 45: 191–207.

Roseghini, M., G. F. Erspamer, and C. Severini. 1988. Biogenic amines and active peptides in the skin of fifty-two African amphibian species other than bufonids. *Comp. Biochem. Physiol.* 91C: 281–6.

Salvador, A. 1996. Amphibians of northwest Africa. *Smith. Herpetol. Info. Serv.* 109: 1–43.

Sanderson, I. T. 1936a. The giant frog problem. *Nigerian Field* 5: 161–70.

——. 1936b. The amphibians of the Mamfe Division, Cameroons.—II. Ecology of the frogs. *Proc. Zool. Soc.* 1936: 165–208, pl I.

Scheel, J. J. 1971. The seven-chromosome karyotype of the African frog *Arthroleptis*, a probable derivative of the thirteen-chromosome karyotype of *Rana* (Ranidae, Anura). *Hereditas* 67: 287–90.

——. 1973. The chromosomes of some African Anuran species. In J. H. Schröder (Ed.), *Genetics and Mutagenesis of Fish*, i-vi, 1–381. Berlin: Springer-Verlag.

Schiøtz, A. 1963. The amphibians of Nigeria. *Vidensk. Medd. Dansk Naturh. Foren.* 125: 1–92.

——. 1971. The superspecies *Hyperolius viridiflavus* (Anura). *Vidensk. Medd. Dansk Naturh. Foren.* 134: 21–76.

——. 1974. Revision of the genus *Afrixalus* (Anura) in eastern Africa. *Vidensk. Medd. Dansk Naturh. Foren.* 137: 9–18.

——. 1975. *The Treefrogs of Eastern Africa.* Copenhagen: Steenstrupia.

——. 1981. The amphibia in the forested basement hills of Tanzania: A biogeographical indicator group. *Afr. J. Ecol.* 19: 205–7.

——. 1982. Two new *Hyperolius* (Anura) from Tanzania. *Steenstrupia* 8: 269–76.

——. 1999. *Treefrogs of Africa.* Frankfurt am Main: Edition Chimaira.

Schiøtz, A., and M. M. Westergaard. 2000. Notes on some *Hyperolius* (Anura: Hyperoliidae) from Tanzania with supplementary information on two recently described species. *Steenstrupia* 25: 1–9.

Schmidt, K. P., and R. F. Inger. 1959. Amphibians. *Explor. Parc. Nat. Upemba* 56: 1–90, pls 1–6.

Schmuck, R., F. Kobelt, and K. E. Linsenmair. 1988. Adaptations of the reed frog *Hyperolius viridiflavus* (Amphibia, Anura, Hyperoliidae) to its arid environment. V. Iridiophores and nitrogen metabolism. *J. Comp. Physiol. [B]* 158: 537–46.

Schmuck, R., and K. E. Linsenmair. 1988. Adaptations of the reed frog *Hyperolius viridiflavus* (Amphibia, Anura, Hyperoliidae) to its arid environment. III. Aspects of nitrogen metabolism and osmoregulation in the reed frog, *Hyperolius viridiflavus taeniatus*, with special reference to the role of iridiophores. *Oecologia* 75: 354–61.

Schneichel, W., and H. Schneider. 1988. Hearing and calls of the banana frog, *Afrixalus fornasinii* (Bianconi) (Anura: Rhacophoridae). *Amphibia-Reptilia* 9: 251–63.

Seymour, R. S., and J. P. Loveridge. 1994. Embryonic and larval respiration in the arboreal foam nests of the African frog *Chiromantis xerampelina*. *J. Exp. Biol.* 197: 31–46.

Shoemaker, V. H., L. L. McClanahan, P. C. Withers, S. S. Hillman, and R. C. Drewes. 1987. Thermoregulatory responses to heat in the waterproof frogs *Phyllomedusa* and *Chiromantis*. *Physiol. Zool.* 60: 365–72.

Skelton-Bourgeois, M. 1961. Reptiles et Batraciens du Stanley Pool. *Ann. Mus. R. Afr. Centr. Tervuren Ser. 8° Sci. Zool.* 103: 169–83.

Spieler, M., and K. E. Linsenmair. 1997. Choice of optimal oviposition site by *Hoplo-*

batrachus occipitalis (Anura: Ranidae) in an unpredictable and patchy environment. *Oecologia* 109: 184–99.

——. 1998. Migration patterns and diurnal use of shelter in a ranid frog of a West African savannah: A telemetric study. *Amphibia-Reptilia* 19: 43–64.

——. 1999. Aggregation behavior of *Bufo maculatus* tadpoles as an antipredator mechanism. *Ethology* 105: 665–86.

Spinar, Z. V. 1980. The discovery of a new species of pipid frog (Anura, Pipidae) in the Oligocene of central Libya. In M. J. Salem and M. T. Busrewil (Eds.), *The Geology of Libya*, Vol. 1, 327–48. London: Academic Press.

Stevenson, T., and J. Fanshawe. 2002. *Field Guide to the Birds of East Africa.* London: T. & A. D. Poyser.

Stewart, M. M. 1967. *Amphibians of Malawi.* Albany: State University of New York Press.

——. 1974. Parallel pattern polymorphism in the genus *Phrynobatrachus* (Amphibia: Ranidae). *Copeia* 1974: 823–32.

Stewart, M. M., and V. J. Wilson. 1966. Herpetofauna of the Nyika Plateau (Malawi and Zambia). *Ann. Natal Mus.* 18: 287–313.

Stuart, C. T. 1981. Bufotoxins and bufogenins—their effectivity in protecting their producer and their potential as aversive conditioning agents. *J. Herpetol. Assoc. Afr.* 26: 3–6.

Stuart, S. N., J. S. Chanson, N. A. Cox, B. E. Young, A. S. L. Rodrigues, D. L. Fischman, and R. W. Waller. 2004. Status and Trends of Amphibian Declines and Extinctions Worldwide. *Science* 306: 1783–6.

Tandy, M., and R. C. Drewes. 1985. Mating calls of the "kassinoid" genera *Kassina, Kassinula, Phlyctimantis* and *Tornierella. S. Afr. J. Sci.* 81: 191–5.

Tandy, M., and D. J. Feener. 1985. Geographic variation in species of the *Bufo blandfordi* group (Amphibia: Anura: Bufonidae) and description of a new species. *Proc. Int. Symp. Afr. Vert. Bonn* 549–85.

Tandy, M., and R. Keith. 1972. *Bufo* of Africa. In W. F. Blair (Ed.), *Evolution in the Genus Bufo*, 119–70. Austin: University of Texas Press.

Tandy, M., and J. Tandy. 1976. Evolution of acoustic behavior of African *Bufo. Zool. Afr.* 11: 349–68.

Tandy, M., J. Tandy, R. Keith, and A. Duff-McKay. 1976. A new species of *Bufo* (Anura: Bufonidae) from Africa's dry savannas. *The Pierce-Sellards Series.* 24: 1–20. Austin: Texas Memorial Museum.

Tandy, R. M. 1972. The evolution of African *Bufo.* Ph.D. dissertation, University of Texas at Austin.

Taylor, P. 1971. Observations on the breeding habits of *Chiromantis xerampelina* Peters. *J. Herpetol. Assoc. Afr.* 8: 7–8.

——. 1973. A note on colour changes in tadpoles of *Bufo garmani* Meek. *J. Herpetol. Assoc. Afr.* 11: 12–13.

——. 1982. Notes on the ecology and life history of the light nosed toad *Bufo garmani* Meek in the lowveld. *Zimb. Sci. News* 16: 60–2.

Telford, S. R., and J. van Sickle. 1989. Sexual selection in an African toad (*Bufo gutturalis*): The roles of morphology, amplexus displacement and chorus participation. *Behaviour* 110: 62–75.

Tinsley, R. C. 1973. Studies on the ecology and systematics of a new species of clawed toad, the genus *Xenopus* from western Uganda. *J. Zool.* (Lond.) 169: 1–27.

——. 1981a. Interaction between *Xenopus* species (Anura: Pipidae). *Monit. Zool. Ital. (N. S.) Suppl.* 15: 133–50.

——. 1981b. The evidence from parasite relationships for the evolutionary status of *Xenopus* (Anura Pipidae). *Monit. Zool. Ital. (N. S.) Suppl.* 15: 367–85.

——. 1996. Evolutionary inferences from host and parasite co-speciation. In R. C. Tinsley and H. R. Kobel (Eds.), *The Biology of Xenopus*, 403–20. Zoological Society of London. Oxford: Clarendon Press.

Tinsley, R. C., and H. R. Kobel (Eds.). 1996. *The Biology of* Xenopus. Zoological Society of London. Oxford: Clarendon Press.

Tornier, G. 1898. Die Reptilien und Amphibien Deutsch-Ost-Afrika's. In C. W. Werther (Ed.), *Wissenschaftliche Ergebnisse der Ivangi-Expedition 1896–1897 nebst kurzer Reisebeschreibung*, 281–304. Berlin: Dietrich Reimer.

Trueb, L. 1996. Historical constraints and morphological novelties in the evolution of the skeletal system of pipid frogs (Anura: Pipidae). In R. C. Tinsley and H. R. Kobel (Eds.), *The Biology of Xenopus*, 349–77. Zoological Society of London. Oxford: Clarendon Press.

Van Berkom, W. A. 1975. Die Zucht von *Afrixalus dorsalis* im Paludarium. *D. Aquar. Terr. Z. Stuttgart* 28: 282–4.

Van der Elzen, P., and D. Kreulen. 1979. Notes on the vocalisations of some amphibians from the Serengeti National Park, Tanzania. *Bonn. Zool. Beitr.* 30: 385–403.

Van Dijk, D. E. 1985. *Hemisus marmoratum* adults reported to carry tadpoles. *S. Afr. J. Sci.* 81: 209–10.

Van Wyk, J. C. P., and D. J. Kok. 1992. Life history notes. *Pyxicephalus adspersus adspersus* predation on tadpoles. *J. Herpetol. Assoc. Afr.* 41: 40.

Van Wyk, J. C. P., D. J. Kok, and L. H. du Preez. 1991. Growth and behaviour of *Pyxicephalus adspersus* tadpoles. *Abstr. Herpetol. Assoc. Afr. Symp.* 2: 17–18.

Vences, M., F. Glaw, J. Kosuch, I. Das, and M. Veith. 2000. Polyphyly of *Tomopterna* (Amphibia: Ranidae) based on sequences of the mitochondrial 16S and 12S rRNA genes, and ecological biogeography of Malagasy relict amphibian groups. In W. R. Lourenço and S. M. Goodman (Eds.), *Diversité et endémisme à Madagascar*, 229–42. Paris: Mémoires de la Société de Biogéographie.

Vestergaard, M. M. 1994. *An Annotated and Illustrated Checklist of the Amphibians of the Usambara Mountains; with a Tentative Key and the Description of Two New Taxa.* Copenhagen: Zoological Museum, University of Copenhagen.

Vigny, C. 1979a. The mating calls of 12 species and sub-species of the genus *Xenopus* (Amphibia: Anura). *J. Zool. (Lond.)* 168: 103–22.

——. 1979b. Morphologie larvaire de 12 espèces et sous-espèces du genre *Xenopus*. *Rev. Suisse Zool.* 86: 877–91.

Vonesh, J. R. 1998. The amphibians and reptiles of the Kibale National Park, Uganda. M.Sc. thesis, University of Florida.

——. 2000. Dipteran predation on the arboreal eggs of four *Hyperolius* frog species in western Uganda. *Copeia* 2000: 560–6.

Wasser, S. K., and J. C. Lovett. 1993. Introduction to the biogeography and ecology of the rain forests of eastern Africa. In J. C. Lovett and S. K. Wasser (Eds.), *Biogeography and Ecology of the Rain Forests of Eastern Africa*, 3–7. Cambridge: Cambridge University Press.

Wickler, W., and U. Seibt. 1974. Rufen und Antworten bei *Kassina senegalensis*, *Bufo regularis* und anderen Anuren. *Z. Tierpsychol.* 34: 524–37.

Wieczorek, A. M., A. Channing, and R. C. Drewes. 1998. A review of the taxonomy of the *Hyperolius viridiflavus* complex. *Herpetol. J.* 8: 29–34.

——. 2000. Biogeography and evolutionary history of *Hyperolius* species: Application of molecular phylogeny. *J. Biogeogr.* 27: 1231–43.

——. 2001. Phylogenetic relationships within the *Hyperolius viridiflavus* complex (Anura: Hyperoliidae), and comments on taxonomic status. *Amphibia-Reptilia* 22: 155–66.

Wilkinson, M., S. P. Loader, D. J. Gower, J. A. Sheps, and B. L. Cohen. 2003. Phylogenetic relationships of African caecilians (Amphibia: Gymnophiona): Insights from mitochondrial rRNA gene sequences. *Afr. J. Herpetol.* 52: 83–92.

Withers, P. C., S. S. Hillman, R. C. Drewes, and O. M. Sokol. 1982. Water loss and nitrogen excretion in sharp-nosed frogs (*Hyperolius nasutus*: Anura, Hyperoliidae). *J. Exp. Biol.* 97: 335–43.

Zasloff, M. 1987. Magainins, a class of antimicrobial peptides from *Xenopus* skin: Isolation, characterization of two active forms, and partial cDNA sequence of a precursor. *Proc. Natl. Acad. Sci. USA* 84: 5449–53.

Zimmerman, H. 1975. Nachzucht des Marmorriedfrosches *Hyperolius marmoratus marmoratus. Aquar. Wupperthal* 9: 261–5.

——. 1979. Durch Nachzucht erhalten: Marmorriedfrösche, *Hyperolius marmoratus. Aquarien. Mag.* 13: 472–7.

Systematic Index

Alphabetical Index